AF443229

Isotope Labeling in Biomolecular NMR

Advances in Experimental Medicine and Biology

For further volumes:
http://www.springer.com/series/5584

Preface

One of the major advancements in NMR spectroscopy of biomolecules has been the development of new methods for isotope labeling. These methods have contributed to increase in both sensitivity and resolution of NMR experiments. At the same time, there has been an explosion in the number of NMR experiments that utilize such isotope labeled samples. Thus, a combination of isotopic labeling and multidimensional, multinuclear experiments has significantly expanded the range of problems in structural biology amenable to NMR.

Isotope labeling in proteins can be classified into different categories: uniform, amino acid type selective, segmental, site-specific and random/fractional labeling. In addition, different expression systems such as *E. coli*, insect, mammalian or cell-free methods are used for expressing the biomolecule of interest. This book is intended to serve as a compendium of isotope labeling for the biomolecular NMR community providing comprehensive coverage of the existing methods and latest developments along with protocols and practical hints on the various experimental aspects. It is our hope that students/ researchers in the field will find it very useful to have a single book covering a wide range of topics including emerging areas of large proteins, metabolonomics and solid state NMR.

The book has been divided into four parts: proteins, nucleic acids, metabolomics and expression systems with each part containing chapters that describe with different isotope labeling methods relevant to the part. The first part contains chapters that cover structural studies of large globular/ multidomain proteins as well as membrane proteins in solution and solid state. Three chapters in this part are dedicated to more specialized aspects such as study of dynamics, site-specific labeling and methods for sequence specific resonance assignments. The second part on nucleic acids has a chapter covering the latest developments in the field of isotope labeling of RNA. The third part on metabolomics covers the methods used in the rapidly expanding field of metabolomics. An important advancement in recent years has been the development of cell-free methods and strategies for selective labeling of proteins expressed in higher organisms such as insect cells and mammalian cells. The last part of the book contains chapters focusing on each of these aspects.

The chapters have been contributed by experts who have extensively worked in the areas covered. It was a pleasure to work with these colleagues. I would like to thank them for their contributions. I also thank Sara Huisman and Thjis from Springer for their continuous support throughout this endeavor.

Hanudatta S. Atreya

Contents

Part IV   Expression Systems

Part I
Proteins

Chapter 1
Isotope Labeling Methods for Large Systems

Patrik Lundström, Alexandra Ahlner, and Annica Theresia Blissing

Abstract A major drawback of nuclear magnetic resonance (NMR) spectroscopy compared to other methods is that the technique has been limited to relatively small molecules. However, in the last two decades the size limit has been pushed upwards considerably and it is now possible to use NMR spectroscopy for structure calculations of proteins of molecular weights approaching 100 kDa and to probe dynamics for supramolecular complexes of molecular weights in excess of 500 kDa. Instrumental for this progress has been development in instrumentation and pulse sequence design but also improved isotopic labeling schemes that lead to increased sensitivity as well as improved spectral resolution and simplification. These are described and discussed in this chapter, focusing on labeling schemes for amide proton and methyl proton detected experiments. We also discuss labeling methods for other potentially useful positions in proteins.

1.1 Introduction

In the first few years following the first successful nuclear magnetic resonance (NMR) spectroscopy experiments [1, 2], the technique was primarily the physicist's tool. After all, it was a nuclear phenomenon that apparently was of little relevance for chemistry. This dramatically changed when chemical shifts of spectral lines due to chemical environment were demonstrated [3, 4]. Now, molecules could be identified based on their NMR spectra and this is the major reason why NMR has been the most important spectroscopic method ever since. Around the same time, two almost as important nuclear magnetic phenomena were discovered. One was the scalar coupling between nuclear spins mediated by electrons in the covalent bonds separating the nuclei. In contrast to the direct dipolar coupling, this coupling leads to splitting of spectral lines, and equivalently provides a way to transfer magnetization, in isotropic samples. This is important as it provides a way of correlating two nuclei. The other important discovery was the Overhauser and nuclear Overhauser effect (NOE) that correctly predicted that nuclear polarization depends on the spin state of nearby (in space) unpaired electrons or nuclei [5–7] and enables measurements of internuclear distances. The chemical shift, the scalar coupling and the NOE in principle provide the tools needed for structure calculations of complex molecules such as proteins. The scalar coupling is used to establish correlations between nuclei and thus to assign the

P. Lundström (✉) • A. Ahlner • A.T. Blissing
Division of Molecular Biotechnology, Department of Physics, Chemistry and Biology, Linköping University,
SE-58183 Linköping, Sweden
e-mail: patlu@ifm.liu.se

H.S. Atreya (ed.), *Isotope Labeling in Biomolecular NMR*, Advances in Experimental Medicine and Biology 992,
DOI 10.1007/978-94-007-4954-2_1, © Springer Science+Business Media Dordrecht 2012

resonances and the NOE is used to measure their separation. Using protocols such as distance geometry and simulated annealing, the three-dimensional structure can then be calculated.

Because of signal overlap, one dimensional NMR experiments are not feasible for structure calculations or other high resolution applications even for small proteins. Luckily, methods for recording two dimensional homonuclear experiments to establish correlations between nuclei were developed in the late 1970s and in 1982 Kurt Wüthrich and coworkers had completed the resonance assignments of the 6.5 kDa protein basic pancreatic trypsin inhibitor using ^{1}H homonuclear experiments [8]. In 1985 the same group were able to calculate the solution structure of the protein bull seminal protease inhibitor (6 kDa) primarily from distance restraints derived from NOESY experiments [9] and in the years that followed, the solution structures of several other proteins of similar size were calculated. However, the process of obtaining resonance assignments was labor intensive and the signal overlap was too severe for the method to be practical for proteins larger than approximately 10 kDa. A partial remedy that did not require isotopic labeling was to generalize the experiments to three dimensions and thus to correlate three different protons using two different mixing sequences, such as one NOE and one Hartmann-Hahn transfer period [10]. While these experiments were useful to increase spectral resolution and to establish many correlations between nuclei in a single experiment they suffered from the shortcoming of small scalar couplings between protons separated by three covalent bonds. Many potential experiments would thus need prohibitively long transfer times with concomitant reduction in sensitivity. In contrast, many heteronuclear scalar couplings are significantly stronger and it was recognized that if a heteronucleus such as ^{15}N or ^{13}C is used in the third dimension, superior sensitivity could be achieved [11–13]. Today, the most important of these heteronuclear experiments are NOESY-HSQC and TOCSY-HSQC and by using these types of experiments proteins up to 20 kDa can often be assigned and their structures calculated. Using this approach, the assignment process is performed by first identifying spin systems using the TOCSY-HSQC experiments and connecting them sequentially by the aid of the NOESY-HSQC experiments.

When triple-resonance experiments were developed [14], a new, more effective way of obtaining resonance assignments was possible by correlating the amide proton and nitrogen of one residue with one or two carbon nuclei of the same residue and of the preceding residue. Side-chain assignments could then be completed with ease using TOCSY experiments and chemical shift information of one or more carbon nuclei obtained by the triple-resonance experiments. The structures themselves were still mainly calculated from distance restraints obtained from NOESY-HSQC experiments. With these methods and with the addition of residual dipolar coupling (RDC) restraints [15], structure calculations by NMR could be completed in less time and with higher precision than before. A necessary price that had to be paid for using these conceptually simpler and less cumbersome experiments was that the proteins must be simultaneously labeled with ^{13}C and ^{15}N.

Another major breakthrough came with the advent of transverse relaxation optimized spectroscopy (TROSY) pulse sequences [16]. For these to work well, most non-labile protons must be replaced by deuterons. Using these methods it is possible to increase the size limit significantly and in favorable cases it is possible to perform backbone resonance assignments and structure calculations of proteins approaching 100 kDa [17]. A significant drawback of these methods is that most protons that are used as distance restraints, including in the protein core, have been removed and a NOE driven structure calculation then has to rely only on distances between amide protons leading to very few restraints for each residue. A way of improving the situation is to add side-chain protons at strategic places in an otherwise deuterated background [18]. For large proteins this would be at side-chain methyl groups, because of their favorable relaxation properties and because of their numerous contacts. They can be assigned using experiments that correlate the methyl groups with the protein backbone [19] and add crucial information about the protein core.

The progress in NMR spectroscopy applied to large systems has been extremely rapid in recent years and the applications have been impressive. One reason for this is the development of high-field instruments and cryogenically cooled probes with superior sensitivity compared to standard instruments two decades ago. Another important aspect is the development of new pulse sequences that allow magnetization transfer in a spin state selective way so that relaxation losses are minimized. However, neither sensitive instruments nor clever pulse sequences suffice for recording spectra of high sensitivity for large systems. An equally important requirement is labeling schemes designed to enhance resolution and to reduce relaxation rates and spectral crowding. These labeling schemes are the focus of this chapter. The definition I will use for a large system is that the usual combination of pulse sequences and isotopic labeling schemes, i.e. non-TROSY pulse sequences and fully protonated uniformly ^{15}N and/or ^{13}C labeled samples, will fail. It is not possible to provide a certain number for what this means in terms of molecular weight but a biomolecule or a complex of biomolecules larger than 30 kDa fulfills this criterion for most applications.

1.2   Spin Relaxation and TROSY

There are two main challenges with NMR spectroscopy applied to large systems. One is signal overlap due to spectral crowding. While this is a concern, many applications have focused on oligomeric complexes so that although the complex tumbles as a large unit, the number of signals is manageable. Also for a monomeric protein as large as the 82 kDa *E. coli* malate synthase G, Tugarinov et al. recorded a beautiful ^{15}N-^{1}H correlation spectrum that, while crowded, had most peaks resolved [17]. Furthermore, if necessary it is possible to selectively label a subset of the amino acid residues as will be discussed below. The more serious problem is line-broadening and concomitant reduced sensitivity for large systems due to rapid transverse relaxation. This, in practice, sets the limit for how large systems that can be studied. A useful strategy for the study of large systems is thus to try to increase the tumbling rate by increasing the temperature. For proteins this can of course not be done indefinitely since stability is decreased at high temperature. Proteins from thermophilic organisms are useful in this regard. However, it is usually not possible to improve spectral quality sufficiently for high molecular weight systems by only increasing the temperature. The method that has opened the door to NMR studies of larger proteins is line narrowing by TROSY [20].

To see how line narrowing is achieved in TROSY type experiments, we consider a spin-pair IS that is scalar coupled with coupling constant J_{IS} and evaluate the time evolution of the two components of I transverse magnetization. Without loss of generality we will disregard from chemical shift evolution. If the two components are relaxed by the dipole-dipole interaction with spin S and by I chemical shift anisotropy they evolve as [21]

$$\frac{d}{dt}\begin{pmatrix} I^+S^\alpha(t) \\ I^+S^\beta(t) \end{pmatrix} = \begin{pmatrix} i\pi J_{IS} + \bar{R}_2 + \eta_{xy} & \left(R_{2I}^{DD} - R_{2IS}^{DD}\right)/2 \\ \left(R_{2I}^{DD} - R_{2IS}^{DD}\right)/2 & -i\pi J_{IS} + \bar{R}_2 - \eta_{xy} \end{pmatrix}\begin{pmatrix} I^+S^\alpha(t) \\ I^+S^\beta(t) \end{pmatrix} \tag{1.1}$$

where

$$\bar{R}_2 = \left(R_{2I}^{DD} + R_{2IS}^{DD}\right)/2 + R_{2I}^{CSA} \tag{1.2}$$

$$R_{2I}^{DD} = \frac{d^2}{8}\left[4J(0) + 3J(\omega_I) + 6J(\omega_S) + 6J(\omega_I + \omega_S) + J(\omega_I - \omega_S)\right] \tag{1.3}$$

$$R_{2IS}^{DD} = \frac{d^2}{8}\left[4J(0) + 3J(\omega_I) + 6J(\omega_I + \omega_S) + J(\omega_I - \omega_S)\right] \tag{1.4}$$

$$R_{2I}^{CSA} = \frac{c^2}{6}\left[4J(0)+3J\left(\omega_I\right)\right]$$

(1.5)

$$\eta_{xy} = \frac{\sqrt{3}}{6}cdP_2\left(\cos\theta\right)\left[4J(0)+3J\left(\omega_I\right)\right]$$

(1.6)

where $d = \mu_0\hbar\gamma_I\gamma_S\left\langle r_{IS}\right\rangle^{-3}/4\pi; c = \gamma_I B_0\Delta\sigma/\sqrt{3}; \mu_0$ is the permeability of vacuum; $\hbar$ is the reduced Planck constant; γ_I and γ_S are the magnetogyric ratios; r_{IS} is the internuclear distance; B_0 is the static magnetic field strength, $\Delta\sigma$ is the anisotropy of the (axially symmetric) chemical shift tensor and $P_2\left(\cos\theta\right)$ is the second order Legendre polynomial of the cosine of the angle between the principal frames of the dipolar and chemical shift anisotropy interactions. $J\left(\omega\right)$ is the spectral density that is usually modeled by the model-free formalism [22–24]

$$J\left(\omega\right)=\frac{2}{5}\left[\frac{S^2\tau_c}{1+\left(\omega\tau_c\right)^2}+\frac{\left(1-S_f^2\right)\tau_f'}{1+\left(\omega\tau_f'\right)^2}+\frac{\left(S_f^2-S^2\right)\tau_s'}{1+\left(\omega\tau_s'\right)^2}\right]$$

(1.7)

where $S^2 = S_f^2 S_s^2; S_f^2$ and S_s^2 are the generalized order parameters for fast and slow internal motions, respectively; $\tau_f' = \tau_f\tau_c/(\tau_f+\tau_c), \tau_s' = \tau_s\tau_c/(\tau_s+\tau_c); \tau_f$ and τ_s are the correlation times for the fast and slow internal motions respectively and τ_c is the correlation time for molecular tumbling.

If $2\pi J_{IS}^2 \gg \left(R_{2I}^{DD}-R_{2IS}^{DD}\right)^2$ the off-diagonal elements are unimportant and cross-relaxation between the two components can be neglected. Equation 1.1 shows that the relaxation rate of one component is reduced while the relaxation rate of the other is enhanced by the cross-correlation relaxation rate between the dipole-dipole and chemical shift anisotropy tensors $\left(\eta_{xy}\right)$ and consequently one component will be broad while the other will be narrow. TROSY experiments select the narrow component in NMR spectra [20] and are designed not to mix slowly and rapidly relaxing components. The auto-correlated and cross-correlated contributions to relaxation have different field dependence which means that complete cancellation occurs at certain field strength. This argument is however based on the assumption that the spin-pair is isolated from dipole-dipole interactions with remote spins. In proteins this is not the case and especially for protonated samples there is additional broadening. The line-narrowing effect by the TROSY principle for an isolated ^{15}N-^{1}H spin-pair is shown in Fig. 1.1.

For IS_2 and IS_3 spin systems there are additional cross-correlations between different dipolar inter-actions. The TROSY principle can be applied in these cases as well and line narrowing can be achieved by selecting the appropriate lines in such multiplets. TROSY experiments have been described for methyl [25] and also methylene [26] groups. The theoretical description in these cases is rather involved and the interested reader is referred to the original works.

The original TROSY experiment for recording ^{15}N-^{1}HN correlation maps is shown in Fig. 1.2a [20]. In this experiment the slowly relaxing component is selected in both dimensions and it is note-worthy that polarization originating from ^{15}N contributes to the signal. Gradient-selected sensitivity enhanced versions [28, 29] as well as versions that actively suppress the rapidly relaxing components without increasing the phase cycle have subsequently been described [30]. TROSY detection for ^{15}N and ^{1}HN is also readily incorporated into triple-resonance pulse sequences [31, 32]. This has been instrumental for resonance assignments of large systems.

In the case of applications to methyl groups the heteronuclear multiple-quantum coherence (HMQC) experiment is used to record TROSY correlation maps. Slowly relaxing single quantum ^{1}H coherence, produced by the only 90° ^{1}H pulse, is converted to slowly relaxing heteronuclear double and zero quantum coherence by a 90° ^{13}C pulse. It is then converted back to slowly relaxing single quantum ^{1}H coherence by a second 90° ^{13}C pulse. Because ^{13}C pulses are used exclusively to change the

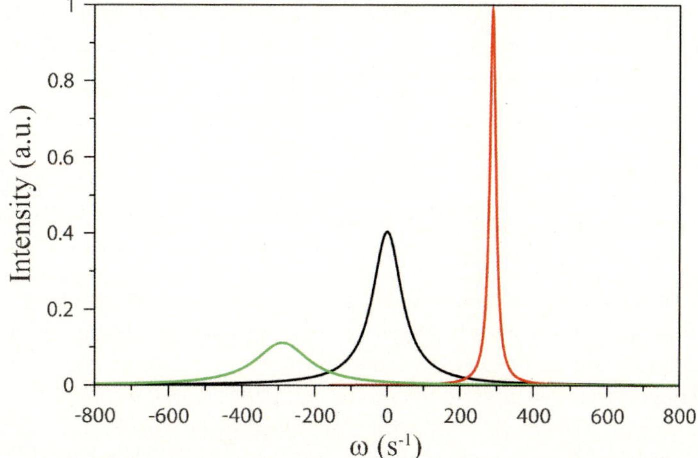

Fig. 1.1 Illustration of the TROSY effect on the line shape. The figure was prepared by solving Eq. 1.1 for an isolated ^{15}N-^{1}H spin-pair, tumbling with a correlation time of 30 ns at a static magnetic field of 18.8 T. The scalar coupling constant is 92 Hz. The *red* and *green lines* correspond to the narrow and broad components of the doublet, respectively, whereas the *black line* represents the line shape in a decoupled spectrum

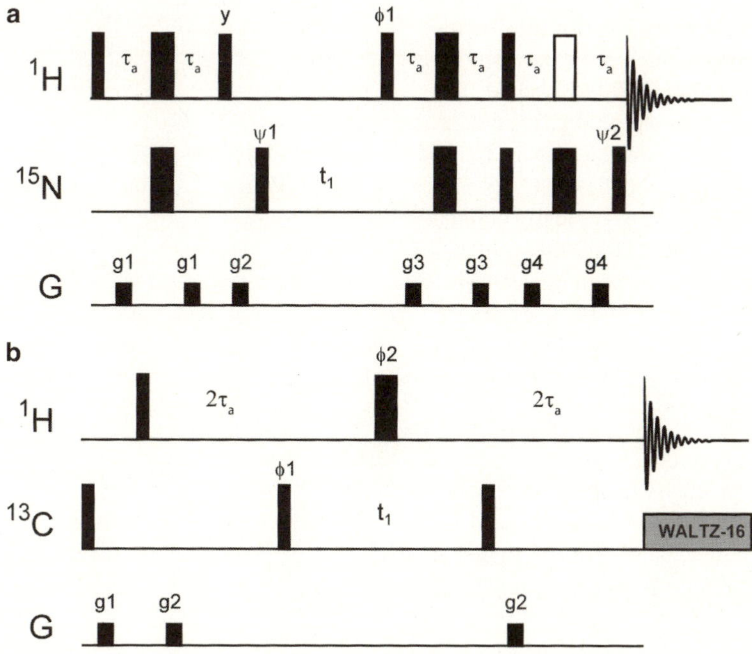

Fig. 1.2 (**a**) The original ^{15}N-^{1}HN TROSY pulse sequence [20]. *Narrow* and *wide rectangles* depict 90° and 180° pulses, respectively. The *open rectangle* represents a WATERGATE element. The delay τ_a is equal to $1/(4J_{NH})$. The phase cycling is $\psi 1 = y,-y,-x,x$; $\psi 2 = 4(x),4(-x)$; $\phi 1 = 4(y),4(-y)$ and receiver = y,-y,-x,x. (**b**) The basic HMQC pulse sequence [27]. In this case the delay τ_a is equal to $1/(4J_{CH})$. The phase cycling is $\phi 1 = x,-x$; $\phi 2 = x,x,y,y,-x,-x,-y,-y$; receiver = x,-x,-x,x. Quadrature detection is achieved by States-TPPI phase incrementation in both cases

coherence level after the initial ^{1}H pulse, rapidly and slowly relaxing coherences are never mixed [25]. The HMQC pulse sequence is shown in Fig. 1.2b. A way of achieving enhanced resolution with only modest (10%) loss in sensitivity is to use heteronuclear zero-quantum (HZQC) instead of HMQC experiments [33].

1.3   Labeling Protocol for Large Proteins

Isotopically enriched proteins are usually expressed in M9 minimal medium [34], referred to as M9 medium in the following. M9 medium is primarily composed of 6 g/L Na_2HPO_4, 3 g/L KH_2PO_4 and 0.5 g/L NaCl. These salts are dissolved in D_2O and the medium is supplemented with 1 mM $MgSO_4$, 0.1 mM $CaCl_2$, 10 mg/L biotin, 10 mg/L thiamine and antibiotics. Stock solutions for these components must be dissolved in D_2O. 0.5–1 g/L $^{15}NH_4Cl$ is also added to the medium. Depending on whether simultaneous labeling with ^{13}C is required or not the carbon source is either 2–3 g/L $[^{13}C_6, {}^2H_7]$-glucose or 2–3 g/L $[^{12}C_6, {}^2H_7]$-glucose. The use of protonated glucose in combination with 100% D_2O as solvent is strongly discouraged since many hydrogen positions are derived from both these sources. In addition to lowering the sensitivity because of introduction of a high magnetogyric ratio nucleus it leads to a mixture of different isotope shifts and further broadened lines. It is imperative to not auto-clave deuterated growth media since this leads to partial protonation due to $H_2O \leftrightarrow D_2O$ exchange. Instead, sterile filtration should be performed. Below is an expression protocol that has frequently been used to produce highly deuterated samples.

1. Transfer one or more freshly transformed *E. coli* colonies of BL21(DE3) strain to 30 mL LB (in H_2O) supplemented with the appropriate antibiotic(s) and grow cells at 37 °C in a shaking incubator until $OD_{600} = 1.0$ is reached.
2. Spin down the cells at 1,200 g, 15 min. at room temperature (25 °C).
3. Resuspend a fraction of the cells in 10% of the isotopically labeled M9 medium to achieve OD_{600} of 0.1–0.2. Grow the cells at 37 °C until $OD_{600} = 1.0$. Pour the starter culture directly into the remaining 90% of the isotopically enriched M9 medium.
4. Grow the cells at 37 °C until $OD_{600} = 0.6$–1.0.
5. Induce over expression with 0.5–1 mM IPTG. Perform over expression either at 37 °C for 2–5 h or at room temperature or 16 °C overnight. The final OD_{600} will depend on the growth medium and on the duration of over expression. This step should be modified if selectively labeled precursors or amino acids are used. These compounds are added to the growth medium 1 h before over expression is induced.

A purification protocol has to be developed for each protein. Denaturation and refolding is usually necessary if complete exchange of deuterons to protons, or vice versa, at amide positions is required. Typically, the protein is unfolded in 6 M GdnCl or 8 M urea after cell lysis. Refolding is done by exchange to a buffer that favors the folded state and can be performed by several different methods including dialysis, on column or by rapid dilution [35].

1.4   Labeling Methods for Amide Proton Detected Experiments

Generally, structural investigations of large proteins require perdeuterated samples although methods for dealing with quite large systems using proteins only $^{13}C/^{15}N$ labeled have been described. For instance, Xu et al. have developed a strategy for assigning spectra of large uniformly ^{13}C, ^{15}N labeled proteins [36]. By using TROSY-HNCA, $^{13}C,^{15}N$-NOESY, $^{13}C,^{13}C$-NOESY and MQ-HCCH-TOCSY

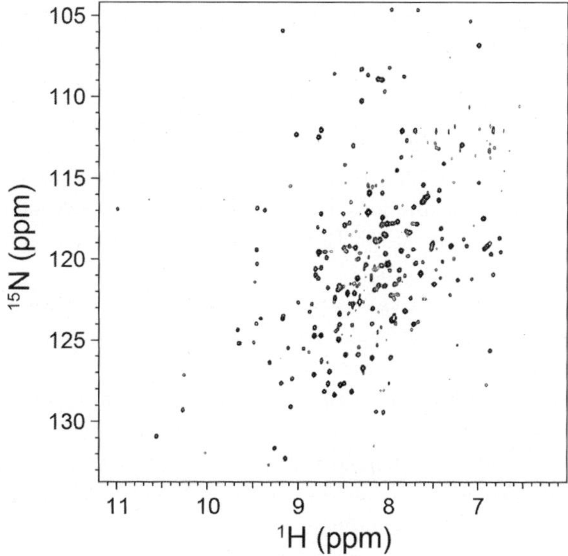

Fig. 1.3 ^{15}N-^{1}HN correlation map of a perdeuterated homodimer of the kinase domain from EphB2, 2×289 residues, 65 kDa, recorded with a TROSY pulse sequence at 25 °C and a static magnetic field of 18.8 T

spectra, clusters with the same HN shifts were identified, and individual spins systems were determined, creating dipeptide fragments that were pieced together and assigned to the sequence. They were able to assign the 42 kDa maltose binding protein and the 65 kDa protein hemoglobin [36].

The reason why perdeuteration is beneficial in TROSY experiments is because the amide proton is not only relaxed by the dipolar interaction with its attached ^{15}N spin and its chemical shift anisotropy but also by dipolar interactions with (especially) other protons that are close in space. Since this means that the total dipolar interaction increases while the chemical shift anisotropy interaction stays the same, the TROSY effect is diminished. It also means that $2\pi J_{IS}^{\,2} \gg \left(R_{2I}^{DD} - R_{2IS}^{DD}\right)^2$ no longer holds which leads to cross-relaxation between the two components and a further reduction of the TROSY effect. For large proteins it is thus essential to reduce the pool of protons. This can be done by perdeuteration, replacing aliphatic and aromatic protons with deuterons. By reducing the number of possible interactions with other protons, the TROSY effect is retained.

As is evident from Fig. 1.3, which shows a TROSY ^{15}N-^{1}HN correlation map of a perdeuterated ^{15}N/^{13}C labeled sample of dimeric kinase domain of the protein EphB2 (2×289 residue, 65 kDa), spectra of excellent quality are possible even for quite large systems. The spectrum was recorded at 25 °C at a static magnetic field strength 18.8 T and the protein was purified from inclusion bodies by solubilization in 6 M GdnHCl and subsequently refolded by dialysis. TROSY pulse sequences in combination with perdeuteration are quite sensitive even for a monomeric protein as large as the *E. coli* malate synthase G that has a molecular weight of 82 kDa (723 residues). The protein was assigned to 95% at backbone positions and 97% at Cα, Cβ, CO positions [17]. In experiments involving transverse magnetization of a nucleus that is primarily relaxed by chemical shift anisotropy neither the TROSY effect nor perdeuteration is of much help. Furthermore, since the chemical shift anisotropy mediated relaxation scales with the square of the external magnetic field going from 600 to 800 MHz in fact reduces sensitivity [17]. Another cause of sensitivity loss in many of these experiments is due to imperfections in the large number of pulses in these sequences. It is thus imperative to keep the number of applied pulses to a minimum and to use optimal pulses. In this regard recent developments in pulse sequences promise much for the future. For instance, a combination of TROSY and multiple-quantum evolution elements on average leads to gains in sensitivity of a factor 1.8 when compared to the conventional TROSY-HN(CO)CA experiment for the membrane protein-detergent complex KcsA with a rotational correlation time of around 60 ns [37]. The reason for the gain in sensitivity is a combination of these relaxation-optimized elements and that ten less pulses on ^{13}C are required. Another complication for highly deuterated samples of high molecular weight proteins relates to slow proton longitudinal relaxation and consequently the need for long inter scan delays in NMR experiments. Acceleration of longitudinal proton relaxation between scans can be achieved by the band-selective excitation short-transient (BEST) technique where band-selective pulse centered on the amide region are substituted for hard pulses on proton [38]. This means that experiments can be repeated very rapidly, resulting in higher sensitivity per unit time.

A drawback with perdeuteration is that experiments that require interactions with and between aliphatic and aromatic protons are not possible. The most important example is the NOESY experiment that is used to obtain distance restraints for structure calculations. For very large systems, like malate synthase G described above, the most viable strategy is to introduce protons at strategic positions, where relaxation properties are favorable, such as methyl groups. Selective labeling schemes for methyl groups are described later in this text. Indeed, when the global fold of this protein was solved by NMR spectroscopy, the relatively few distance restraints obtained from ^{1}HN and ^{1}H^{methyl} had to be complemented by other measured parameters such as residual dipolar couplings and secondary chemical shifts [39].

Amide proton detected experiments for even larger systems, comprising homo-oligomeric supramolecular complexes, have also been described. These complexes benefit from having the spectral appearance of monomers and from the fact that the intensity of the NMR signal is multiplied with the number of monomers per oligomer. Wüthrich and coworkers have recorded ^{15}N-^{1}H correlation maps of the heptameric co-chaperonin GroES (72 kDa) bound to either SR1 (400 kDa) or GroEL (800 kDa). While GroES was perdeuterated it did not matter whether the other proteins were perdeuterated or not. Because of significant relaxation losses during transfer periods TROSY experiments perform significantly worse than ones incorporating cross-relaxation induced polarization transfer (CRIPT) for systems of this size. In CRIPT spectra of GroES bound to GroEL, the slowly relaxing component for most of the expected 94 resonances were observed [40].

To decrease spectral crowding for large proteins and for assignment purposes, selective ^{15}N labeling of only one or several residue types can be employed. For small to medium-sized proteins this can be achieved by growth and expression in rich medium supplemented with ^{15}N labeled amino acids for the selected residues and high levels of all other amino acids in unlabeled form [41]. For large proteins a similar strategy is growth in M9 medium in 100% D$_2$O supplemented with 1 g/L glucose, 1 g/L natural abundance NH$_4$Cl, 1–3 g/L deuterated algal lysate and 4 mg/L deuterated and ^{15}N labeled forms of each of the selected amino acids [42]. Since the algal lysate contains amino acids it is necessary to carefully optimize the amount of labeled amino acids to avoid cross-labeling due to transamination. Completely clean labeling of the desired residue types was not possible unless bacterial strains deficient in transaminases were used [42].

1.5   Labeling Methods for Aromatic Side-Chain Positions

TROSY line-narrowing can also be realized for other positions than backbone amides. Two examples of this are ^{1}H^{15}Nε2 moieties of the tryptophan side-chain and ^{1}H^{13}C groups of the side-chains of Phe and Tyr.

1.5.1   Labeling of Tryptophan Side-Chains

Tryptophan side-chains frequently provide important interactions in the early stages of protein folding [43] and participate in interactions with ligands or other proteins [44]. Also, tryptophan residues have been shown to be sensitive to solvation and temperature [45]. These characteristics combined make this amino acid particularly useful as a biophysical probe for NMR spectroscopy.

Since tryptophan residues report on several biophysical events, there are many ways to study these occurrences. Löhr et al. have written a three-dimensional experiment to correlate tryptophan side-chain with $^{13}C\beta$ [46]. Monitoring perturbations of the chemical shifts upon interaction with various ligands provide information on the nature of the association, without the need for complete structure determination. Similarly, cross-saturation experiments can be employed to characterize large protein interaction complexes [47].

A simple and cost-effective way of selectively labeling tryptophan side-chains is to add perdeuterated $[2,4-^{13}C_2]$-indole, which is a precursor of tryptophan, to the culture medium prior to induction of protein expression. This method results in excellent incorporation with virtually no scrambling of label [45].

1.5.2   Labeling of Aromatic Side-Chain Positions

Because of the large chemical shift anisotropy for the carbon positions of aromatic side-chains, substantial line narrowing by the TROSY principle can also be achieved in this case. TROSY experiments applied to aromatic side-chains have been described [16]. In this case, maximum line-narrowing is achieved at static magnetic fields between 11.7 and 18.8 T. For uniformly ^{13}C labeled proteins these experiments require constant-time evolution in the indirect dimension and since it is not feasible to selectively protonate a protein at certain aromatic side-chain positions by using glucose or related molecules as carbon sources, these experiments are not widely used although one application is HCCH-TOCSY experiments with TROSY detection in both carbon dimensions [48]. It should be noted however that specifically labeled aromatic amino acids can be added to a deuterated background. The amino acids should optimally be $^{13}C^{1}H$ labeled at the positions that are detected and $^{12}C^{2}H$ elsewhere. The in vitro stereo-array isotope labeling (SAIL) protocol produces aromatic amino acid side-chains with alternating $^{13}C^{1}H$ and $^{12}C^{2}H$ isotopomers and could thus be useful in this regard [49].

1.6   Labeling Methods for Methyl Groups

Methyl groups provide the most sensitive probes in NMR experiments because of their rapid rotation around a three-fold axis. Additional line-narrowing required for applications to large systems is achieved by methyl TROSY experiments. For applications involving high molecular weight systems, non-methyl positions must be deuterated and scalar couplings to adjacent carbons must not be present to avoid using constant-time evolutions periods. The use of methyl groups as probes for supra-molecular structure and dynamics has been reviewed extensively by Ruschak and Kay [50]. In this section, we will sometimes use the notation H for ^{1}H and D for ^{2}H for increased clarity.

A clean way of selectively labeling certain methyl side-chains is to add the commercially available compounds α-ketobutyrate (precursor of Ile) and/or α-ketoisovalerate (precursor of Leu, Val) that are specifically labeled with ^{13}C at the methyl groups to the growth medium [18]. This is usually referred to as ILV labeling and is shown in Fig. 1.4. Although these compounds can potentially be degraded into precursors for other amino acids the authors noticed essentially no labeling at other positions. A very useful feature of this method is that labeling of the methyl groups can be customized. For instance, in addition to having all methyl groups $^{13}CH_3$ it is also possible to have them $^{13}CH_2D$ or $^{13}CHD_2$. One can also label different methyl groups differently. By using α-ketoisovalerate labeled with $^{13}CH_3$ at one methyl group and with $^{12}CD_3$ at the other only the *proR* or *proS* methyl group of Leu and Val is detected. For applications involving high molecular weight systems, the non-methyl position

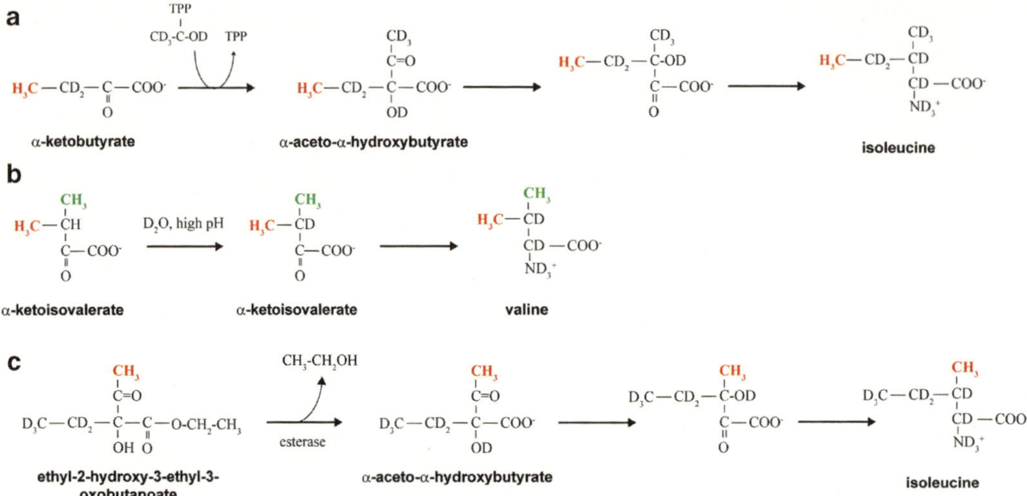

Fig. 1.4 Synthesis of the amino acids Ile and Val from the precursors α-ketobutyrate and α-ketoisovalerate, respectively for high molecular weight NMR applications. The methyl group originally at α-ketobutyrate is highlighted in *red* and the two methyl groups originally at α-ketoisovalerate are shown in *red* and *green*. The labeling pattern of the highlighted methyl groups can be customized independently. (**a**) Synthesis of Ile, selectively labeled at the δ1 position. (**b**) Synthesis of Val from the precursor α-ketoisovalerate. The initial step is deuteration of position 3 that can be performed in-house to cut costs. The methyl groups of Leu will be labeled in the same manner if α-ketoisovalerate is added to the growth medium. (**c**) Selective labeling of Ileγ2, achieved by adding α-aceto-α-hydroxybutyrate to the growth medium. This precursor is purchased in its ethyl ester form and deesterification is performed by treatment with esterase as illustrated by the first step

of α-ketoisovalerate can be deuterated in-house by incubation of 25 mM α-ketoisovalerate in D_2O at pH 12.5 for 2–3 h at 45 °C [18]. The completeness of the reaction can be followed by NMR.

To produce these samples, a similar protocol for bacterial growth and over expression as given above is used. The methyl groups of Ileδ1, Leu and Val are labeled to 90% if the precursors are supplied in concentrations of 50 mg/L for α-ketobutyrate and 100 mg/L for α-ketoisovalerate 1 h prior to induction of protein expression [18]. Using this approach it has for instance been possible to measure dynamics for systems as large as the proteasome 20S core particle of 670 kDa [51] and to reveal the structural basis for signal-sequence recognition by the translocase motor SecA [52].

To compare $^{13}CHD_2$ and $^{13}CH_3$ labeled methyl groups in NMR applications involving supra-molecules, Religa and Kay compared the relative sensitivity of experiments involving these isotopomers. The sensitivity of Ile, Leu, Val $^{13}CH_3$-labeled samples was found to be between 1.5 and 2 fold higher than the corresponding data sets obtained with $^{13}CHD_2$ probes. For supra-molecules, labeling with $^{13}CH_3$ isotopomers is thus the method of choice if maximum sensitivity is desired. However, applications that require $^{13}CHD_2$ moieties can still be performed with high sensitivity [53].

Recently, Ruschak et al. suggested a method to instead label Ileγ2 to enable measurements at this position in large proteins [54]. The protocol is based on adding the precursor α-aceto-α-hydroxybutyrate that is $^{13}C^1H_3$ labeled at the relevant methyl group and $^{12}C/^2H$ labeled elsewhere. For reasons of stability the compound is purchased in its ethyl ester form and de-esterified by incubation with esterase. An amount corresponding to 100 mg/L of the acid form is added to the growth medium. Contrary to what is observed for the ILV labeling scheme, scrambling leads to the presence of weak correlations of *proR* Valγ and Leuδ [54]. However, these do not complicate the interpretation of the spectra considerably.

Methyl groups in Met residues are also most useful probes. For applications involving large proteins it is, of course, essential that all positions except the Met methyl groups are deuterated. This is conveniently achieved by supplying this residue, selectively ^{13}C, ^{1}H labeled at the methyl group, to a deuterated growth medium at a concentration of 100 mg/L 1 h prior to induction [55].

1.7 Cell-Free Protein Synthesis

Some proteins cannot be expressed successfully in *E coli*. Proteins that are involved in apoptosis, are rapidly degraded, have low solubility or are toxic can instead be expressed in a cell-free system. Cell-free expression is an *in vitro* method to produce proteins by isolating the protein translational machinery of eukaryotic or prokaryotic cells, allowing protein synthesis to take place *in vitro* and has since long been used for NMR applications [56]. Since the protein production takes place in an artificial environment, specific labels can be incorporated by supplying amino acids with the desired labeling pattern suitable for NMR studies. It is also possible to produce perdeuterated protein samples [57] and introduce non-natural amino acids [58].

Cell-free protein synthesis can be achieved either by batch-mode or continuous-exchange. The batch-mode protocol includes mixing the reactants (cell extract, template DNA, RNA polymerase, NTPs) in a buffered solution and incubating. However, build-up of by-products and depletion of NTPs are limiting factors. The continuous-exchange protocol involves incubating the reaction in a dialysis membrane, allowing exchange of solutes and thereby removing by-products and supplying NTPs to the synthesis reaction across the membrane. The continuous-flow cell-free system is a variation of this concept. Here, NTPs are continuously added to the reaction mixture while by-products are simultaneously removed through a membrane that retains the translational components.

1.8 Concluding Remarks

We have herein described isotopic labeling schemes and experimental methods for characterization of high molecular weight systems by NMR spectroscopy. Current methodology allows resonance assignments and structural characterization of single chain proteins approaching 100 kDa [17, 39] as well as relating structure and dynamics to function for supra-molecular systems as large as the 20S core particle of the ribosome [51, 55, 59]. Key to most labeling techniques is substitution of deuterons for protons at most positions. Protons are only retained at strategic positions, such as at backbone amide positions or at certain methyl groups, to gain as much sensitivity and resolution enhancement as possible in TROSY type experiments. The continuous development of new labeling schemes in combination with improved pulse sequences, more powerful magnets and more sensitive probes ensures that even larger and more complex systems can be studied with NMR spectroscopy in the future.

References

1. Purcell EM, Torrey HC, Pound RV (1946) Resonance absorption of nuclear magnetic moments in a solid. Phys Rev 69:37–38
2. Bloch F, Hansen WW, Packard ME (1946) Nuclear induction. Phys Rev 69:680
3. Proctor WG, Yu FC (1950) The dependence of a nuclear magnetic resonance frequency upon chemical compounds. Phys Rev 77:717
4. Dickinson WC (1950) Dependence of the ^{19}F nuclear resonance position on chemical compound. Phys Rev 77: 736–737
5. Overhauser AW (1953) Polarization of nuclei in materials. Phys Rev 92:411–415
6. Carver TR, Slichter CP (1953) Polarization of nuclear spins in metals. Phys Rev 92:212–213

7. Solomon I (1955) Relaxation processes in a system of two spins. Phys Rev 99:559–566
8. Wagner G, Wüthrich K (1982) Sequential resonance assignments in protein ^{1}H nuclear magnetic resonance spectra: basic pancreatic trypsin inhibitor. J Mol Biol 155:347–366
9. Williamson MP, Havel TF, Wüthrich K (1985) Solution conformation of proteinase inhibitor IIa from bull seminal plasma by ^{1}H nuclear magnetic resonance and distance geometry. J Mol Biol 182:295–315
10. Oschkinat H, Griesinger C, Kraulis PJ, Sorensen OW, Ernst RR, Gronenborn AM, Clore GM (1988) 3-dimensional NMR spectroscopy of a protein in solution. Nature 332:374–376
11. Fesik SW, Zuiderweg ERP (1988) Heteronuclear 3-dimensional NMR spectroscopy – a strategy for the simplification of homonuclear two-dimensional NMR spectra. J Magn Reson 78:588–593
12. Marion D, Driscoll PC, Kay LE, Wingfield PT, Bax A, Gronenborn AM, Clore GM (1989) Overcoming the overlap problem in the assignment of ^{1}H NMR spectra of larger proteins by use of 3-dimensional heteronuclear ^{1}H-^{15}N Hartmann-Hahn multiple quantum coherence and nuclear Overhauser multiple quantum coherence spectroscopy – application to interleukin-1 beta. Biochemistry 28:6150–6156
13. Ikura M, Kay LE, Tschudin R, Bax A (1990) 3-dimensional NOESY-HMQC spectroscopy of a ^{13}C labeled protein. J Magn Reson 86:204–209
14. Kay LE, Ikura M, Tschudin R, Bax A (1990) Three-dimensional triple-resonance NMR spectroscopy of isotopically enriched proteins. J Magn Reson 89:496–514
15. Tjandra N, Bax A (1997) Direct measurement of distances and angles in biomolecules by NMR in a dilute liquid crystalline medium. Science 278:1111–1114
16. Pervushin K, Riek R, Wider G, Wüthrich K (1998) Transverse relaxation-optimized spectroscopy (TROSY) for NMR studies of aromatic spin systems in ^{13}C-labeled proteins. J Am Chem Soc 120:6394–6400
17. Tugarinov V, Muhandiram R, Ayed A, Kay LE (2002) Four-dimensional NMR spectroscopy of a 723-residue protein: chemical shift assignments and secondary structure of malate synthase G. J Am Chem Soc 124:10025–10035
18. Goto NK, Gardner KH, Mueller GA, Willis RC, Kay LE (1999) A robust and cost-effective method for the production of Val, Leu, Ile (δ1) methyl-protonated ^{15}N-, ^{13}C-, ^{2}H-labeled proteins. J Biomol NMR 13:369–374
19. Tugarinov V, Kay LE (2003) Ile, Leu, and Val methyl assignments of the 723-residue malate synthase G using a new labeling strategy and novel NMR methods. J Am Chem Soc 125:13868–13878
20. Pervushin K, Riek R, Wider G, Wüthrich K (1997) Attenuated T_2 relaxation by mutual cancellation of dipole-dipole coupling and chemical shift anisotropy indicates an avenue to NMR structures of very large biological macromolecules in solution. Proc Natl Acad Sci USA 94:12366–12371
21. Cavanagh J, Fairbrother WJ, Palmer AG 3rd, Rance M, Skelton NJ (2007) Protein NMR spectroscopy: principles and practice. Elsevier Academic Press, Burlington
22. Lipari G, Szabo A (1982) Model-free approach to the interpretation of nuclear magnetic-resonance relaxation in macromolecules.1. Theory and range of validity. J Am Chem Soc 104:4546–4559
23. Lipari G, Szabo A (1982) Model-free approach to the interpretation of nuclear magnetic-resonance relaxation in macromolecules. 2. Analysis of experimental results. J Am Chem Soc 104:4559–4570
24. Clore GM, Szabo A, Bax A, Kay LE, Driscoll PC, Gronenborn AM (1990) Deviations from the simple two-parameter model-free approach to the interpretation of ^{15}N nuclear magnetic relaxation of proteins. J Am Chem Soc 112:4989–4991
25. Ollerenshaw JE, Tugarinov V, Kay LE (2003) Methyl TROSY: explanation and experimental verification. Magn Reson Chem 41:843–852
26. Miclet E, Williams DC Jr, Clore GM, Bryce DL, Boisbouvier J, Bax A (2004) Relaxation-optimized NMR spectroscopy of methylene groups in proteins and nucleic acids. J Am Chem Soc 126:10560–10570
27. Tugarinov V, Hwang PM, Ollerenshaw JE, Kay LE (2003) Cross-correlated relaxation enhanced H-1-C-13 NMR spectroscopy of methyl groups in very high molecular weight proteins and protein complexes. J Am Chem Soc 125:10420–10428
28. Czisch M, Boelens R (1998) Sensitivity enhancement in the TROSY experiment. J Magn Reson 134:158–160
29. Weigelt J (1998) Single scan, sensitivity- and gradient-enhanced TROSY for multidimensional NMR experiments. J Am Chem Soc 120:10778–10779
30. Nietlispach D (2005) Suppression of anti-TROSY lines in a sensitivity enhanced gradient selection TROSY scheme. J Biomol NMR 31:161–166
31. Salzmann M, Pervushin K, Wider G, Senn H, Wüthrich K (1998) TROSY in triple-resonance experiments: new perspectives for sequential NMR assignment of large proteins. Proc Natl Acad Sci USA 95:13585–13590
32. Yang DW, Kay LE (1999) Improved ^{1}HN-detected triple resonance TROSY-based experiments. J Biomol NMR 13:3–10
33. Tugarinov V, Sprangers R, Kay LE (2004) Line narrowing in methyl-TROSY using zero-quantum ^{1}H-^{13}C NMR spectroscopy. J Am Chem Soc 126:4921–4925
34. Maniatis T, Sambrook J, Fritsch EF (1982) Molecular cloning: a laboratory manual. Cold Spring Harbor Laboratory Press, Cold Spring Harbor, pp 68–69

35. Middelberg APJ (2002) Preparative protein refolding. Trends Biotechnol 20:437–443
36. Xu YQ, Zheng Y, Fan JS, Yang DW (2006) A new strategy for structure determination of large proteins in solution without deuteration. Nat Methods 3:931–937
37. Bayrhuber M, Riek R (2011) Very simple combination of TROSY, CRINEPT and multiple quantum coherence for signal enhancement in an HN(CO)CA experiment for large proteins. J Magn Reson 209:310–314
38. Schanda P, Van Melckebeke H, Brutscher B (2006) Speeding up three-dimensional protein NMR experiments to a few minutes. J Am Chem Soc 128:9042–9043
39. Tugarinov V, Choy WY, Orekhov VY, Kay LE (2005) Solution NMR-derived global fold of a monomeric 82-kDa enzyme. Proc Natl Acad Sci USA 102:622–627
40. Fiaux J, Bertelsen EB, Horwich AL, Wüthrich K (2002) NMR analysis of a 900 K GroEL GroES complex. Nature 418:207–211
41. McIntosh LP, Dahlquist FW (1990) Biosynthetic incorporation of ^{15}N and ^{13}C for assignment and interpretation of nuclear magnetic resonance spectra of proteins. Q Rev Biophys 23:1–38
42. Fiaux J, Bertelsen EB, Horwich AL, Wüthrich K (2004) Uniform and residue-specific 15N-labeling of proteins on a highly deuterated background. J Biomol NMR 29:289–297
43. Baldwin RL (2002) Making a network of hydrophobic clusters. Science 295:1657–1658
44. Bogan AA, Thorn KS (1998) Anatomy of hot spots in protein interfaces. J Mol Biol 280:1–9
45. Rodriguez-Mias RA, Pellecchia M (2003) Use of selective Trp side chain labeling to characterize protein-protein and protein-ligand interactions by NMR spectroscopy. J Am Chem Soc 125:2892–2893
46. Löhr F, Katsemi V, Betz M, Hartleib J, Rüterjans H (2002) Sequence-specific assignment of histidine and tryptophan ring ^{1}H, ^{13}C and ^{15}N resonances in ^{13}C/^{15}N- and ^{2}H/^{13}C/^{15}N-labelled proteins. J Biomol NMR 22:153–164
47. Takahashi H, Nakanishi T, Kami K, Arata Y, Shimada I (2000) A novel NMR method for determining the interfaces of large protein-protein complexes. Nat Struct Biol 7:220–223
48. Meissner A, Sorensen OW (1999) Optimization of three-dimensional TROSY-type HCCH NMR correlation of aromatic ^{1}H-^{13}C groups in proteins. J Magn Reson 139:447–450
49. Kainosho M, Torizawa T, Iwashita Y, Terauchi T, Ono AM, Guntert P (2006) Optimal isotope labelling for NMR protein structure determinations. Nature 440:52–57
50. Ruschak AM, Kay LE (2010) Methyl groups as probes of supra-molecular structure, dynamics and function. J Biomol NMR 46:75–87
51. Sprangers R, Kay LE (2007) Quantitative dynamics and binding studies of the 20S proteasome by NMR. Nature 445:618–622
52. Gelis I, Bonvin A, Keramisanou D, Koukaki M, Gouridis G, Karamanou S, Economou A, Kalodimos CG (2007) Structural basis for signal-sequence recognition by the translocase motor SecA as determined by NMR. Cell 131: 756–769
53. Religa TL, Kay LE (2010) Optimal methyl labeling for studies of supra-molecular systems. J Biomol NMR 47: 163–169
54. Ruschak AM, Velyvis A, Kay LE (2010) A simple strategy for ^{13}C,^{1}H labeling at the Ile-gamma 2 methyl position in highly deuterated proteins. J Biomol NMR 48:129–135
55. Religa TL, Sprangers R, Kay LE (2010) Dynamic regulation of archaeal proteasome gate opening as studied by TROSY NMR. Science 328:98–102
56. Kigawa T, Muto Y, Yokoyama S (1995) Cell-free synthesis and amino acid-selective stable isotope labeling of proteins for NMR analysis. J Biomol NMR 6:129–134
57. Etezady-Esfarjani T, Hiller S, Villalba C, Wüthrich K (2007) Cell-free protein synthesis of perdeuterated proteins for NMR studies. J Biomol NMR 39:229–238
58. Goerke AR, Swartz JR (2009) High-level cell-free synthesis yields of proteins containing site-specific non-natural amino acids. Biotechnol Bioeng 102:400–416
59. Ruschak AM, Religa TL, Breuer S, Witt S, Kay LE (2010) The proteasome antechamber maintains substrates in an unfolded state. Nature 467:868–871

Chapter 2
Segmental Labeling to Study Multidomain Proteins

Jing Xue, David S. Burz, and Alexander Shekhtman

Abstract This chapter contains a review of methodologies and recent applications of segmental labeling for NMR structural studies of proteins and protein complexes. Segmental labeling is used to specifically label a segment of protein structure with NMR active nuclei, thus reducing NMR spectral complexity and greatly facilitating structural NMR studies of large multi-domain proteins. It can also be used to introduce a synthetic fragment into a protein structure to study post-translationally modified proteins. Detailed protocols describing segmental labeling techniques are also included.

2.1 Introduction

Multi-domain protein complexes are the eukaryotic solution to biogenetic pathways found in prokaryotes. Each domain is a fully functional unit whose activity is necessary for the processivity of the reaction pathway. Interactions between domains are a common means of regulating the overall activity of such a complex. Therefore it is useful to be able to study a single domain within a complex to understand regulation of biological activity.

NMR analyses of protein structure have been traditionally limited by the molecular weight or overall size of the molecule being examined; more residues contribute more complexity to the spectrum. To reduce spectral complexity, several methodological approaches have been used. Among them, partial labeling of the molecule in question by uniformly labeling the backbone or by selectively labeling specific amino acids. In both of these cases the information gleaned represents a subset of all structural information available. Segmental labeling is used to isotopically label one domain or segment within a multi-domain complex and offers the dual advantage of providing complete structural information while minimizing spectral complexity.

Segmental labeling of separate domains within a large multidomain protein became possible after extending methodologies of synthetic protein chemistry to the area of recombinant protein production. Since the 1980s, people have sought to increase the size of high purity, synthetic peptides or proteins available for structural, functional, or physiological study. Solid phase peptide synthesis (SPPS) combined with size-exclusion or ion-exchange chromatography was used to prepare large quantities of high purity peptides [1]. This synthetic technique led to native chemical ligation (NCL)

J. Xue • D.S. Burz • A. Shekhtman (✉)
Department of Chemistry, State University of New York at Albany,
Albany, NY 12222, USA
e-mail: ashekhtman@albany.edu

H.S. Atreya (ed.), *Isotope Labeling in Biomolecular NMR*, Advances in Experimental Medicine and Biology 992,   17
DOI 10.1007/978-94-007-4954-2_2, © Springer Science+Business Media Dordrecht 2012

[2, 3] in which two synthetic peptides are ligated *in vitro* through a C-terminal cysteine on one peptide and an N-terminal thioester group on the other. NCL evolved to expressed protein ligation (EPL) [4] in which cloned, recombinant peptides and proteins, overexpressed in *E. coli,* are utilized to form a native peptide bond, resulting in a macromolecule that is functionally similar to the natural protein.

Segmental isotope labeling combined with modern NMR techniques can identify whether protein domains interact with one another and helps to define the precise interaction interface and orientation between them, or whether they do not interact and can therefore be structurally and biologically characterized independently. It has been used to study enzymatic reactions mechanisms and to structurally characterize proteins that are difficult to obtain in chemically pure form, such as glycosylated proteins.

2.2 Methods and Techniques

2.2.1 *Stepwise Solid-Phase Peptide Synthesis (SPPS)*

Stepwise solid-phase peptide synthesis (SPPS) was established in the 1980s [1, 5] as the most efficient way to prepare peptide fragments. The peptide is immobilized (covalently linked) to a stationary phase resin via a linker. Single amino acids are incorporated one at a time from the C-terminus to the N-terminus. The peptide is constructed through repeated cycles of deprotecting, washing, coupling and washing. The final product is cleaved from resin using HF or trifluoroacetic acid. This process allows chemical modifications to the peptide backbone and the incorporation of unnatural or isotopically-labeled amino acids.

The effective limit of SPPS is 60–70 amino acids, beyond this length the overall yield of product makes this technique impractical. To prevent unintended reactions, the N-terminal amine is protected. SPPS is defined by the nature of the chemical group used to protect the α-amino group during stepwise synthesis. The protecting groups typically employed are di-tert-butyl-dicarbonate (Boc) or 9-fluorenylmethyloxycarbonyl (Fmoc).

Boc is used to reduce aggregation during synthesis and during incorporation of base-sensitive peptide analogs (utilizing non-natural amino acids) [6]. The Boc group is removed using trifluoroacetic acid (TFA), which results in a positively-charged amino group that must be neutralized concomitant with coupling to the next activated amino acid. During the cleavage reaction cresol is added to scavenge t-butyl cations and prevent the formation of undesired products. Exposure to HF is harsh and may result in the degradation of the nascent peptides; this led to the use of a less harsh, base-labile reagent (Fmoc).

Fmoc uses piperidine in DMF to remove the protecting group and results in a neutral exposed amino group [7]. TFA is used to cleave the peptide from the resin. The lack of residual charge may lead to increased aggregation of the peptide. Nonetheless, the use of Fmoc is generally preferred over Boc because of the ease of cleavage, despite the increase in cost of synthesis. Finally, Boc SPPS generates fluoride salts, which are highly soluble, whereas Fmoc SPPS yields a TFA salt, which is less soluble.

SPPS had been optimized to yield high purity peptides by using reverse phase HPLC or ion exchange chromatography, for peptides less than 50 amino acids in length. Due to the limitations of SPPS, it was necessary to develop a better technique that not only produces a high yield and purity of protein, but abundantly synthesizes much larger proteins.

2.2.2 *Native Chemical Ligation (NCL)*

SPPS was combined with native chemical ligation (NCL) [2] to produce longer polypeptides. NCL is based on the reaction between two unprotected synthetic peptides, one of which contains a C-terminal

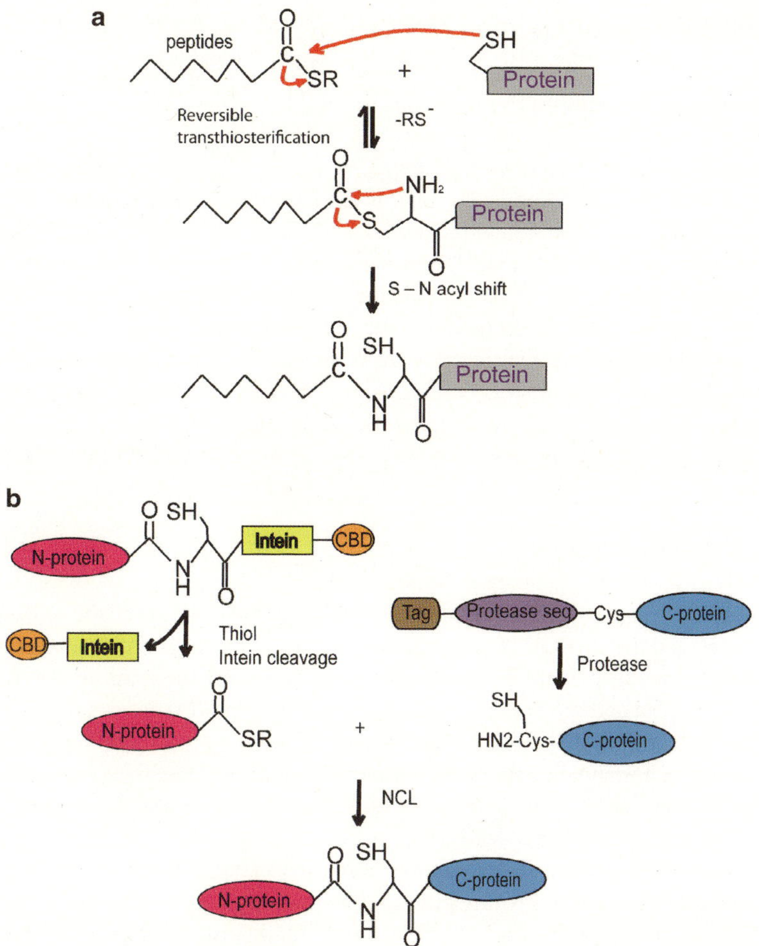

Fig. 2.1 (**a**) Peptide-protein ligation (semisynthesis). An N-terminal synthetic peptide containing a C-terminal thioester reacts with a C-terminal recombinant protein through an N-terminal cysteine via reversible trans-thioesterification and S-N acyl shift to form a native peptide bond. (**b**) Protein-protein ligation. A C-terminal thioester is created when intein-CBD is cleaved from the N-terminal protein fragment and an N-terminal cysteine is formed when protease cleaves the C-terminal fragment at the protease site. A native peptide bond is then formed via native chemical ligation. CBD is a chitin binding domain used for ease of purification

thioester (α-thioester) and the other an N-terminal cysteine residue (α-cysteine), to form a native peptide bond. The essence of NCL is the formation of an intermediate thioester-linked product. The thioester-linked product undergoes spontaneous rearrangement, via an intramolecular nucleophilic attack, to reach the final desired amide-linked product (Fig. 2.1). The result is a native polypeptide chain that functions *in vitro* or *in vivo*. NCL is widely used in conjunction with NMR [7, 8] and mass spectroscopy to observe single domains within a full-length protein by ligating together isotope-labeled and unlabeled peptides. In NMR, the technique alleviates the difficulty of obtaining clear and non-overlapping spectra by reducing spectral complexity [8].

NCL is performed in aqueous solution at neutral pH under denaturing conditions. This highly chemoselective reaction depends strongly on the amino acid present at the C-terminal thioester, with Gly increasing the reaction rate and β-branched amino acids, like Leu or Ile, reducing the reaction rate and producing lower yields [9]. Interestingly, cysteine and histidine, which are among the least sterically hindered amino acids, react at the same speed as glycine, while valine, isoleucine and proline, which are more sterically hindered, react less favorably [10].

2.2.3 Expressed Protein Ligation (EPL)

2.2.3.1 Protein-Peptide Ligation (Semisynthesis)

Because of the length limitation of synthetic peptides, chemical ligation of short synthetic peptides with recombinant proteins expressed in *E.coli* was developed to extend the size of ligated proteins from less than one hundred to several hundred amino acids. This method, called semisynthesis [11, 12], also allows unnatural amino acids [13], fluorescent probes [13], or posttranslational modifications [14] to be introduced into any size protein and can be used to attach a synthetic peptide to either the C-terminus or N-terminus of the recombinant expressed protein (Fig. 2.1a). To introduce a synthetic peptide at the N-terminus, the peptide is synthesized with a C-terminal α-thioester and the recombinant protein must have an N-terminal Cys. To position the peptide at the C-terminus, the protein is expressed as a fusion with an engineered intein. Self-cleavage of the intein results in a C-terminal α-thioester that can be used for the ligation with a synthetic peptide possessing an N-terminal Cys.

2.2.3.2 Protein-Protein Ligation

Protein-protein ligation requires overexpression of protein fragments in which the N-terminal fragment is fused to an intein and the C-terminal fragment has a Cys at the N-terminus. The engineered intein catalyzes its own excision to yield N-terminal protein fragments containing reactive termini for optimum ligation. There are three broad categories of inteins: (1) Maxi-inteins that contain a homing endonuclease domain within the core sequence, which is not required for splicing activity [15]; (2) Mini-inteins that lack a homing endonuclease domain [15–17]; (3) Trans-splicing inteins that have no peptide linkage between the N- and C- terminal halves of the intein resulting in two fragments that must come together for splicing activity to occur.

The basic protein splicing mechanism involves three steps (Fig. 2.1b): First, an $N \rightarrow S$ (or $N \rightarrow O$) acyl shift in which the N-extein is transferred to the –SH or –OH group of a Cys or Ser at the N-terminus of the intein. Second, the entire N-extein is transferred, via thioesterfication, to a second, conserved Cys/Ser/Thr at the +1 position within C-extein. Third, the resulting branched intermediate undergoes cyclization with a conserved asparagine at the C-terminus of the intein, and the intein is excised as a C-terminal succinimide derivative. Spontaneous chemical rearrangement leads to the formation of an amide bond between the two exteins in an intein-independent manner.

Intein-mediated ligation gene products are derived from *Synechocystis* sp. *dnaB* (*Ssp* DnaB) [18], *Mycobacterium xenopi gyrA* (*Mxe* GyrA) [19], and from *Methanobacterium thermoautotrophicum rir1* (*Mth* RIR1) [20]. The *Ssp* DnaB intein has been engineered to undergo pH or temperature dependent cleavage at the C-terminus to generate a fragment containing the desired N-terminal amino acid residue [21]. *Mxe* GyrA and *Mth* RIR1 inteins have been modified to undergo thio-induced cleavage at the N-termini to yield a C-terminal α-thioester on the resulting fragment [22].

To generate a C-terminal protein fragment for EPL, the protein is designed to contain a specific protease cleavage sequence that, after cleavage, will leave an N-terminal Cys. So far, three proteases have proved to be useful for this purpose: Factor Xa, which cleaves immediately after its recognition sequence, Ile-Glu-Gly-Arg; TeV protease, which cleaves within its recognition sequence, Glu-Asn-Leu-Tyr-Phe-Gln-Cys, between Gln and Cys; and thrombin, which cleaves within its recognition sequence, Leu-Val-Pro-Arg-Cys-Ser, between Arg and Cys.

It is important to choose less sterically hindered amino acids at the ligation site to improve the ligation reaction. It is also important to design a functional assay for the ligation product to ensure that the modifications introduced to facilitate EPL do not influence the structure and biological function of the protein [23]. For example, to construct SH32 protein [24], which consists of the Src homology type 3 (SH3) and type 2 (SH2) domains, the SH3 domain contained an α-thioester at the C-terminus, and the

SH2 domain contained a Cys at the N-terminus. The location of the ligation site was chosen to be within the short linker region between the SH3 and SH2 domains and involved two mutations, N120G and S121C. The S-C mutation is required to facilitate the ligation reaction, whereas the N-G mutation is expected to improve the kinetics of ligation reaction. As a result, the NMR spectra of the ligation product, SH32, and recombinant expressed SH32 were quite similar, which means that the ligation reaction did not affect protein folding, even though a few expected chemical shift changes are observed in the amino acids located spatially close to the ligation site.

2.3 Applications

2.3.1 Conformational Changes

The 400-kDa bacterial core RNA polymerase (RNAP) depends on the binding of σ factors for promoter recognition and specific transcription initiation in RNA polymerization. σ^{70} is responsible for the bulk of transcription during exponential growth. Structural studies confirm [25, 26] that the −35 elements of the binding site for RNAP are recognized by amino acid residues of σ^{70} region 4.2. The latent DNA binding activity of σ^{70} is inhibited by N-terminal region 1.1, which directly masks the DNA binding determinants of region 4.2 [27, 28]. This inhibition is relieved by a conformational change when σ^{70} factor binds to the RNAP core. The autoinhibition of σ^{70} was difficult to resolve by using X-ray crystallography due to the flexibility of region 1.1 [29]. Segmental labeling of region 4.2 and isotope edited NMR spectroscopy was used to observe interactions between regions 1.1 and 4.2 required for σ-factor autoinhibition. A thermostable variant of σ^{70}, σ_A from *Thermatoga maritima* [30] was used to facilitate NMR studies. Two constructs were created, σ-factor with [U-^{15}N] region 4.2 and σ-factor with a deletion in region 1.1 and [U-^{15}N] region 4.2.

Segmental isotopic labeling was accomplished by using expressed protein ligation (EPL) [31]. To facilitate the ligation reaction, a Cys was inserted between Gly348 and Lys349. Two different N-terminal fragments, full-length σ_A* (1–348) and Δ1.1-σ_A* (137–348) which lacks region 1.1, were fused with an *Mxe* GyrA intein-CBD fragment. In this construct an α-thioester group is released by thiolysis to ligate with the C-terminal fragment, σA factor region 4.2, [U-^{2}H, ^{13}C, ^{15}N]-CG-σ^A (349–399).

Experimentally, segmentally labeled σ^{A*} and Δ1.1-σ^{A*} are both active *in vitro*. The σ^{A*} and Δ1.1-σ^{A*} constructs are able to bind the −35 promoter DNA with similar affinities in low salt buffer. In high salt buffer, the affinity of σ^{A*} for the −35 promoter DNA is reduced by more than two orders of magnitude as compared to Δ1.1-σ^{A*}. This result proves that deletion of region 1.1 allows the truncated σ^A factor, Δ1.1-σ^{A*}, to make tight and specific interactions with the −35 promoter DNA, confirming that region 1.1 is involved in the previously observed autoinhibition of σ^A.

^{1}H{^{15}N}HSQC-TROSY and ^{1}H{^{13}C}HSQC experiments were performed to present the spectra of region 4.2 in the context of Δ1.1-σ^{A*} and in σ^{A*} (Fig. 2.2). The well-dispersed signals indicate that region 4.2 assumes a defined tertiary fold in Δ1.1-σ^{A*} and in σ^{A*}. In contrast, the isolated region 4.2 lacks a defined fold in solution and many peaks are overlapping. Further experiments demonstrated that adding T4Asia, a known ligand of *E. coli* σ70 region 4.2, results in significant conformational changes in the NMR spectrum of Δ1.1-σ^{A*} (Fig. 2.3), implying that the presence of region 1.1 inhibits binding of T4Asia to region 4.2. There are only minor differences between the NMR spectra of region 4.2 in the context of Δ1.1-σ^{A*} and in σ^{A*}. It was concluded that region 1.1 indirectly inhibits σA binding to the promoter DNA, possibly by electrostatic interaction [32].

To further understand the nature of autoinhibition, region 1.1 was segmentally labeled and characterized by NMR spectroscopy in the context of full-length σ^A [33]. Region 1.1 (residues 25–120) was expressed in and purified from *E.coli* [32]. The standard set of double and triple resonance 2D and 3D experiments and restraints generated from a series of multi-dimensional NMR experiments were

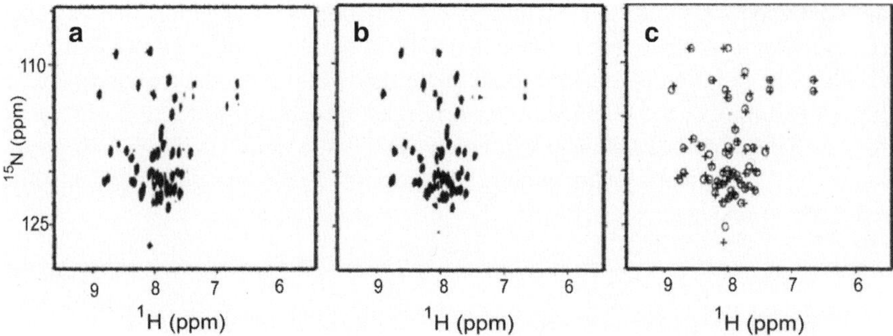

Fig. 2.2 Effect of context on the solution structure of σ^A region 4.2. (**a**) $^1H\{^{15}N\}$ HSQC-TROSY spectrum of [U-^{15}N] region 4.2 in context of σ^{A*}. (**b**) $^1H\{^{15}N\}$ HSQC-TROSY spectrum of [U-^{15}N] region 4.2 in context of $\Delta1.1$-σ^{A*}. (**c**) The overlay of *panel* (**a**) and (**b**). *Panel* (**a**) peaks are represented by *circles*, and *panel* (**b**) peaks are shown as *crosses* (This figure is reproduced from Camarero et al. [32])

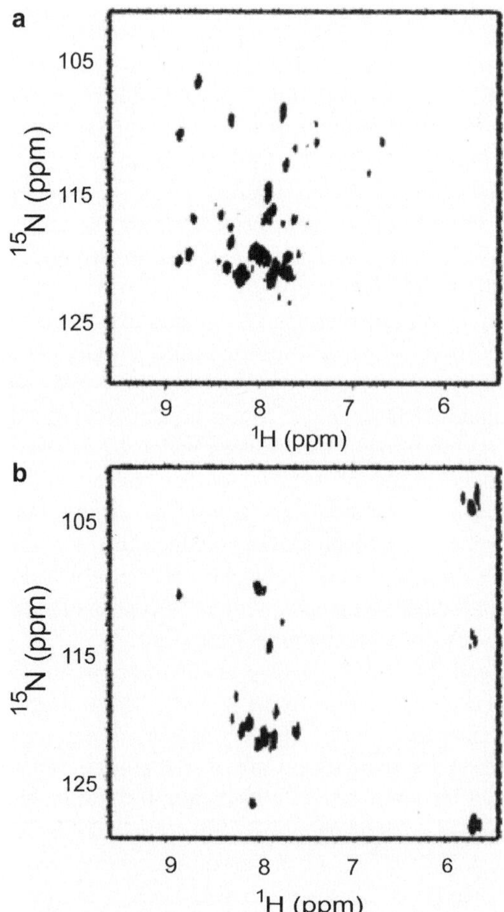

Fig. 2.3 Binding of T4 AsiA and promoter DNA to $\sigma^{1.1}$-σ^{A*}. $^1H\{^{15}N\}$HSQC-TROSY spectra of [U-^{15}N] region 4.2 in context of $\Delta1.1$-σ^{A*} with 1.2 M equivalents of purified AsiA (**a**) and promoter DNA (**b**) (This figure is reproduced from Camarero et al. [32])

performed for the chemical shifts assignments and the solution structure determination of region 1.1 [27, 28, 32]. The ^{1}H{^{15}N]-HSQC NMR spectrum of region 1.1 in the context of full-length σ^A was sufficiently broadened to allow the interaction surface of σ^A to be mapped. These results, in combination with cross-linking experiments [34], clearly indicate that region 1.1 interacts with regions 3 and 4.1 of the full length σ^A.

Importantly, this work shows that segmental isotopic labeling does not interfere with the folding of full-length proteins and allows us to observe domain-domain interactions without the absolute requirement for assignments. It can be used to observe the effects of ligands on a segmentally labeled domain in the context of full-length protein by using isotope edited NMR experiments.

2.3.2 *Interdomain Interactions*

Advances in the study of multidomain proteins by using solution NMR have been made possible in recent years by the development of new segmental isotope labeling methods that identify and map interdomain interactions and allow structural characterization in the absence and presence of such interactions. Skrisovska and Allain [35] developed a technique to segmentally isotope label multidomain proteins and to provide a high yield recovery of ligated product. The protocol employs an on-column expressed protein ligation (EPL) step and permits ligation of insoluble, non-interacting and improperly folded domains. The technique was successfully demonstrated by using two multidomain proteins, heterogenous nuclear ribonucleoprotein L (hnRNP L)and Npl3p, each of which contain RNA recognition motifs (RRMs).

hnRNP L is an abundant RNA-binding protein involved in alternative splicing and mRNA degradation [36]. hnRNP L contains four RRMs and evidence suggested that RRM3 and RRM4, which are connected by a long linker region (417–461), may interact with one another. Two constructs were prepared, the first, RRM3-Mxe GyrA-CBD, consists of RRM3 fused at the C-terminus to the *Mxe* GyrA intein and a chitin binding domain (CBD), the second, CBD-Ssp DnaB-RRM4, consists of RRM4 fused at the N-terminus to CBD and the *Ssp* DnaB intein. To ligate the constructs following cleavage of the intein-CBD moieties requires a cysteine residue at the N-terminus of RRM4 that can react with a thioester at the C-terminus of RRM3. The ligation site lies in the linker region at position 452. Since the linker does not contain any cysteine residues, serine 452 was mutated to cysteine (S452C) at the N-terminus of RRM4.

Each construct was overexpressed in *E. coli* under non-labeling and [U-^{15}N] or [U-^{15}N, ^{13}C] labeling growth conditions and resulted in the formation of inclusion bodies. Inclusion bodies from labeling and non-labeling overexpression were combined, solubilized in 8 M urea, refolded by fast dilution and bound to a chitin column. Intein cleavage and subsequent ligation was induced by the addition of sodium 2-mercaptoethanesulfonate (MESNA); the reaction was allowed to proceed for 24 h at 37°C on the column before eluting the final product, hnRNP RRM34. Ligation efficiency was ~90%, hnRNP L RRM34 ran as a single band on SDS-PAGE and its molecular weight was confirmed by mass spectrometry. The cleavage reaction, estimated to be 80–90% complete for *Mxe* GyrA [19] and 60–70% for *Ssp* DnaB [18], proved to be the limiting step in preventing an even higher ligation efficiency.

^{1}H{^{15}N}-HSQC and ^{1}H{^{13}C}-HSQC spectra were collected for hnRNP L RRM34 containing [U-^{13}C, ^{15}N] RRM3 and [U-^{15}N] RRM4 and compared to ^{1}H{^{15}N}-HSQC and ^{1}H{^{13}C}-HSQC spectra collected for [U-^{13}C, ^{15}N]- hnRNP L RRM34 to confirm that the ligated product was properly folded. 2D homonuclear and 3D 13C-edited NOESY half-filter spectra acquired using segmentally labeled hnRNP L RRM34 identified 101 NOE crosspeaks between the two domains, confirming that the domains interact. The interaction interface was defined and the structure of the ligated structure is currently being characterized.

Npl3p (nuclear protein localization) is a yeast RNA binding protein. It is a member of the serine/arginine-rich (SR) protein family that selects and regulates splice sites in eukaryotic mRNA. Npl3p contains two RRMs (RRM1 and RRM2) connected by a short linker and a C-terminal glycine/arginine-rich domain. Two constructs were prepared: RRM1-*Mxe* GyrA-CBD and CBD-*Ssp* DnaB-RRM2 to yield Npl3p RRM12. A cysteine, introduced into the short linker by mutating serine 193 (S193C) was used as the N-terminal residue of RRM2.

Each construct was overexpressed in *E. coli* under non-labeling and [U-^{15}N] or [U-^{15}N, ^{13}C] labeling growth conditions and was soluble. The on-column cleavage efficiency of both proteins was 80–90%, but the ligation efficiency was only ~10% as estimated from SDS-PAGE. The chitin column elution containing ligated and non-ligated product was concentrated to ~1 mM and further incubated at 42°C for 24 h; this additional step improved the ligation efficiency to 80–90%. RRM12 was separated from the non-ligated species by using gel filtration chromatography. To show that ligation is independent of the folded state of the proteins, segmentally labeled protein was successfully prepared by concentrating and ligating protein domains in the presence of 6 M guanidinium chloride. Ligation using a naturally occurring cysteine (C211) located within RRM2 left the domain unstructured but capable of undergoing ligation, albeit at a lower level than obtained when the domains were properly folded.

^{1}H{^{15}N}-HSQC spectra acquired for each domain indicates that both fold properly after cleavage and during the ligation reaction. There were no significant differences between the ^{1}H{^{15}N}-HSQC spectra of segmentally and uniformly labeled Npl3p RRM12 indicating that the S193C mutation has no effect on the protein fold of ligated RRM12. An overlay of the spectra of each domain is very similar to the ^{1}H{^{15}N}-HSQC spectrum of Npl3p RRM12, suggesting that there is no interaction between the two domains. No NOE crosspeaks were observed between RRM1 and RRM2 in 3D ^{13}C-edited NOESY half-filtered data collected using a segmentally labeled sample in which only one of the RRMs was ^{13}C-labeled, further indicating that RRM1 and RRM2 do not interact. Minor changes in chemical shifts indicate that there may be small changes in the conformation and dynamics of the domains or weak interactions for which NOEs were not detected. Because RRM1 and RRM2 do not interact, the structure of each domain was determined individually, greatly simplifying the analysis.

In sum, the on-column ligation technique is very robust, interacting domains ligate more efficiently than non-interacting domains, but this is not necessary for successful ligation. Low level ligation efficiency is improved by concentrating the eluted fragments, thereby increasing the concentration of protein termini available for the ligation reaction. Successful ligation of two protein fragments is largely independent of their solubility and folding state. This technique will be broadly applicable in future solution NMR studies on large, multidomain proteins.

2.3.3 Glycoproteins

Glycosylation is a common post translational modification (PTM) that facilitates a variety of biological processes involving primarily inter- and intra-cellular communication. Understanding the molecular basis for these processes is limited by the dearth of structural information available for glycoproteins. The use of NMR to acquire structural information is hampered by the inability to generate sufficient quantities of uniformly glycosylated protein and the spectral complexity arising from overlapping signals attributed to carbohydrate and protein moieties. Segmental labeling can help overcome these problems.

Slynko et al. [37] used *in vitro* glycosylation to attach an unlabeled glycan to [U-^{13}C, ^{15}N]-labeled protein and NMR spectroscopy to deduce the structures of the N-linked oligosaccharide PTM and the corresponding modified protein (Fig. 2.4). The protein, AcrA$^{61-210ΔΔ}$ from *Campylobacter jejuni*, is a drug efflux pump protein with broad substrate specificity that is easily glycosylated *in vivo* and *in vitro*. The *in vitro* glycosylation method requires three components, each isolated from separate, dedicated strains of *E. coli*: Oligosaccharyltransferase, PglB, which is purified from solubilized membrane fractions [38]; a lipid-linked oligosaccharide (LLO) prepared from cells containing an inactive

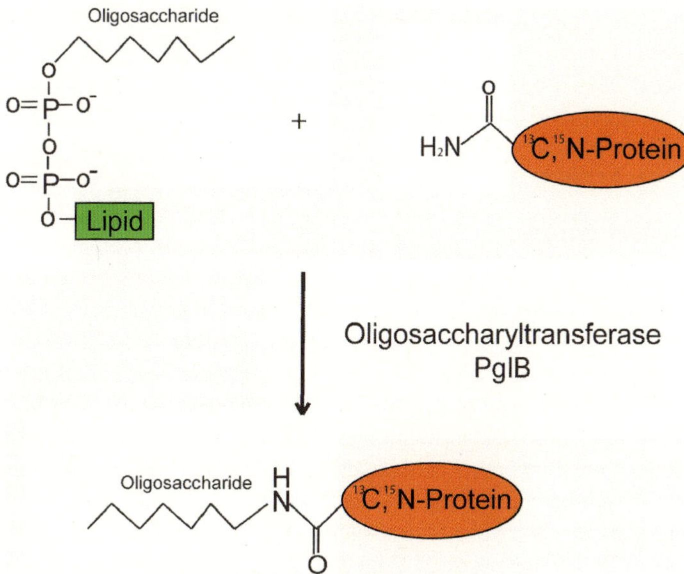

Fig. 2.4 *In vitro* synthesis of glycosylated protein. Oligosaccharyltransferase PglB mediates the reaction between an oligosaccharide attached to a lipid and an Asn of a labeled protein

pgl ORF, and the [U-^{13}C, ^{15}N] target protein, AcrA$^{61-210\Delta\Delta}$. The small (13 kDa) protein contains a lipoyl domain and an extended loop that includes the glycosylation site. In addition, the model protein contains two deletions (ΔF97-N117 and ΔF146-D166) and two point mutations (K96Q and K131Q). Purified AcrA$^{61-210\Delta\Delta}$ was ~90% glycosylated *in vitro* and further purified by Ni-NTA chromatography before using in NMR experiments. The protein yields good dispersion and line widths in NMR.

By using [U-^{13}C, ^{15}N] protein, chemical shift overlap between the protein and the carbohydrate is resolved using specific filtered/edited 2D NOESY-type experiments that suppress resonances from the labeled protein and allow the non-labeled glycan to be assigned. In these experiments NOE transfer becomes very efficient due to increased overall tumbling rate. The use of a very high magnetic field (900 MHz) also maximizes the sensitivity and resolution of the NOESY spectra, enabling long-range distance restraints to be extended to 6 Å. Amino acid type-selective ^{1}H-^{15}N correlation and standard triple resonance experiments were used to identify N42 as the glycosylation site [39]. The observed ^{1}H and ^{15}N resonances are consistent with those previously observed for glycosylated asparagines [40]. The segmentally labeled complex contains a lipoyl domain fold with four N-terminal and four C-terminal intertwined β-strands forming a β-sandwich. The glycosylation site is contained within a flexible loop that lies between the two half-lipoyl domains motifs. Flexibility was assessed by acquiring steady-state heteronuclear ^{1}H{^{15}N} NOE spectra. The NH of the glycosylated N42 side chain is less flexible than the backbone of the surrounding amino acid residues most likely due to the reduced degree of conformational freedom imparted by the attached glycan.

By acquiring 2D ^{13}C-filtered-filtered NOESY and natural abundance ^{13}C-HSQC spectra, assignment of all residues in the attached glycan was accomplished. 2D ^{13}C-filtered-filtered NOESY suppresses all protein resonances, resulting in a NOESY spectrum of the unlabeled glycan. 2D ^{15}N-filtered-filtered NOESY, containing NOEs to the carbohydrate amides while suppressing protein amide signals, confirmed the assignment of the linked polysaccharide structure. In total, 125 interproton distance constraints within the attached glycan were obtained. Half of which are inter-residue NOEs (~11 per glycosidic linkage). The final structure had an RMSD of 0.56±0.10 Å. The glycan structure was calculated exclusively by using torsion angle dynamics from experimentally derived upper limit NOE distance restraints.

Segmental labeling of a glycosylated target protein is a new strategy for studying glycoproteins by NMR [41]. Problems associated with the spectral overlap between protein and oligosacharide resonances are avoided. This technique also generates chemically pure glycosylated protein, which is difficult to achieve by enzymatic glycosylation.

2.3.4 Reaction Mechanisms

To gain insight into the mechanistic details of autocatalysis, Romanelli et al. (2004) [12] used a combination of NMR spectroscopy and segmental isotopic labeling to study the structure of an active protein splicing precursor. In the first step of the splicing reaction an $N \rightarrow S$ (or $N \rightarrow O$) acyl shift occurs in which the N-extein is transferred to the –SH or –OH group of a Cys or Ser at the N-terminus of the intein. In the ground state destabilization model, it is believed that the scissile (-1) peptide bond is distorted from its most favorable conformation rendering it susceptible to nucleophilic attack. To examine the validity of this model in the autocatalytic splicing reaction mechanism, semisynthesis was employed to prepare an N-extein fusion of the *Mxe* GyrA intein in which the protein was uniformly labeled with ^{15}N, and the peptide ^{13}C labeled at the (-1) phenylalanine residue. This ligation scheme allowed the scissile (-1) peptide bond to be doubly labeled and examined in detail.

The peptide, H-AAMR[$^{13}C'$]F-SR, was synthesized by Boc-N^a-SPPS using the *in situ* neutralization/2-(1H-benzotriazole-1-yl)-1,1,3,3-tetramethyluronium hexafluorophosphate (HBTU) activation protocol [6] on 3-mercaptopropionamide-4-methylbenzhydrylamine (MNHA) resin. The peptide was cleaved from the resin using anhydrous HF contining 4% (v/v) *p*-cresol, purified by reverse phase chromatography (RP-HPLC) and characterized by using electrospray mass spectrometry (ESMS). DNA coding for GyrA protein was amplified by PCR off plasmid pTXB1 (NEBiolabs), which encodes a GyrA intein with a single mutation (N198A) that prevents cleavage of the intein-C-extein bond, but does not affect N-terminal splicing [42]. The PCR product was cloned into pTrcHisA (Invitrogen) to generate pTrcHis-Xa-GyrA$_{WT}$, in which a factor Xa cleavage site is inserted between the His-tag and the GyrA coding region. pTrcHis-Xa-GyrA$_{WT}$ was used as a template to prepare a host of mutant GyrA proteins, including GyrA(H75A), which contains a H75A mutation that abolishes N-terminal splicing activity. GyrA$_{WT}$ was overexpressed from plasmid pTrcHis-SH3-GyrA$_{WT}$, which contains a Src homology domain 3 between the His-tag and the intein.

To uniformly label the inteins, pTrcHis-Xa-GyrA(H75A) and pTrcHis-SH3-GyrA$_{WT}$ were transformed into *E. coli* strain BL21(DE3), grown in minimal (M9) medium containing 0.2% (w/v) ^{15}N-NH$_4$Cl as the sole nitrogen source. The cells were induced with IPTG and the fusion proteins purified by using Ni^{2+}-NTA affinity chromatography. To prepare [U-^{15}N]GyrA, His-SH3-GyrA was incubated overnight in DTT cleavage buffer (50 mM Tris, pH 8, 100 mM DTT). To prepare [U-^{15}N]GyrA(H75A), His-Xa-GyrA(H75A) was incubated in proteolysis buffer (50 mM Tris, pH 8, 100 mM NaCl, 1 mM CaCl$_2$) in the presence of factor Xa for 10 h at room temperature. Both proteins were further purified by using RP-HPLC and characterized by ESMS.

Two segmentally labeled constructs, AAMR[$^{13}C'$]F-[U-^{15}N]GyrA and AAMR[$^{13}C'$]F-[U-^{15}N]GyrA(H75A), were prepared by chemically ligating H-AAMR[$^{13}C'$]F-SR with either [U-^{15}N]GyrA or [U-^{15}N]GyrA(H75A). Reactions were carried out using ~1 mM protein and ~10 mM peptide in 0.1 M NaP$_i$ buffer, pH 8, containing 6 M guanidinium chloride, 3% MESNA and 2% ethanethiol, for 5 h at room tempertaure. The ligation products were purified by using RP-HPLC and characterized by using ESMS. Purified protein was re-folded by stepwise dialysis into NMR buffer (20 mM KP$_i$, pH 6.6, 100 mM NaCl) for spectroscopy.

The uniformly labeled N-exteins have similar ^{1}H{^{15}N}HSQC spectra, only a small number of signals are unique to one construct or other. The spectral similarity factor [32] between the constructs is 16 Hz, indicating that the H75A mutation does not result in a major structural rearrangements.

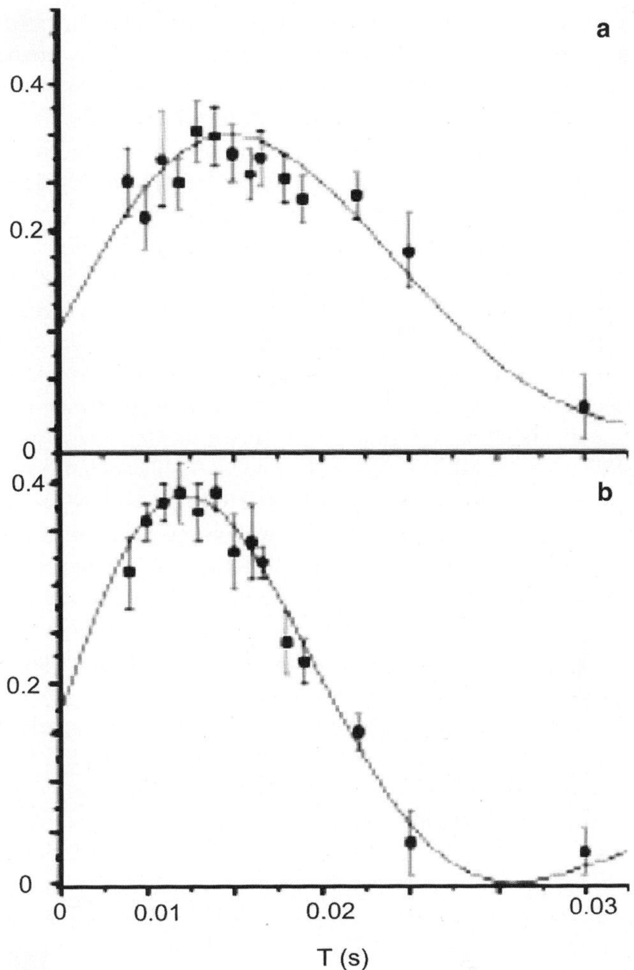

Fig. 2.5 Determination of the $^1J_{NC'}$ coupling constant for the scissile (−1)amide in AAMR[^{13}C′]F-[U-^{15}N]GyrA and AAMR[^{13}C′]F-[U-^{15}N]GyrA(H75A), respectively. The time evolution of the normalized peak intensities extracted from a series of H(N)CO experiments was nonlinearly fit to Eq. 2.1 to give the single-bond coupling constants. (**a**) Fit obtained for AAMR[^{13}C′]F-[U-^{15}N]GyrA, $^1J_{NC'}$ = 12.3±0.3 Hz and R_2 = 17.3±0.3 s^{-1}. (**b**) Fit obtained for AAMR[^{13}C′]F-[U-^{15}N]GyrA(H75A), $^1J_{NC'}$ = 16.2±0.2 Hz and R_2 = 17.5±0.4 s^{-1} (This figure is reproduced from Romanelli et al. [12])

The small differences are attributed to localized changes in structure and/or a magnetic shielding effect due to the loss of the imidazole ring in the H75A construct. Since the H75A mutant renders the intein inactive, these results imply that H75 plays an essential, catalytic role in the splicing reaction.

The dual-isotopic labeling pattern allowed unequivocal assignment of the scissile (−1) amide resonances. The most dramatic effect is on the amide proton, which shifts from 6.61 ppm in AAMR[^{13}C′]F-[U-^{15}N]GyrA to 10.01 ppm in AAMR[^{13}C′]F-[U-^{15}N]GyrA(H75A). The chemical shift of the (−1) amide proton in AAMR[^{13}C′]F-[U-^{15}N]GyrA lies upfield of the typical value observed for cysteine within a random coil [8], however, neither the ^{13}C nor the ^{15}N chemical shifts of the (−1) amide are unusually shifted. For AAMR[^{13}C′]F-[U-^{15}N]GyrA(H75A), all of the resonances are shifted downfield.

One-bond dipolar couplings were evaluated from the time evolution of normalized peak intensities derived from HNCO-type experiments [43]. The $^1J_{NC'}$ values obtained for the scissile (−1) amide were 12.3±0.3 and 16.2±0.2 Hz, for AAMR[^{13}C′]F-[U-^{15}N]GyrA and AAMR[^{13}C′]F-[U-^{15}N]GyrA(H75A), respectively (Fig. 2.5). Typical $^1J_{NC'}$ values reported for proteins are 13–17 Hz [44, 45].

Amide $^1J_{NC'}$ values correlate with hydrogen bonding: Hydrogen bonding to the amide carbonyl increases the coupling constant while hydrogen bonding to the amide hydrogen decreases the coupling constant [45–47]. The authors speculate that the low $^1J_{NC'}$ coupling observed for the scissile amide in the active construct is evidence of a backbone distortion primarily due to the fact that the H75A mutant has a significantly higher $^1J_{NC'}$, value similar to what is typically observed in proteins. The conclusion is that the first step in protein splicing is facilitated in part by destabilizing the scissile amide bond, in agreement with the proposed destabilization theory.

Another study [48] investigated the next step in the autocatalytic splicing reaction: the excision of the intein. They showed that intern-succinimide formation, which follows branched intermediate formation, is the rate-limiting step in protein splicing, and that this helps regulate the overall fidelity of the reaction. To examine the hypothesis that structural changes during the splicing reaction reflects a re-organization of the catalytic apparatus to accelerate succinimide formation at the C-terminal splice junction, branched intermediates of the *Mxe* GyrA intein were prepared using semi-synthesis.

Branched peptides, corresponding to residues 185–198 of the *Mxe* GyrA intein, were synthesized by using Fmoc/tBu SPPS using the HBTU activation protocol for linear chain assembly on Rink-amide ChemMatrix resin. Branched chain assembly utilized HOBt and (N,N'-diisopropylcarbodiimide) DICP activation. The peptides were cleaved from the resin using a cocktail comprised of TFA, triisopropylsilane (TIS) ethandiol and water, purified by RP-HPLC and characterized by using ESI-MS. DNA coding for *Mxe* GyrA intein residues 1–184 was amplified from pTXB1 to incorporate a factor Xa site and His-tag and cloned back into pTXB1, which also contains a chitin binding domain (CBD). The first intein residue, Cys1, was mutated to Ser (C1S) to generate the final fusion product, His-Xa-Ser1-(2–184)-GyrA- CBD, used to prepare branched constructs.

To uniformly [U-, ^{15}N] label the fusion protein, the appropriate plasmid was transformed into *E. coli* strain BL21(DE3), grown in minimal (M9) medium containing 0.2% (w/v) ^{15}N-NH$_4$Cl as the sole nitrogen source. Overexpressed protein was purified from inclusion bodies by using Ni-NTA affinity chromatography under denaturing conditions and renatured by stepwise dialysis. Thiolysis was performed in 50 mM Tris–HCl, pH 7.6, 100 mM NaCl, 1 mM EDTA and 200 mM MESNA for 2 days at 25°C to yield the protein α-thioester. Products were purified by using RP-HPLC on a C4 column. Purified thioester was lyophilized and refolded by stepwise dialysis into 50 mM Tris–HCl, pH 7.6, 100 mM NaCl and 100 mM MESNA. The N-terminal tag was removed with factor Xa and the final products purified by C4 RP-HPLC and characterized by ESI-MS.

EPL was performed using a 3:1 equivalent ratio of peptide to protein α-thioester in ligation buffer (100 mM NaPi, pH 7.8, 6 M guanidinium chloride and 100 mM NaCl) containing 100 mM MESNA and 10 mM Tris[2-carboxyethyl] phosphine (TCEP) for 5 days at 4 °C. Semisynthetic protein was separated from unreacted material on a Ni-NTA column and the ligated product purified by C4 RP-HPLC and characterized by using EMI-MS. Purified constructs were refolded by stepwise dialysis into NMR buffer (50 mM Tris–HCl, pH 7.5, 100 mM NaCl, 1 mM TCEP) for analysis.

NMR spectroscopy was used to compare the local structure around the scissile +1 peptide bond in the context of the linear precursor and the branched intermediate. The linear construct showed a signal around 8 ppm at pH 7.5 and 4°C; in contrast, no signal was obtained for the branched construct under the same condition, however, when the branched construct was denatured, a clear HNCO signal was obtained. The lack of signals in the HNCO spectrum of the branched construct reflects an exchange process, either chemical and/or conformational, around the labeled amide. A new construct was prepared in which two peptide bonds, the scissile +1 amide and the amide connecting Phe 194 and Val 195, were labeled with ^{13}C and ^{15}N. A single resonance was observed at 7.89 ppm at pH 7.5, while two peaks were detected, at 7.89 and 8.29 ppm, at pH 4.5. This observation was reversible, demonstrating that the signal from the scissile +1 amide is highly sensitive to pH, whereas that of the Phe 194-Val195 amide is not. The $^1J_{NC'}$ coupling constants for the scissile +1 and Phe 194-Val 195 amides at pH 4.5 were found to be 15.4±0.5 and 15.5±0.2 Hz, respectively, which indicates a normal trans-planar conformation.

Segmental labeling proved to be a viable technique to analyze the mechanism of intein mediated protein splicing reaction at atomic resolution and defined key conformations of the protein backbone preceding the enzymatic catalysis.

2.4 Protocols

2.4.1 Semisynthesis of a Segmental Isotopically Labeled Protein Splicing Precursor

To determine the structure of an active N-extein-intein splicing precursor, [12] used semisynthesis to prepare segmentally isotopic labeled constructs in which a short N-extein peptide α-thioester, H-AAMR[^{13}C′]F-SR, is ligated to an intein sequence derived from the *Mycobacterium xenopi* DNA gyrase A (*Mxe* GyrA) intein. The peptide is prepared by Boc-Na-SPPS and contains a single ^{13}C isotope at the C′ position of the phenylalanine. The intein protein is overexpressed as a uniformly ^{15}N labeled [*U*-, ^{15}N] polyhistidine-cleavage site-intein fusion product. This approach results in only the scissile (−1) amide being dual labeled with ^{13}C and ^{15}N.

2.4.1.1 Peptide Synthesis

1. The peptide, H-AAMR[^{13}C′]F-SR, is synthesized on 3-mercaptopropionamide-4-methylbenzhydrylamine (MBHA) resin by using the *in situ* neutralization/2-(H-benzotriazole-1-yl)-1,1,3,3-tetramethyluronium hexafluorophosphate (HBTU) activation protocol for Boc-SPPS.
2. The peptide is cleaved off the resin by using anhydrous HF containing 4% (v/v) *p*-cresol for 1 h at 4°C.
3. The peptide is purified by preparative reverse phase HPLC (Vydac C18 resin) using a linear gradient of 13.5–31.5% solution B (9:1 MeCN:water, 0.1% trifluroacetic acid) over 60 min at a flow rate of 3 mL/min. The final product, ~40 mg of purified peptide, is characterized by using electrospray mass spectrometry (ESMS).

2.4.1.2 Cloning and Protein Expression

1. Plasmid pTXB1 (NEBiolabs) encodes the GyrA intein with a single mutation, N198A, which prevents cleavage of the intein–C-extein peptide bond but does not affect the N-terminal splicing reaction. DNA encoding the *Mxe* GyrA intein (residues 1–198) is PCR amplified using the pTXB1 vector as a template.
2. To construct the plasmid pTrc-His-Xa-GyrA$_{WT}$, which encodes a factor Xa cleavage sequence between a poly(His) tag and the wild type intein, the PCR product is cloned into the *Bam*HI and *Hind*III restriction sites of pTrcHisA (Invitrogen).
3. pTrc-His-Xa-GyrA$_{WT}$ is used as a template for site-directed mutagenesis using the QuikChange site-directed mutagenesis kit (Stratagene) to generate single site mutations in the GyrA intein, in particular H75A, which abolishes N-terminal splicing activity.
4. To overexpress the fusion proteins for segmental labeling, *E. coli* strain BL21 (DE3), is transformed with either pTrc-His-Xa-GyrA$_{H75A}$ or pTrc-His-SH3-GyrA$_{WT}$, and grown to mid-log phase at 37°C in LB medium. pTrc-His-SH3-GyrA$_{WT}$ (V. Muralidharan, Rockefeller University), which encodes

the Src homology 3 (SH3) domain of murine c-Crk-II between the poly(His) tag and the intein, is used to overexpress the wild type fusion protein.

5. For [U-^{15}N]-labeling, LB medium is replaced by M9 minimal medium containing 0.2% (w/v) ^{15}NH$_4$Cl as the sole nitrogen source.
6. Overexpression is induced with 0.4 mM isopropyl β-D-thiogalactoside (IPTG) at 37°C for 5 h, after which cells are harvested and lysed.
7. Fusion protein is purified by affinity chromatography using a Ni^{2+} high-trap column (Amersham) and dialyzed into 50 mM Tris–HCl, pH 8.0, 1 mM EDTA.

2.4.1.3 Generation of [U-^{15}N]-GyrA$_{H75A}$ and [U-^{15}N]-GyrA$_{WT}$

1. [U-^{15}N]-GyrA$_{WT}$ is generated by incubating purified, labeled His–SH3–GyrA$_{WT}$ overnight in DTT cleavage buffer (50 mM Tris–HCl, pH 8.0, 100 mM DTT).
2. [U-^{15}N]-GyrA$_{H75A}$ is generated by incubating purified, labeled His–Xa–GyrA$_{H75A}$ in proteolysis buffer (50 mM Tris–HCl, pH 8.0, 0.1 M NaCl, 1 mM CaCl$_2$) with factor Xa for 10 h at room temperature.
3. Both proteins are purified to >95% homogeneity by preparative RP-HPLC and characterized by ESMS.

2.4.1.4 Preparation of AAMR[^{13}C′]F-[U-^{15}N]-GyrA$_{WT}$ and AAMR[^{13}C′]F-[U-^{15}N]GyrA$_{H75A}$

1. Ligation reactions between H-AAMR[^{13}C′]F-SR and either -[U-^{15}N]-GyrA$_{WT}$ or -[U-^{15}N]-GyrA$_{H75A}$, are initiated by dissolving purified, lyophilized peptide (10 eq) and protein (1 eq) in ligation buffer (6 M Gdm-HCl, pH 8.0, 0.1 M NaPi containing 3% MESNA and 2% ethanethiol). The final concentration of peptide is ~10 mM and that of protein is ~1 mM. The reaction is complete after 5 h at room temperature.
2. The ligation products are purified by preparative RP-HPLC and characterized by ESMS.
3. Purified proteins (1 mg/mL) are folded by stepwise dialysis at 4 °C from denaturing buffer (6 M Gdm-HCl, pH 6.6, 0.1 M NaCl, 20 mM KPi, 1 mM DTT) into NMR sample buffer (20 mM KPi, pH 6.6, 0.1 M NaCl).

2.4.1.5 NMR Spectroscopy

1. NMR samples are prepared by concentrating labeled proteins to 100 μM in NMR sample buffer containing 10% D$_2$O and 0.01% NaN$_3$.
2. ^{1}H{^{15}N} HSQC (heteronuclear single quantum correlation) and 2D planes of HNCO spectra are collected at 4 °C on a spectrometer equipped with a cryoprobe.
3. For the HSQC experiments, 512 complex points are collected in the ^{1}H and ^{15}N dimensions. In the 2D H{N}CO experiments, 512 complex points are collected in the 1H dimension and 40 complex points in the 13C dimension. Data sets are multiplied by a cosine-bell window function and zero-filled to 1,000 points using XWINNMR (Bruker) before Fourier transformation. The corresponding sweep widths are 12.5, 12, and 30 ppm in the ^{1}H, ^{13}C, and ^{15}N dimensions, respectively.
4. Experimental amide ^{1}J$_{NC'}$ coupling constants are obtained by fitting the time evolution of the normalized peak intensities extracted from a series of HNCO-type experiments using the expression:

$$I_k = \exp\left(-4t_1 R_{2k}\right)\sin^2\left(2\pi\, ^1J_{NC'}t_1\right) \tag{2.1}$$

where I_k is the normalized peak intensity for peak k, R_{2k} is the transverse relaxation time for peak k, and t_1 is the indirect dimension delay.

2.4.2 Expressed Protein Ligation of a Segmentally Labeled Bacterial σ Factor

To examine the effect of σ factor region 1.1 interactions with promoter DNA, and, in particular, intermolecular interactions between regions 1.1 and 4.2, segmentally labeled σ factor was prepared containing σ^A(1–348)-GyrA or Δ1.1-σ^A(137–348)-GyrA ligated to [U-^{2}H, ^{13}C, ^{15}N]-CG-σ^A(349–399) [12].

2.4.2.1 Cloning and Protein Expression

1. Intein proteins σ^A[1–348]-GyrA-CBD and Δ1.1-σ^A[137–348]-GyrA-CBD are expressed as a chitin binding domain fusions (CBD) in LB medium off pTXB1 (NEBiolabs) in *E. coli* strain BL21(DE3)pLysS.
2. The proteins are purified on chitin-agarose beads (NEBiolabs).
3. Region 4.2 of σ^A is expressed off pET28 (Novagen)in *E. coli* strain BL21(DE3) as a His-tagged fusion containing a factor Xa cleavage site between the poly(His) tag and region 4.2. [U-^{2}H, ^{13}C, ^{15}N]-CG- σ^A(349–399) protein is prepared by growing the cells in M9 minimal medium in ^{2}H$_2$O containing 0.2% [U-^{13}C]glucose and 0.1% ^{15}NH$_4$Cl. Introducing a Gly residue immediately after the Cys was found to greatly improve the yield of the cleavage reaction.
4. His-tagged protein is purified by affinity chromatography on Ni-nitrilotriacetate (NTA) beads (Qiagen) followed by preparative RP-HPLC on a Vydac C18 column.

2.4.2.2 Generation of Labeled Species

1. [U-^{2}H, ^{13}C, ^{15}N]-CG-σ^A(349–399) is generated by incubating purified, triple-labeled His–Xa–CG-σ^A(349–399) in proteolysis buffer (50 mM Tris–HCl, pH 8.0, 0.1 M NaCl, 1 mM CaCl$_2$) with factor Xa for 10 h at room temperature.
2. The protein is purified to >95% homogeneity by preparative RP-HPLC.

2.4.2.3 Preparation of Segmentally Labeled Species

1. Ethyl α-thioester derivatives of σ^A[1–348]-GyrA-CBD and Δ1.1-σ^A[137–348]-GyrA-CBD are generated *in situ* in the ligation mixture by thiolysis of chitin beads. Equal molar amounts of each of the two fragments ([U-^{2}H, ^{13}C, ^{15}N]-CG- σ^A(349–399) and σ^A[1–348]-GyrA-CBD or Δ1.1-σ^A[137–348]-GyrA-CBD) are used at a concentration of ~50 µM each. The ligation reaction is carried out in ligation buffer (25 mM KPi, pH 7.2, 200 mM Gdm-HCl, 250 mM NaCl, 1 mM EDTA containing 0.2% octyl glucoside and 3% (v/v) ethanethiol) overnight at room temperature.
2. The slurry is filtered and the beads washed several times with ligation buffer. All washes are combined with the supernatant.
3. The ligated protein is concentrated and exchanged into storage buffer (30 mM Tris–HCl, pH 7.6, 100 mM NaCl, 20 mM CHAPSO, 20 mM DTT). *N.b.* CHAPSO is 3-[(3-cholamidopropyl) dimethylammonio]-2-hydroxy-1-propanesulfonate.

2.4.2.4 NMR Spectroscopy

1. NMR samples are prepared by exchanging pure protein (final concentration 100–200 µM) into 30 mM Tris–HCl, pH 7.6, containing 100 mM NaCl, 20 mM CHAPSO, 20 mM [^{2}H$_{10}$]DTT, 0.1% NaN$_3$, and 10% (v/v) ^{2}H$_2$O.

2. ^{1}H{^{15}N} HSQC-TROSY and ^{1}H{^{13}C} constant –time HSQC spectra are collected at 35°C with 1,000 scans per transient. Five hundred and twelve complex points are collected in the ^{1}H, ^{13}C and ^{15}N dimensions and multiplied by a cosine-bell window function and zero-filled to 1,000 points using XWINNMR (Bruker)before Fourier transformation. The corresponding sweep widths are 12.5, 70, and 30 ppm in the ^{1}H, ^{13}C, and ^{15}N dimensions, respectively. The triple-labeling procedure affords a low proton density in region 4.2.

3. Unlabeled –35 element promoter DNA or T4 AsiA, which binds to region 4 to inhibit gene expression, is added to the NMR samples to a final molar ration of 1.2:1.

References

1. Heath WF, Merrifield RB (1986) A synthetic approach to structure-function relationships in the murine epidermal growth factor molecule. Proc Natl Acad Sci USA 83:6367–6371

2. Dawson PE, Muir TW, Clark-Lewis I, Kent SB (1994) Synthesis of proteins by native chemical ligation. Science 266:776–779

3. Tam JP, Lu YA, Liu CF, Shao J (1995) Peptide synthesis using unprotected peptides through orthogonal coupling methods. Proc Natl Acad Sci USA 92:12485–12489

4. Dawson PE, Kent SB (2000) Synthesis of native proteins by chemical ligation. Annu Rev Biochem 69:923–960

5. Kent SB (1988) Chemical synthesis of peptides and proteins. Annu Rev Biochem 57:957–989

6. Schnolzer M, Alewood P, Jones A, Alewood D, Kent SB (1992) In situ neutralization in Boc-chemistry solid phase peptide synthesis. Rapid, high yield assembly of difficult sequences. Int J Pept Protein Res 40:180–193

7. Camarero JA, Hackel BJ, de Yoreo JJ, Mitchell AR (2004) Fmoc-based synthesis of peptide alpha-thioesters using an aryl hydrazine support. J Org Chem 69:4145–4151

8. Wishart DS, Sykes BD, Richards FM (1991) Relationship between nuclear magnetic resonance chemical shift and protein secondary structure. J Mol Biol 222:311–333

9. Hackeng TM, Mounier CM, Bon C, Dawson PE, Griffin JH, Kent SB (1997) Total chemical synthesis of enzymatically active human type II secretory phospholipase A2. Proc Natl Acad Sci USA 94:7845–7850

10. Hackeng TM, Griffin JH, Dawson PE (1999) Protein synthesis by native chemical ligation: expanded scope by using straightforward methodology. Proc Natl Acad Sci USA 96:10068–10073

11. Severinov K, Muir TW (1998) Expressed protein ligation, a novel method for studying protein-protein interactions in transcription. J Biol Chem 273:16205–16209

12. Romanelli A, Shekhtman A, Cowburn D, Muir TW (2004) Semisynthesis of a segmental isotopically labeled protein splicing precursor: NMR evidence for an unusual peptide bond at the N-extein-intein junction. Proc Natl Acad Sci USA 101:6397–6402

13. Muir TW, Sondhi D, Cole PA (1998) Expressed protein ligation: a general method for protein engineering. Proc Natl Acad Sci USA 95:6705–6710

14. Tolbert TJ, Franke D, Wong CH (2005) A new strategy for glycoprotein synthesis: ligation of synthetic glycopeptides with truncated proteins expressed in E. coli as TEV protease cleavable fusion protein. Bioorg Med Chem 13:909–915

15. Chong S, Xu MQ (1997) Protein splicing of the Saccharomyces cerevisiae VMA intein without the endonuclease motifs. J Biol Chem 272:15587–15590

16. Derbyshire V, Belfort M (1998) Lightning strikes twice: intron-intein coincidence. Proc Natl Acad Sci USA 95:1356–1357

17. Shingledecker K, Jiang SQ, Paulus H (1998) Molecular dissection of the Mycobacterium tuberculosis RecA intein: design of a minimal intein and of a trans-splicing system involving two intein fragments. Gene 207:187–195

18. Wu H, Xu MQ, Liu XQ (1998) Protein trans-splicing and functional mini-inteins of a cyanobacterial dnaB intein. Biochim Biophys Acta 1387:422–432

19. Telenti A, Southworth M, Alcaide F, Daugelat S, Jacobs WR Jr, Perler FB (1997) The Mycobacterium xenopi GyrA protein splicing element: characterization of a minimal intein. J Bacteriol 179:6378–6382

20. Smith DR, Doucette-Stamm LA, Deloughery C, Lee H, Dubois J, Aldredge T, Bashirzadeh R, Blakely D, Cook R, Gilbert K, Harrison D, Hoang L, Keagle P, Lumm W, Pothier B, Qiu D, Spadafora R, Vicaire R, Wang Y, Wierzbowski J, Gibson R, Jiwani N, Caruso A, Bush D, Reeve JN et al (1997) Complete genome sequence of Methanobacterium thermoautotrophicum deltaH: functional analysis and comparative genomics. J Bacteriol 179: 7135–7155

21. Mathys S, Evans TC, Chute IC, Wu H, Chong S, Benner J, Liu XQ, Xu MQ (1999) Characterization of a self-splicing mini-intein and its conversion into autocatalytic N- and C-terminal cleavage elements: facile production of protein building blocks for protein ligation. Gene 231:1–13

22. Evans TC Jr, Benner J, Xu MQ (1999) The in vitro ligation of bacterially expressed proteins using an intein from Methanobacterium thermoautotrophicum. J Biol Chem 274:3923–3926
23. Hauser PS, Ryan RO (2007) Expressed protein ligation using an N-terminal cysteine containing fragment generated in vivo from a pelB fusion protein. Protein Expr Purif 54:227–233
24. Xu R, Ayers B, Cowburn D, Muir TW (1999) Chemical ligation of folded recombinant proteins: segmental isotopic labeling of domains for NMR studies. Proc Natl Acad Sci USA 96:388–393
25. Hawley DK, McClure WR (1983) Compilation and analysis of Escherichia coli promoter DNA sequences. Nucleic Acids Res 11:2237–2255
26. Harley CB, Reynolds RP (1987) Analysis of E. coli promoter sequences. Nucleic Acids Res 15:2343–2361
27. Dombroski AJ, Walter WA, Record MT Jr, Siegele DA, Gross CA (1992) Polypeptides containing highly conserved regions of transcription initiation factor sigma 70 exhibit specificity of binding to promoter DNA. Cell 70:501–512
28. Dombroski AJ, Walter WA, Gross CA (1993) Amino-terminal amino acids modulate sigma-factor DNA-binding activity. Genes Dev 7:2446–2455
29. Patikoglou GA, Westblade LF, Campbell EA, Lamour V, Lane WJ, Darst SA (2007) Crystal structure of the Escherichia coli regulator of sigma70, Rsd, in complex with sigma70 domain 4. J Mol Biol 372:649–659
30. Gruber TM, Bryant DA (1997) Molecular systematic studies of eubacteria, using sigma70-type sigma factors of group 1 and group 2. J Bacteriol 179:1734–1747
31. Camarero JA, Muir TW (2001) Native chemical ligation of polypeptides. Curr Protoc Protein Sci Chapter 18: Unit18.14
32. Camarero JA, Shekhtman A, Campbell EA, Chlenov M, Gruber TM, Bryant DA, Darst SA, Cowburn D, Muir TW (2002) Autoregulation of a bacterial sigma factor explored by using segmental isotopic labeling and NMR. Proc Natl Acad Sci USA 99:8536–8541
33. Vuthoori S, Bowers CW, McCracken A, Dombroski AJ, Hinton DM (2001) Domain 1.1 of the sigma(70) subunit of Escherichia coli RNA polymerase modulates the formation of stable polymerase/promoter complexes. J Mol Biol 309:561–572
34. Sorenson MK, Darst SA (2006) Disulfide cross-linking indicates that FlgM-bound and free sigma28 adopt similar conformations. Proc Natl Acad Sci USA 103:16722–16727
35. Skrisovska L, Allain FH (2008) Improved segmental isotope labeling methods for the NMR study of multidomain or large proteins: application to the RRMs of Npl3p and hnRNP L. J Mol Biol 375:151–164
36. Hui J, Bindereif A (2005) Alternative pre-mRNA splicing in the human system: unexpected role of repetitive sequences as regulatory elements. Biol Chem 386:1265–1271
37. Slynko V, Schubert M, Numao S, Kowarik M, Aebi M, Allain FH (2009) NMR structure determination of a segmentally labeled glycoprotein using in vitro glycosylation. J Am Chem Soc 131:1274–1281
38. Kowarik M, Numao S, Feldman MF, Schulz BL, Callewaert N, Kiermaier E, Catrein I, Aebi M (2006) N-linked glycosylation of folded proteins by the bacterial oligosaccharyltransferase. Science 314:1148–1150
39. Schubert M, Oschkinat H, Schmieder P (2001) MUSIC, selective pulses, and tuned delays: amino acid type-selective (1)H-(15)N correlations, II. J Magn Reson 148:61–72
40. Wood MJ, Sampoli Benitez BA, Komives EA (2000) Solution structure of the smallest cofactor-active fragment of thrombomodulin. Nat Struct Biol 7:200–204
41. Skrisovska L, Schubert M, Allain FH (2010) Recent advances in segmental isotope labeling of proteins: NMR applications to large proteins and glycoproteins. J Biomol NMR 46:51–65
42. Southworth MW, Amaya K, Evans TC, Xu MQ, Perler FB (1999) Purification of proteins fused to either the amino or carboxy terminus of the Mycobacterium xenopi gyrase A intein. Biotechniques 27:110–114, 116, 118–120
43. Cornilescu G, Delaglio F, Bax A (1999) Protein backbone angle restraints from searching a database for chemical shift and sequence homology. J Biomol NMR 13:289–302
44. Ritt S, Boschitz ET, Meier R, Tacik R, Wessler M, Junker K, Konter JA, Mango S, Renker D, van den Brandt B, Efimovyhk VV, Kovaljov A, Prokofiev A, Mach R, Chaumette P, Deregel J, Durand G, Fabre J, Thiel W (1991) Measurement of the vector analyzing power iT11 in pi +−6Li. Phys Rev C Nucl Phys 43:745–760
45. Juranic N, Moncrieffe MC, Likic VA, Prendergast FG, Macura S (2002) Structural dependencies of h3JNC′ scalar coupling in protein H-bond chains. J Am Chem Soc 124:14221–14226
46. Juranic N, Vuk-Pavlovic S, Nikolic AT, Chen TB, Macura S (1996) Nitrogen-15 NMR chemical shifts in oligopeptides coordinated to cobalt(III). J Inorg Biochem 62:117–126
47. Juranic N, Macura S (2001) Correlations among (1)J(NC)′ and (h3)J(NC)′ coupling constants in the hydrogen-bonding network of human ubiquitin. J Am Chem Soc 123:4099–4100
48. Frutos S, Goger M, Giovani B, Cowburn D, Muir TW (2010) Branched intermediate formation stimulates peptide bond cleavage in protein splicing. Nat Chem Biol 6:527–533

Chapter 3
Isotope Labeling for Solution and Solid-State NMR Spectroscopy of Membrane Proteins

Raffaello Verardi, Nathaniel J. Traaseth, Larry R. Masterson, Vitaly V. Vostrikov, and Gianluigi Veglia

Abstract In this chapter, we summarize the isotopic labeling strategies used to obtain high-quality solution and solid-state NMR spectra of biological samples, with emphasis on integral membrane proteins (IMPs). While solution NMR is used to study IMPs under fast tumbling conditions, such as in the presence of detergent micelles or isotropic bicelles, solid-state NMR is used to study the structure and orientation of IMPs in lipid vesicles and bilayers. In spite of the tremendous progress in biomolecular NMR spectroscopy, the homogeneity and overall quality of the sample is still a substantial obstacle to overcome. Isotopic labeling is a major avenue to simplify *overlapped spectra* by either diluting the NMR active nuclei or allowing the resonances to be separated in multiple dimensions. In the following we will discuss isotopic labeling approaches that have been successfully used in the study of IMPs by solution and solid-state NMR spectroscopy.

Abbreviations

IMP	Integral Membrane Protein
SSNMR	Solid-State NMR
O-SSNMR	Oriented SSNMR
MAS-SSNMR	Magic-Angle-Spinning SSNMR
PISEMA	Polarization Inversion Spin Exchange at Magic Angle

R. Verardi
Department of Biochemistry, Molecular Biology, and Biophysics,
University of Minnesota, Minneapolis, MN 55455, USA

N.J. Traaseth
Department of Chemistry, New York University, New York, NY 10003, USA

L.R. Masterson • V.V. Vostrikov
Department of Chemistry, University of Minnesota, 6-155 Jackson Hall, 321 Church St SE,
Minneapolis, MN 55455, USA

G. Veglia (⊠)
Department of Biochemistry, Molecular Biology, and Biophysics, University of Minnesota,
Minneapolis, MN 55455, USA
vegli001@umn.edu

Department of Chemistry, University of Minnesota,
6-155 Jackson Hall, 321 Church St SE, Minneapolis, MN 55455, USA

H.S. Atreya (ed.), *Isotope Labeling in Biomolecular NMR*, Advances in Experimental Medicine and Biology 992,
DOI 10.1007/978-94-007-4954-2_3, © Springer Science+Business Media Dordrecht 2012

3.1   Introduction

Isotopic enrichment has been an integral part of the advancements made by nuclear magnetic resonance (NMR) spectroscopy for the characterization of biomacromolecules at atomic resolution. The first pioneering studies on isotopically labeled proteins were carried out in the late 1960s, resulting in the production of isotopically labeled proteins extracted from organisms (bacteria and plants) cultured in media containing isotopically labeled nutrients [1–4]. In the past few years, there has been a true explosion of labeling schemes and production techniques that has enabled NMR spectroscopic studies of proteins and protein complexes larger than 100 kDa [5–7].

While most of the structural biology has been focusing on soluble proteins, outstanding progress is being made both in liquid and solid-state NMR for the structural analysis of membrane-bound proteins. In fact, an estimated 30% of all proteins synthesized in most organisms are integral membrane proteins [8, 9], which necessitate lipid environments to properly fold and function. IMPs are involved in signal transduction, transport of molecules across the membrane, conduction of ions and many other vital cellular processes [10–13]. Despite their importance, only 308 IMPs (http://blanco. biomol.uci.edu/mpstruc/listAll/list) have been deposited in the protein data bank (PDB) as of 2011, which is a rather exiguous number compared to the thousands of high-resolution structures determined for their soluble counterparts. There are several reasons for the paucity of high-resolution IMP structures. First of all, IMPs are difficult to express and purify in large amounts (tens of milligrams) and with the proper folding. Second, IMPs need lipids or detergents for structural and functional studies. The membrane mimetic environments coat the proteins forming large and slowly tumbling complexes that complicate NMR analysis. In recent years however, improvements in protein production systems, NMR hardware, pulse sequences and isotopic labeling strategies have made possible a number of successes in the study of IMPs [6, 14].

This chapter highlights the recent progress in isotopic labeling technologies to aid solution and solid-state NMR studies of IMPs. Although only four isotopes (^{1}H, ^{15}N, ^{13}C, ^{2}H) are routinely used in biomolecular NMR, there are several ways for introducing them along the amino acid sequence (see Fig. 3.1). We focus on the recent progress from our laboratory and other research groups in the production of isotopically labeled IMPs for both liquid- and solid-state NMR studies. In addition, we review how isotopic labeling schemes can be exploited for studying protein-protein interactions in micelles and lipid vesicles. Finally, we will discuss some of the most common techniques to engineer spin-labels and isotopically labeled chemical groups to image large mammalian membrane proteins.

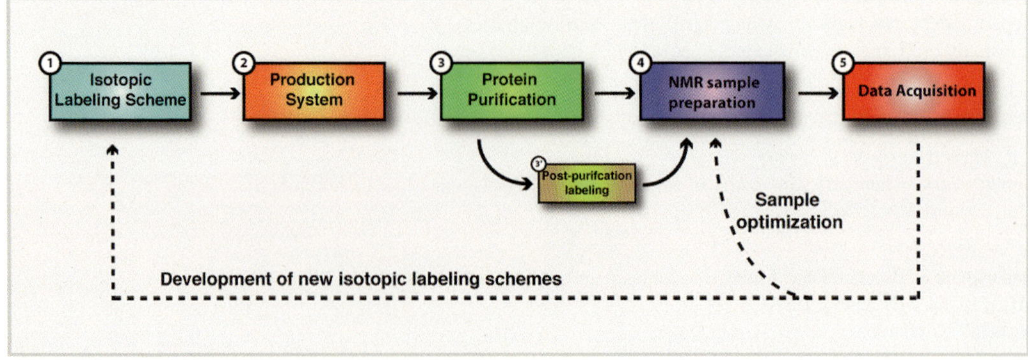

Fig. 3.1 Production of isotopically labeled membrane proteins for NMR spectroscopy

3.2 Recent Advances in the Production of IMPs

The main isotopes routinely used in protein NMR spectroscopy are ^1H, ^2H, ^{13}C and ^{15}N, with a more sparse use of ^{31}P, ^{19}F and ^{17}O. Among the main isotopes, only ^1H is found naturally at high abundance (>99.9%), whereas the others must be artificially introduced in proteins. Isotopic labeling schemes can be divided into two broad categories: uniform and selective labeling. In the first category, we list all methods that produce a protein with uniform incorporation of NMR active isotope (i.e., uniformly ^{13}C labeled or U-^{13}C). Conversely, if a protein is enriched with an isotope only at particular sites, the protein will be defined as selective labeled.

Because of the inherent insensitivity of NMR, it is generally necessary to have an production system capable of yielding milligram amounts of IMPs properly folded and biologically active. There are three well-established approaches: (1) heterologous overexpression, (2) total chemical synthesis and (3) cell-free expression. Depending on the protein under investigation each one of these approaches can be a viable choice. However, each system has advantages and drawbacks that need to be evaluated on a case-by-case basis.

3.2.1 Heterologous Overexpression Systems for Membrane Proteins

Heterologous overexpression consists of the use of living cells to synthesize proteins. It involves manipulation of the host DNA in such a way that the foreign gene is transcribed and translated at high levels. There are several heterologous systems for the expression and purification of IMPs [15–17], but the most widely used for isotopic labeling are: bacteria, yeasts, and insect cells. Each system has its own advantages and drawbacks, nonetheless a number of IMPs have been successfully produced for NMR studies [6]. When choosing an expression system, there are at least three important parameters to consider and eventually optimize: (1) the amount of final product (pure protein) per liter of growth medium, (2) whether the protein is properly folded and (3) whether biological activity of the expressed protein is retained.

3.2.1.1 Bacteria

The use of bacteria (especially *Escherichia coli* strains) for heterologous expression of proteins was established in the 1980s when molecular cloning techniques became widely available [18]. Bacteria offer a number of advantages over other expression systems: they can grow at high densities in a variety of synthetic media, foreign genes can be inserted in their genome using simple molecular cloning techniques, and growth rates are fast (doubling time is on the order of 30 min). *E. coli* strains can be grown in fermenter vessels, where important parameters such as pH, temperature and dissolved oxygen are monitored to increase biomass and protein expression levels. Several strategies for efficient isotopic labeling of recombinant proteins in *E. coli* have been proposed [4, 19–21]. All these methods focus on obtaining high cell densities using inexpensive unlabeled media and subsequent transfer in labeled medium immediately before expression. High expression levels for IMPs have also been obtained using a clever manipulation of the common T7 expression system, which cause autoinduction of the recombinant gene [22].

A promising new strategy for the efficient expression of labeled proteins in *E. coli* is the single protein production system [23, 24]. By expression of an mRNA interferase (MazF) that cleaves RNA at ACA nucleotide sequences, it is possible to stop cellular growth. If the mRNA of the gene of interest is engineered so that no ACA sequences are present, MazF will not cleave it and translation will continue undisturbed. By using this expression system, it has been estimated that up to 30% of total cellular content is comprised of the recombinant protein, making it possible to acquire NMR spectra without substantial purification. When such a system is used for the production of isotopically labeled poteins the savings in terms of materials could be substantial. Indeed its success has been demonstrated by producing several IMPs [23, 24].

Although *E. coli* is a robust and reliable host cell, it presents a number of problems for the expression of IMPs. Overexpression of IMPs is often toxic to the cell, thereby decreasing the viability of the cell itself. When IMPs are expressed at high levels they often tend to aggregate into inclusion bodies [25] which require unfolding and refolding strategies in order to extract the target protein. Although these problems can be circumvented by expressing the IMPs at lower temperature, or using soluble fusion tags, the IMPs might not be in an active form since *E. coli* bacteria do not possess post-translational modification machinery.

In addition to *E. coli* bacteria, other prokaryotes have been investigated for the overexpression of IMPs. The two most promising organisms are *Pseudomonas Aeruginosa* and *Lactococcus lactis*. *P. Aeruginosa* is a gram-negative bacterium that breaks down glucose using the Entner-Doudoroff pathway rather than glycolysis, producing alternative labeling patterns. McDermott and coworkers produced Pf1 coat protein labeled with ^{13}C only at the carbonyl position by feeding *P. aeruginosa* with 1-^{13}C-glucose [26]. Although this labeling scheme was used for solid-state NMR investigation of Pf1, it has great potential for studying the dynamics of IMPs by solution NMR as well.

The second promising prokaryote for the production of IMPs is *L. lactis*. This gram-positive bacterium offers several attractive features: (1) it has a single cellular membrane, which facilitates the insertion of heterologous IMPs and reduces the formation of inclusion bodies, (2) it can grow at high cell densities in the absence of oxygen and (3) it possesses a tightly regulated inducible expression system that uses the peptide nisin for induction [27]. IMPs have been successfully produced using this system [28], although the use for isotopic labeling in NMR studies has yet to be demonstrated.

3.2.1.2 Yeasts

The inability to introduce complex post-translational modifications and obtain properly folded and functional proteins are among the most significant drawbacks for the expression of IMPs in bacteria. A solution to these problems is to use more sophisticated expression systems, such as eukaryotic cells. The simplest and most studied eukaryotic system for the expression of IMPs are the yeasts *Saccharomyces cerevisiae* and *Pichia pastoris*.

Both systems have been used to produce many IMPs for NMR and X-ray studies [29, 30]. As for *E. coli*, yeast can be cultured in completely defined media composed of simple sugars and salts. Moreover, molecular biology techniques for the recombinant expression of foreign genes are available and readily applicable for the isotopic labeling of IMPs.

3.2.1.3 Higher Eukaryotes

Other eukaryotic organisms have been used for the production of IMPs. The major advantage of using higher eukaryotes over simpler systems is the presence of more complex folding machinery and post-translational patterns. Some of the most promising systems for the isotopic labeling of IMPs are baculovirus-infected insect cells and transfected mammalian cells. Recently, a simple and inexpensive protocol for the selective isotopic labeling of proteins in insect cells has been proposed [31]. Despite their utility, insect cells suffer from some important drawbacks: (1) cost of labeled media can

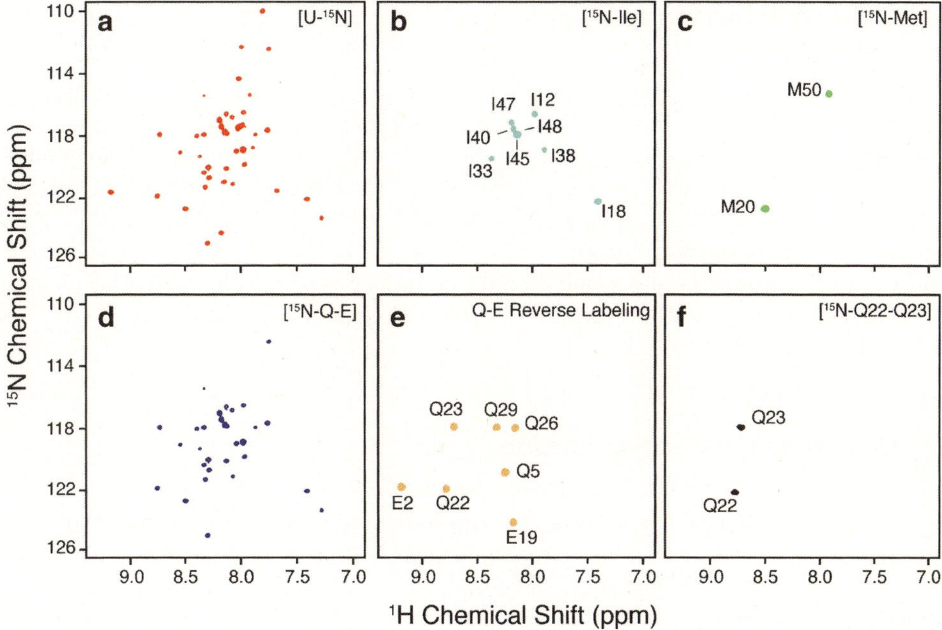

Fig. 3.2 Examples of ^{15}N uniform and selective labeling of the membrane protein PLN. (**a**) ^{15}N-^{1}H HSQC of [U-^{15}N] recombinant PLN in 300 mM DPC. (**b–c**) Selective ^{15}N-Ile and ^{15}N Met labeled recombinant PLN. Notice the absence of isotopic scrambling. (**d**) An attempt to label PLN at Gln and Glu residues using ^{15}N-Gln and ^{15}N-Glu labeled amino acids in the growth medium resulted in significant isotopic scrambling. (**e**) Labeling of Glu and Gln in PLN using the reverse labeling approach. No isotopic scrambling is present. (**f**) PLN selective labeled at Q22-Q23 produced by peptide synthesis

be prohibitive, (2) deuteration has not yet been reported and (3) the yield of pure protein can be substantially lower than other systems.

Transfected mammalian cells are another useful system to express active and properly folded IMPs. Isotopically labeled IMPs have been produced with CHO and HEK293 cells at levels comparable to simpler systems [32]. Moreover, growth media for the incorporation of ^{15}N and ^{13}C are commercially available.

3.2.2   Total Chemical Synthesis

All the production systems described so far involve the use of living cells from different organisms. There are, however, chemical methods for the synthesis of proteins of up to 100 amino acids, which can be easily adapted for isotopic labeling purposes. Chemical synthesis is usually carried out using the standard solid-phase peptide synthesis (SPPS) developed by Merrifield and coworkers [33]. SPPS uses solid resins composed of amino acid covalently linked to polystyrene beads. Protected amino acids are added to the reaction vessel where they form peptide bonds through a series of couplings and deprotection reactions. Thanks to microwave-assisted technologies which increase yields during difficult couplings and make more efficient use of isotopically labeled reagents during synthesis it is now possible to routinely produce IMPs isotopically labeled at single sites in the primary sequence (Figs. 3.2f and 3.4c).

3.2.3 Cell-Free Expression Systems

Cell-free systems are *in vitro* transcription/translation systems extracted from a variety of cells (bacteria, wheat germ, insect cells etc.) [34–36]. For cell-free systems to work, a mixture of all the 20 amino acids must be added in the reaction vessel. Because of the absence of other enzymes other than those necessary for transcription and translation, isotopic scrambling is nearly eliminated for most amino acids. In addition, this approach provides an alternative avenue to obtain IMPs that may be toxic to host cells during overexpression.

Cell-free systems can be used not only to produce residue-type selectively labeled proteins, but also for some ingenious applications such as combinatorial labeling [37–40] and stereo array isotopic labeling (SAIL) [41].

3.2.4 Membrane Protein Purification

So far, we reviewed biological and chemical systems to introduce isotopes in different positions of a protein. However, once the protein has been recombinantly expressed or chemically synthesized it must be purified to high levels (generally more than 90% purity) before NMR experiments can be undertaken. For solid-phase peptide synthesis, purification involves cleavage of the peptide from the resin and subsequent precipitation of the peptide in organic solvents. A final step of reverse-phase chromatography usually yields pure protein suitable for structural studies.

For heterologous expression of IMPs, the purification process is more involved and usually requires the use of fusion tags [42, 43].

A fusion tag is a protein or short peptide included in the same reading frame as the gene of the target protein. When the gene is transcribed and translated, the final protein will be fused to the tag through a peptide bond. Fusion tags are engineered either at the C-terminus or N-terminus and are usually separated from the protein of interest by a flexible loop.

Two important classes of fusion tags in this context are: (1) solubility tags and (2) affinity tags. To the first category belong all those tags that are used to improve solubility of the target protein. The most widely used solubility tags are: maltose binding protein (MBP), glutathione S-transferase (GST), N-utilization substance A (NusA), and Thioredoxin [43, 44].

Affinity tags are used to aid the purification of the target protein. The most common affinity tags for IMPs are: hexahistidine, GST, biotin acceptor peptide, MISTIC (acronym for membrane-integrating sequence for translation of IM protein constructs), and streptavidin binding peptide. Affinity tags bind strongly to solid supports (usually resins or gels) together with their fusion partners. The bound fusion protein can be subsequently eluted off the resin and the affinity tag removed by chemical or proteolytic cleavage [44].

Removal of the fusion tag by proteases requires the presence of specific recognition sequences that must be engineered in the gene. Tobacco etch virus (TEV) protease, factor Xa, thrombin, and enterokinase are the most commonly used enzymes to cleave off fusion tags from the target protein [45–47]. Factor Xa has a four amino acid recognition sequence (IEGR), while TEV has a more stringent seven amino acid recognition sequence (ENLYFQ/G). TEV, however, leaves one amino acid at the C-terminal side of the cleavage site that in most cases can be constructed to coincide with a native N-terminal residue in the protein sequence [48].

Some fusion tags such as MBP and GST act as both solubility and affinity tags. The MBP system is one of the most versatile systems for the expression and purification of IMPs.

In the commercially available pMal plasmid (New England BioLabs Inc.), the gene of interest is inserted upstream of the MBP gene. A recognition sequence for TEV or Factor Xa proteases can also be engineered between the two fusion partners. The plasmid is transformed into *E. coli* BL21(DE3) competent cells and the protein is expressed under the control of the inducible Ptac promoter. Upon expression, the cells are lysed and loaded onto an amylose resin [49], which binds MBP at high affinity. After washing the resin with buffer, the fusion protein is eluted off the resin by addition of maltose, which competes with amylose to bind MBP. Purified fusion protein is cleaved using TEV protease. Following cleavage, the target protein can be separated by reverse-phase HPLC or gel filtration to the desired purity. Alternatively, solvent extraction has been successfully used in some cases [50].

3.3   Labeling Strategies in Solution State NMR

3.3.1   Uniform Isotopic Labeling

Uniform isotopic labeling consists of replacing all nuclei of a certain element with its respective isotope. As of today, the only cost-effective way to produce uniformly labeled proteins is to make use of recombinant expression in heterologous systems (see previous section). The isotope of interest is incorporated into the polypeptide by providing the organism with labeled substrates, which are then converted to labeled amino acids in the metabolic pathways [51, 52]

In the 1980s and 1990s, the development of multidimensional NMR techniques for structure and dynamics studies required proteins to be uniformly enriched in ^{15}N and/or ^{13}C. In general, ^{15}N and ^{13}C are easily introduced in the polypeptide by growing cells in minimum media containing ^{15}N ammonium salts and ^{13}C glucose as the sole nitrogen and carbon sources, respectively [51]. New media containing algal lysate have been recently used to produce uniformly labeled proteins in bacteria, achieving higher yields at lower costs [53, 54]. ^{15}N uniform labeling has become a standard strategy to enable NMR studies. Figure 3.2a shows an example of well-dispersed and homogenous correlation spectrum for a uniformly ^{15}N labeled membrane protein.

For large IMPs, the strong ^1H-^1H dipolar and heteronuclear (^1H-^{13}C or ^1H-^{15}N) relaxation pathways introduced with uniform ^{13}C and ^{15}N labeling, becomes a source of sensitivity loss. To circumvent this problem, partial and complete deuteration of proteins has been introduced [55–57]. Deuterium is a quadrupolar nucleus with a significantly lower gyromagnetic ratio compared to proton, therefore the previous relaxation pathways are largely eliminated [56]. Triple labeled proteins (U-^2H-^{13}C-^{15}N) are now routinely produced and used for resonance assignment purposes [57]. However, complete deuteration has some inconveniences. First, the absence of ^1H sites does not allow the detection of the structurally important ^1H-^1H NOE connectivities. Second, most pulse sequences terminate with detection of the proton resonances to increase sensitivity; therefore they would be useless with a completely deuterated protein. Fortunately, amide deuterons are readily exchanged with water protons and for most soluble proteins ^1H amide exchange is achieved during the purification steps. However, for IMPs the back exchange of amides might be more difficult due to the reduced accessibility and strong hydrogen bonding of the hydrophobic domains buried in the interior of the detergent micelle [58, 59]. In such cases, the protein must be unfolded and refolded in the presence of protonated buffers, which may generate misfolded proteins [60]. For the detection of short-range NOE contacts in large proteins, deuteration can still be useful if it is carried out at lower levels (60–70%). It has been demonstrated that partial deuteration can improve resolution and sensitivity, while enabling the detection of NOE contacts with the remaining protons [56].

As for the other isotopes, uniform deuteration is accomplished by growing cells in media containing only deuterated water as solvent and deuterated carbon sources [1]. Historically, the first isotopic labeling strategy used in protein NMR was selective deuteration in order to simplify the spectra (by dilution of the natural abundance ^1H signals) and decrease the linewidths (by removing the broadening effect of dipolar spin relaxation) [2, 4]. Proteins were enriched in ^2H by growing cells in media containing deuterated carbon sources (^2H-amino acid mixtures derived from algae grown in deuterated water or

^{2}H glucose) and deuterated water [2, 4]. Crespi and coworkers demonstrated how completely deuterated organisms were still able to survive and reproduce, although plant and mammalian cells could only be enriched at 20–60% with ^{2}H [61]. However, extensive deuteration can alter the structure and activity of proteins [62, 63]. Although uniform isotopic labeling still represents the first step for most protein NMR studies, this strategy does not provide the same gain for very large helical IMPs. The main obstacle when using uniform isotopic labeling of IMPs is spectral overlap, which is caused by different factors: (1) increase in the rotational correlation times, which causes line broadening, (2) degenerate chemical shifts due to the presence of only a small number of residue types (mostly Ile, Leu, Val) in transmembrane regions and (3) high occurrence of α-helical secondary structures, which decrease the breath of chemical shifts. These problems can be alleviated by using selective isotopic labeling schemes.

3.3.2 Selective Isotopic Labeling

By selective isotopic labeling, we indicate any labeling strategy that results in the incorporation of isotopes at specific sites along the polypeptide sequence. This results in NMR spectra of particular residue types in a protein sequence. An alternative approach, introduced by Oschkinat and co-workers [64], involves spectroscopic identification of individual or groups of residue types such as Gly, Ala, Thr, Val, Ile, Asn, and Gln. This approach is based on the clever use of INEPT transfer steps. However, the easiest and most widespread approach is the isotopic labeling of specific residue types using ^{15}N (and more recently ^{13}C) labeled amino acids. Traditionally, the ^{15}N and/or ^{13}C labeled amino acids are included in the growth media along with all the other "unlabeled" (^{14}N/^{12}C) amino acids. Residue-type selective labeling is extensively used to simplify spectra for assignment purposes. Not all 20 amino acids can be labeled using this strategy. In fact, the use of some amino acids results in isotopic dilution or scrambling [65]. Scrambling occurs for those amino acids that serve as precursors for the synthesis of other amino acids and results in isotopic dilution and/or distribution of the labels among other amino acids. A classic example is the amino acid glutamate, which is a central precursor for most of the other residues [66]. If ^{15}N-glutamate is used in the growth medium, the protein synthesized will have most of the other residues labeled as well. In the case of ^{15}N-labeling in heterologous expression systems, there are two ways to overcome this problem: (1) use of mutated strains (auxotrophs) and (2) reverse labeling. In the first case, libraries of *E. coli* bacteria strains have been engineered so that the metabolic pathways leading to the synthesis of each amino acid are altered through mutations [66–68]. For the amino acids Arg, Cys, Gln, Gly, His, Ile, Lys, Met, Pro and Thr, a single lesion is sufficient to eliminate isotopic scrambling [66]. This is because all of these amino acids (except Thr and Ile) are located at the end of metabolic pathways and are not used as precursors for other residues [52]. For the other amino acids, more than one genetic deletion is necessary [66]. An alternative approach is reverse labeling, which does not require mutant strains of *E. coli*. With this approach, all of the amino acids are included in the growth medium in the unlabeled (^{14}N) form, whereas the amino acid(s) of interest is omitted. ^{15}N-ammonium chloride is also included in the medium [69]. When cells grow, they will use the unlabeled amino acids for protein synthesis, but they will use ^{15}N-ammonium chloride to make up the missing amino acid(s). The result will be identical to the traditional method, but isotope scrambling can be significantly reduced. Figure 3.2 shows the comparison between an attempt to label Glu and Gln in a membrane protein using the traditional selective labeling method, resulting in severe isotopic scrambling, (Fig. 3.2d) and the reverse labeling method (Fig. 3.2e).

The use of cell-free expression systems has also been applied to a number of membrane proteins, alleviating the scrambling encountered in protein expression with bacterial host cells. In this manner, high resolution spectra of membrane proteins have been obtained from *in vitro* protein synthesis [36, 70]. A number of labeling strategies, including combinatorial, sequence-optimized, or SAIL approaches, have been used in cell-free protein synthesis to aid in resonance assignment and improve the spectral quality of membrane proteins [71–73]. These approaches are different variations of selective-labeling of amino acids into a target protein sequence during cell-free protein expression. However, since *in vitro* expression is not complicated by various catabolic and metabolic pathways, unique protein labeling patterns can be obtained.

Another promising approach for studying large proteins is to incorporate isotopically labeled unnatural amino acids such as p-methoxy-phenylalanine (p-OMePhe), o-nitrobenzyl-tyrosine (oNBTyr), 2-amino-3-(4-(trifluoromethoxy)phenyl)propanoic acid (OCF$_3$Phe), trifluoromethyl-l-phenylalanine [74–76] into specific single positions along the primary sequence of a protein. This is possible by using orthogonal tRNA/tRNA synthetase pairs, which generates tRNA charged with the unnatural amino acid [75, 77, 78]. The validity of this approach was demonstrated by incorporating three unnatural amino acid in the 33 kDa thioesterase domain of human fatty acid synthase without perturbation of the protein structure [74].

Fluorine can also be selectively introduced in proteins by using fluorinated tryptophan, tyrosine or phenylalanine amino acids in *E. coli* strains auxotrophic for those amino acids [79]. Fluorine labeled amino acids have been used extensively to study protein folding, ligand binding, dynamics [79, 80], membrane immersion depth [81] and more recently solvent accessibility [82].

Finally, a new method for the labeling of specific domains of proteins has been proposed with the name "segmental labeling". This method exploits the post-translational modification, known as splicing, performed by inteins [78]. For a detailed description of this technique see previous reviews [83]. The main point of this approach is that it is possible to label (with ^{15}N and/or ^{13}C) only specific domains, while the rest of the protein remains unlabeled. This has important consequences in NMR, since the spectra are considerably simplified while retaining important inter-residue information for the labeled domain. Although useful, this technique has not been extensively applied for the production of IMPs.

3.3.3   Methyl Labeling

In highly deuterated proteins, it is advantageous to reintroduce some of the protons at specific positions [84]. For the methyl groups of isoleucine, leucine and valine, this is achieved by adding protonated precursors to the deuterated growth medium just before induction [84]. The most common of these precursors are α-ketobutyrate (yielding isoleucine) and α-ketosovalerate (yielding leucine and valine) (Fig. 3.3a, b). Due to the high degree of sensitivity via TROSY NMR of deuterated, methyl labeled proteins, a number of commercially available precursors with specific labeling patterns have been developed. For the methyl labeling of methionine, alfa-oxomethionine is added as precursor in the presence of glucose (Fig. 3.3c), whereas labeling of the methyl group of threonine can be achieved by growing cells in a medium containing a mixture of 2-^{13}C-glycerol and NaH^{13}CO$_3$ [7] (Fig. 3.3d). Slightly more involving is the ^{13}C labeling of alanine, which requires the addition of ^{13}C-labeled alanine supplemented with unlabeled succinate, α-ketoisovalerate and isoleucine to reduce isotopic scrambling (Fig. 3.3e) [7].

Methyl group labeling has proven to be a very useful strategy for membrane proteins since hydrophobic amino acids Ile, Leu and Val occur at high frequency in transmembrane domains and they are often involved in the packing of those domains [85–87]. Selective methyl labeling has been successfully applied to the study of several IMPs by solution NMR in the past few years [88–92].

3.4   Labeling Strategies in Solid State NMR

Unlike in solution NMR where rapid reorientation leads to isotropic chemical shifts and averaging of dipolar interactions, SSNMR spectra are dominated by anisotropic interactions such as anisotropic chemical shifts, quadrupolar, and dipolar couplings. The two primary classes of SSNMR methodology

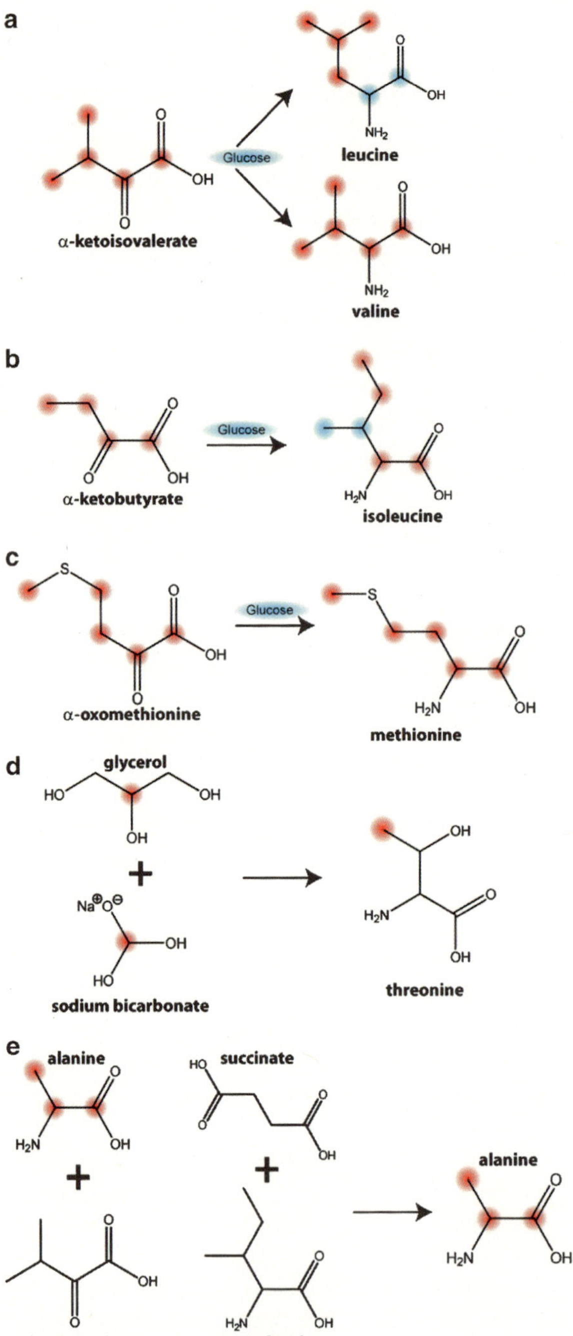

Fig. 3.3 Selective ^{13}C enrichment of methyl containing amino acids using different precursors in the presence of glucose. Carbons derived from the precursors are indicated in *red*. Note that these precursors lead to very high ^{13}C incorporation for all sites (>90%). We did not include other carbon sources (such as ^{13}C-pyruvate) that lead to lower enrichment levels at the methyl sites

are oriented (static) and magic angle spinning (MAS) experiments. MAS experiments most commonly result in solution-like isotropic spectra, whereas oriented solid-state NMR (O-SSNMR) gives orientation dependent parameters, which can be used to determine the orientation of membrane proteins in lipid bilayers or single/liquid crystals such as bicelles. Highly anisotropic systems for MAS or O-SSNMR have primarily utilized detection on ^{15}N or ^{13}C, since ^{1}H observation is hindered due to strong ^{1}H-^{1}H dipolar couplings that give rise to severe line-broadening. Techniques such as fast MAS (>60 kHz) in combination with ^{2}H labeling have made proton detection feasible in biological samples [93]. In addition, stroboscopic detection allows for the detection of signals while *simultaneously* decoupling them in a windowed-fashion [94]. Both windowed PMLG in MAS and PISEMO in O-SSNMR have benefited from these approaches. Advancements in these techniques will play an important role in the future of SSNMR due to the significant gains in sensitivity.

The following section will be broken down into labeling approaches in (1) O-SSNMR and (2) MAS-SSNMR. Subcategories of isotopic labeling strategies will be discussed that (a) reduce spectral complexity and (b) decrease the linewidth of the resonances. These two approaches are often used synergistically for optimal spectral quality.

3.4.1   Labeling Strategies in Magic-Angle-Spinning (MAS)

3.4.1.1   Uniform Isotopic Labeling

While SSNMR lines of the best-behaving samples can approach the quality of solution NMR spectra, the majority of proteins give substantially broader spectra. As an example, consider the following typical backbone ^{15}N and ^{13}C linewidths of the 6 kDa transmembrane protein phospholamban monomer (PLN) at a magnetic field of 14.1 T (600 MHz ^{1}H frequency): (a) solution NMR in detergent micelles ~0.25–0.35 ppm, (b) MAS-SSNMR in lipids ~0.75–1.5 ppm, (c) O-SSNMR in lipid bicelles ~3–6 ppm, and (d) O-SSNMR in mechanically aligned lipid bilayers ~5–10 ppm. As expected from these linewidths, the ability to resolve peaks is substantially reduced in the case of MAS and O-SSNMR. An MAS N-CA 2D correlation spectrum of uniformly labeled ^{13}C, ^{15}N spectra, [U-^{13}C,^{15}N] PLN is shown in Fig. 3.4a. From the known labeling in the sample, 52 peaks are expected. One alternative is to use 3D experiments to improve the resolution by carrying out experiments such as N-CA-C′, N-C′-CX, CA-N-C′, and other *sequential experiments* in SSNMR. However, for redundant primary sequences and helical structures such as membrane proteins, 3D experiments alone are not sufficient to resolve all the peaks. The ^{15}N dimension typically has only ~5–10 ppm in resolution (not including glycine residues). In addition, the sensitivity of multiple magnetization transfers considerably attenuates signal-to-noise, further complicating the scenario. For these reasons, reduction of spectral complexity is needed for unambiguous assignment purposes.

Similar to solution state NMR, deuteration of protein MAS samples eliminates the dipolar interactions involving protons, thus reducing the linewidths of the detected nuclei [95]. A portion of the dipolar network can be reintroduced by back-exchanging the amide protons, while the magnetization transfer to non-exchangeable side chains is achieved by expressing the proteins in the media containing minor amounts of protonated substrates [96, 97]. Since the majority of MAS pulse sequences have cross polarization as an essential block for boosting the sensitivity of low γ nuclei, deuterated samples require either direct polarization of heteronuclei (long T1 values and therefore costly from the experimental time standpoint), but can be shortened by paramagnetic doping [98]. Protein deuteration has also been observed to be beneficial in dynamic nuclear polarization experiments, yielding higher sensitivity relative to the protonated samples [99]. Furthermore, aside from providing line-narrowing of heteronuclear lineshapes (*vide supra*), deuterium itself can be employed for assignment purposes. Utility of ^{2}H in triple uniformly labeled proteins has been demonstrated for the assignment of spin systems in ^{13}C edited spectra [100]. We note that the acquisition of such experiments can be facilitated with the help of DUMAS approach [101].

3.4.1.2 Synthetic Labeling

The simplest strategy that yields the most unambiguous assignment is to label a single residue. In this case, the assignment problem is reduced (or eliminated), and a single broad line does not cause the same resolution problems as when several signals are present. For ^2H or ^{17}O quadrupoles, the inherent linewidths in the spectra are on the order of ~50–100 kHz, with mosaic spread and IMP dynamics further increasing the linewidths, requiring the use of single labeled samples [102, 103]. Interpretation of quadrupolar splitting can give orientation as well as dynamics of peptides and proteins (see Sect. 3.4.2.2) [104]. This approach is very similar to EPR spectroscopy that also utilizes site-specific labeling, often with the methanesulfonothioate (MTSL) spin label if samples are made by single cysteine mutants, or 2,2,6,6-tetramethyl-piperidine-1-oxyl-4-amino-4-carboxylic acid (TOAC), prepared by SPPS.

An extension of single site-specific labeling strategy is the incorporation of two nuclear probes in which distance and dynamics information can be obtained. This is the foundation for a number of rotational-echo double-resonance (REDOR) experiments which have been used extensively in the SSNMR studies of peptides and proteins [105–109].

A further step is to selectively label stretches of residues in the primary sequence in a contiguous fashion. Such an approach has been successfully implemented by a number of MAS research groups for studying fibrils. For example, Jaroniec et al. [110] relied on three samples to assign the chemical shifts from a fragment of transthyretin (residues 105–115) fibrils. In each case the spectra are substantially simplified, since one can avoid overlap from unlike amino acids by carefully choosing the stretches of amino acids to label. Also due to the limited labeling, 2D spectra are usually sufficient to assign the spectra, without the need for longer 3D sequences that can take several weeks to acquire. Many other research groups have used this strategy in the study of amyloid fibrils, where broad lines similar to membrane proteins are present [111, 112]. We recently implemented this strategy for membrane proteins to understand the complicated folding pathways of amphipathic helices at the membrane interface [113]. Figure 3.4c shows an example of the simplification that is expected when solid-phase peptide synthesis is used to introduce a limited number of labeled residues. The main disadvantages of this technique are (a) limited applicability for large proteins (>50–75 residues in length), (b) high costs associated with purchasing some of the isotopically labeled and protected amino acids, and (c) difficulty in measuring long-range distances, since only a limited number of labeled sites are present. Nevertheless, if the protein of interest can be synthesized using SPPS, spectral quality and the ability to unambiguously assign peaks is improved.

3.4.1.3 Residue-Type Labeling

Another potential way to reduce spectral complexity and overlap is to incorporate isotopically labeled amino acids into the growth media. Unfortunately for IMPs this does not reduce a primary problem in the [U-^{15}N,^{13}C] spectra: overlap of peaks of the same residue-type (Fig. 3.4b). However, when multiple [U-^{13}C,^{15}N] amino acids are labeled at the same time, pairwise-selective labeling can be obtained. For example, consider the stretch of six residues Val^1-Ala^2-Ile^3-Ile^4-Asn^5-Ala^6. If all the residues were labeled [U-^{13}C,^{15}N], there would be five $^{13}C'$-^{15}N peptide bonds. Alternatively, residue-type selective labeling with [^{13}C,^{15}N]-Ile and [^{13}C,^{15}N]-Ala would give only two $^{13}C'$-^{15}N *pairwise* peptide bonds (Ala^2-Ile^3 and Ile^3-Ile^4). A 2D N-CO MAS correlation experiment would give five cross-peaks

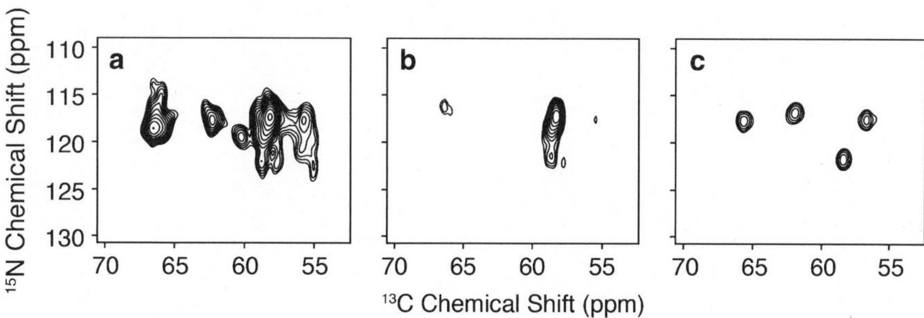

Fig. 3.4 MAS N-CA 2D correlation spectra of PLN in lipid vesicles. (**a**) uniformly labeled, [U-^{13}C,^{15}N] PLN. (**b**) Selective Leu and Val labeled PLN obtained by addition of [Val and Leu-^{13}C,^{15}N] to the growth medium. Notice the severe overlap in both dimensions. (**c**) PLN labeled with ^{13}C,^{15}N at residues Asn^{30}-Leu^{31}-Phe^{32}-Ile^{33} produced by peptide synthesis

for the [U-^{13}C,^{15}N] labeling pattern and only two for the selective labeling, thus improving unambiguous assignment.

Sensitivity of the experiment in connection with the labeling pattern can be improved with new pulse sequences. We recently implemented a complementary approach to the standard backbone experiments that increases the sensitivity of 2D correlation spectra by ~25–40%. Our filtering approach is similar to the spin-echo difference technique developed by Bax and co-workers for solution NMR [114]. This pulse sequence with a schematic and the results are shown in Fig. 3.6b. Broadly, we classify this approach as selective labeling with filtering blocks in pulse sequences to reduce the amount of peaks in the spectrum. This approach incorporates frequency selective REDOR with the N-CA selective CP of Baldus et al. [116]. Recently this approach has been extended to acquire multiple heteronuclear correlation datasets at the same time using afterglow magnetization from the cross-polarization experiment [117].

Residue-type labeling can also be employed in MAS SSNMR with selective amino acids that are not prone to scrambling. For instance, this approach has been utilized with 4-^{19}F-phenylalanine and 4-^{13}C-tyrosine to probe distances in the $\alpha_2\beta_2$ tetrameric enzyme tryptophan synthase using REDOR spectroscopy [118]. An extension of residue-type labeling is achieved using *reverse labeling* or *unlabeling*. These approaches utilize U-^{13}C glucose in the growth medium with isotopically unlabeled amino acids to produce a labeling pattern that labels those amino acids that were not supplied in the growth medium [119, 120]. This can be very advantageous, since several of these amino acids can be quite expensive to purchase, and would scramble in the growth as previously mentioned above.

3.4.1.4   Metabolic Labeling with Precursors in MAS SSNMR

An emerging approach for diluting the spin system in MAS SSNMR is the use of metabolic precursors. This method is beneficial when ^{13}C is the nucleus for direct observation. Since the presence of J-couplings (35-60 Hz) can cause line broadening, removing one-bond J-couplings can substantially improve ^{13}C spectra resolution [121]. For broader resonances > 1 ppm, only minor improvement is expected. The most common ways of diluting the ^{13}C spins is by fractional labeling or use of specifically labeled precursors: glycerol (1,3-^{13}C-glycerol or 2-^{13}C-glycerol) (Fig. 3.5), glucose (1-^{13}C-glucose or 2-^{13}C-glucose), or pyruvate with bicarbonate labeling (Fig. 3.9). Note that there are many other precursors that can be used such as keto-acids (Fig. 3.3), but these labeling patterns are less common and primarily used for methyl group spectroscopy. In the following section, we will focus on obtaining the backbone labels, since these are the foremost challenge to assign crowded SSNMR spectra.

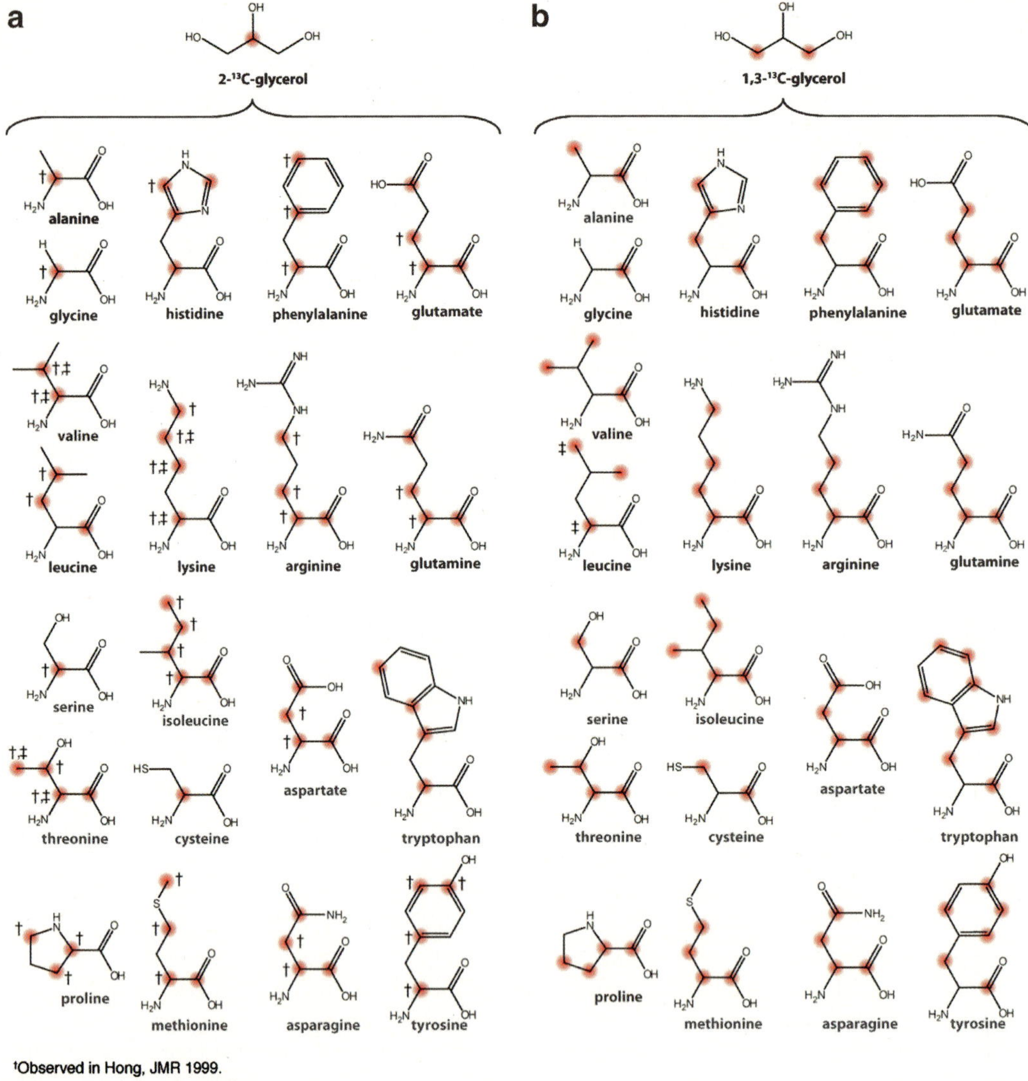

Fig. 3.5 Expected ¹³C distribution using **a**) 2-¹³C-glycerol or **b**) 1,3-¹³C-glycerol as the sole carbon source and *E. coli* BL21(DE3) strain. ¹³C labeled carbons are indicated in red two studies [121, 122] reported different results using 2-¹³C-glycerol therefore both are indicated in the labeling pattern for each amino acid

The original approach to dilute the spin system was simply to fractionally label the protein by using a mixture of unlabeled and labeled carbon source [124]. With this approach, the labels are distributed in a stochastic manner. A significant disadvantage is the lack of pairwise labeling to assign the simplified spectra. To overcome these problems, Hong and Jakes introduced the TEASE approach (<u>te</u>n-<u>a</u>mino acid <u>s</u>elective and <u>e</u>xtensive labeling), which utilizes 2-¹³C-glycerol, ¹⁵NH₄Cl isotopic sources and ten unlabeled amino acids (Asp, Asn, Arg, Gln, Glu, Ile, Lys, Met, Pro and Thr) [124]. This labeling scheme results in 100% ¹³C′ for Gly, Ala, Ser, Cys, Phe, Tyr, Trp, His, Val and 100% incorporation at ¹³Cα for Leu. To avoid or limit the fractional ¹³C or ¹⁵N labeling of these ten amino acids, they are added at natural abundance. Due to the use of unlabeled amino acids such as glutamine and glutamate, a two-fold dilution of ¹⁵N is obtained by this method. Likewise, the 1,3-¹³C-glycerol, gives 100% incorporation for nine amino acids (Gly, Ala, Ser, Cys, Phe, Tyr, Trp, His, Val) at the ¹³Cα

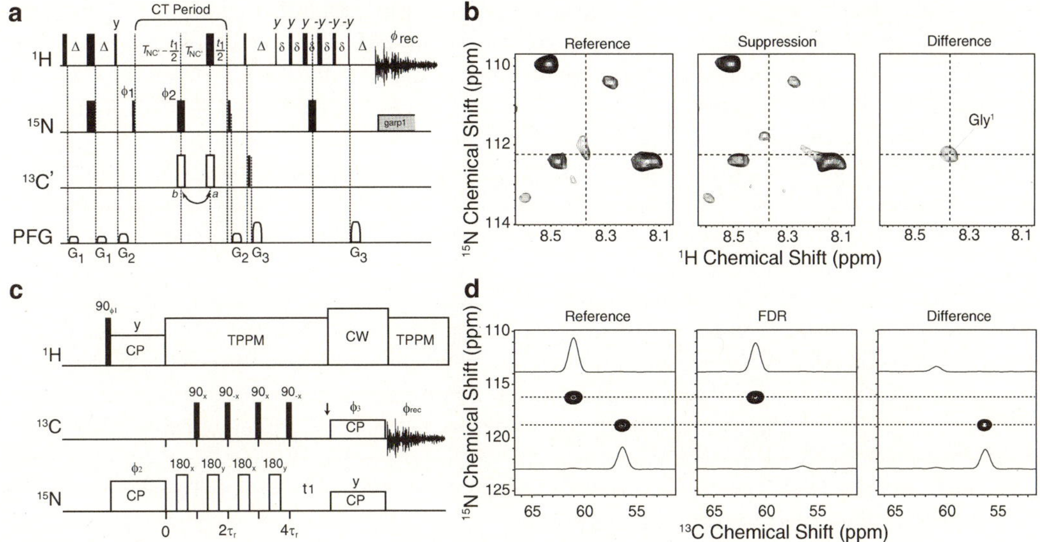

Fig. 3.6 (a–b) CCLS-HSQC. (**a**) Schematic of the CCLS-HSQC pulse sequence. (**b**) The reference spectrum is obtained by executing the pulse sequence with the 180° ^{13}C' pulse (open rectangle) at position a; the ^{13}C' suppressed spectrum is obtained with this pulse at position b. (c–d) Frequency-selective heteronuclear dephasing and selective carbonyl labeling to deconvolute crowded spectra of membrane proteins by magic angle spinning NMR. (**c**) Pulse sequence used to obtain 2D FDR-^{15}N–^{13}Cα. (**d**) FDR-^{15}N–^{13}Cα spectra for N-acetyl-valyl-leucine. Spectra were acquired with (FDR – red spectrum) and without ^{13}C 90° pulses (reference – black spectrum) (Reproduced with permission from Traaseth and Veglia [115])

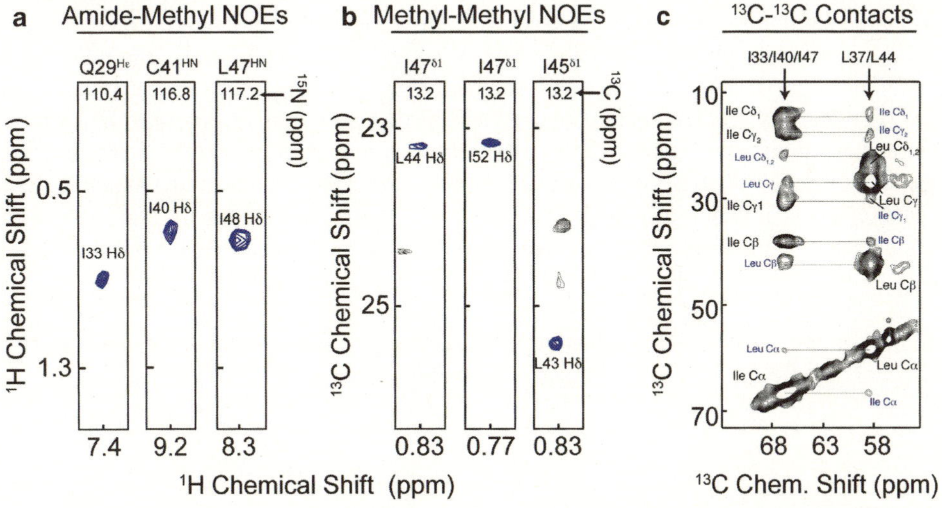

Fig. 3.7 Asymmetric labeling scheme for the detection of inter-protomer contacts in homo-oligomeric membrane proteins using solution and solid-state NMR. (**a**) 2D planes from 3D [^{1}H, ^{1}H, ^{15}N]-NOESY-HSQC (400 ms mixing time) on a mixed PLN sample with 1:1 ratio of [^{2}H-^{15}N] and [^{2}H-^{14}N-^{13}CH$_3$-Ile^{δ1}] PLN. (**b**) 2D planes from 3D [^{1}H, ^{13}C, ^{13}C]-HSQC-NOESY-HSQC experiment performed on a sample containing 1:1 ratio [^{2}H-^{14}N-^{13}CH$_3$-Ile^{δ1}] and [^{2}H-^{14}N-^{13}CH$_3$-Leu^{δ1}/Val^{γ1}] PLN. (**c**) 2D-DARR experiments (200 ms mixing time) on a 50% [U^{13}C]-Leu/ 50% [U^{13}C]-Ile PLN sample. Intra-residue and interprotomer cross-peaks are labeled in black and blue, respectively (Reproduced with permission from Verardi et al. [90])

site. If unlabeled amino acids are not provided as in the TEASE approach, the other ten amino acids are fractionally labeled [126]. In addition to direct J-coupling removal, diluting the ^{13}C spins can also reduce cross-relaxation between ^{13}C nuclei leading to both increased resolution and sensitivity [127].

While 1,3-^{13}C-glycerol and 2-^{13}C-glycerol labeling patterns are not ideal for backbone walk due to non-contiguous ^{13}C labels, improved ^{13}C-^{13}C correlations in spin diffusion experiments have been observed due to the reduction in dipolar truncation effects. Additionally, the 1,3-^{13}C-glycerol labeling scheme is useful to reduce spectral overlap in N-CO correlation spectra (Hiller M. et al. Application note 22, Cambridge Isotope Laboratory, Inc.).

To obtain isolated ^{13}C spins (i.e., non-bonded ^{13}C-^{13}C pairs in the backbone or side chain carbon atoms), it is possible to use 2-^{13}C-glucose. This method generates 20–45% enrichment at the ^{13}Cα position, with virtually no labeling at the ^{13}C′ for all residues with the exception of Leu, which is labeled at the ^{13}C′ position. In addition, all residues are devoid of ^{13}C$_\beta$ labeling with the exception of Leu, Val, and Ile residues. It is also possible to use 1-^{13}C-glucose. This labeling scheme enables the introduction of ^{13}C at the α-position of Leu and Ile, which are very abundant in membrane protein sequences. This labeling scheme also gives stretches of ^{13}C atoms such as ^{13}Cα-^{13}Cβ-^{13}Cγ for many residues, which can be useful for side chain detection. For a detailed summary of labeled atoms in 1-^{13}C-glucose and 2-^{13}C-glucose, see Figure 3 in Lundstrom et al. [128]. A combination of fractional labeling with selectively labeled precursors has also been used to achieve isolated spin systems. Wand et al. [124] used 15% [1-^{13}C-acetate], 15% [2-^{13}C-acetate], and 70% [1,2-^{12}C-acetate] to achieve isolated ^{13}C spins for relaxation experiments on ubiquitin.

3.4.2 Labeling Strategies for Oriented Solid-State NMR (O-SSNMR) Studies

While MAS has been used to study membrane proteins, fibrils, amorphous proteins, and crystalline proteins, O-SSNMR has been primarily used to study membrane protein structure and orientation [129–131]. Complete membrane protein structure determination requires characterization of the orientation of the membrane protein with respect to the lipid bilayer, i.e. topology. Since the energetic penalty for distorting the hydrogen bonding network is high in the lipid bilayer environment with low dielectric permeability [132], the O-SSNMR data has been often successfully interpreted assuming an idealized α-helical environment. Alternatively, O-SSNMR data can be incorporated in a total potential for structure minimization, restraining both protein topology and geometry [133–136]. Furthermore, the analysis of OSS NMR data from multiple isotopes can yield whole body dynamics of the transmembrane segments as well [137]. The O-SSNMR signal is dependent upon the angle θ between the interaction tensor components and the applied magnetic field according to the second order Legendre polynomial, $\frac{1}{2}\left(3\cos^2\theta - 1\right)$. The essential requirement for interpreting the θ angle in scope of the transmembrane domain orientation is that the NMR-active label must be rigidly attached to the polypeptide backbone. Below we discuss three different approaches in O-SSNMR, based on ^{15}N, ^{2}H and ^{19}F labeling.

3.4.2.1 Nitrogen Labeling in O-SSNMR

The most common way to determine the topology of a membrane protein is through separated local field experiments (SLF) such as PISEMA [138]. The PISEMA spectrum is considered the *fingerprint* for oriented membrane proteins, and is the most popular of the SLF class. The PISEMA experiment measures the anisotropic chemical shift of spin S and correlates it to the corresponding I-S dipolar coupling. Typically, the S spin is ^{15}N (although applicability of ^{13}C PISEMA has been illustrated [138]) and the dipolar coupling is ^{1}H-^{15}N, and correspondingly membrane proteins are either uniformly or selectively labeled with ^{15}N.

The PISEMA spectra result in periodic spectral patterns called PISA wheels [140–142]. From these wheels, it is possible to immediately obtain the tilt angles of helices or sheets with respect to the lipid bilayer normal, while determination of the rotation angle requires the assignment of the PISEMA spectrum.

As initial step, the macroscopic alignment of the protein is verified by acquiring a PISEMA spectrum using a U-^{15}N labeled protein. Often, small adjustments to the lipid composition, buffer, and temperature are necessary to find the best (homogenous) alignment. Once conditions are optimal, a high quality U-^{15}N PISEMA can be obtained that can be fit to obtain the global angle of orientation of the helices [140, 141]. One significant challenge that arises is how to assign a labeled PISEMA spectrum. There are several ways this can be done: (1) spin diffusion experiments with a single [U-^{15}N] sample [143, 144]; (2) assignment of isotropic ^{1}H and ^{15}N chemical shifts from solution NMR or MAS SSNMR in conjunction with a pair of flipped and unflipped aligned bicelle SLF spectra, requiring selective labeling [145]; (3) use of periodic assignment algorithms (based on PISA wheel) with uniform and/or selective labeled samples ("shotgun" approach) [146, 147].

Since chemical shifts are anisotropic in O-SSNMR, the orientation of the internuclear NH vector with respect to the magnetic field rather than residue-type or secondary structure determines the resonance position. This is a significant help to resolve spectral overlap in highly degenerate transmembrane helical segments. Nevertheless, uniformly labeled samples can still present severe spectral overlap and are often difficult to assign with selective labeling represents a reliable source for completing the assignments. Fortunately, the majority of the transmembrane helices are enriched with amino acids that have aliphatic side chains, which are not prone to isotopic scrambling. By labeling a protein sample with U-^{15}N-Leu or U-^{15}N-Ile, one can substantially decrease the complexity of the spectra. One can also use residue-specific labeling to determine accessibility as in H/D accessibility or proximity to a spin-label as is commonly done in solution NMR for membrane proteins [148].

Pairwise labeling utilized in solution NMR has not been extensively tested in O-SSNMR. This labeling scheme will be useful to resolve backbone resonances, when triple-resonance experiments will become routine for membrane proteins aligned in bicelles or mechanically aligned bilayers. In addition, isotopic dilution will reduce strong dipolar couplings and enable the acquisition of high quality spectra.

3.4.2.2 Deuterium Labeling in O-SSNMR

While the SLF experiments provide an initial picture of the IMP topology in lipid bilayers, they suffer from an intrinsically low sensitivity due to the orientation of the internuclear ^{15}N-H bond vectors, and in many cases where more precision is required it is often advantageous to employ isotopic labels which axes of interactions are positioned close to the magic angle relative to the helix axis. The combination of Φ-Ψ dihedral angles in a regular α-helix along with the tetrahedral geometry of the C_{α} carbon dictates that the C_{α}-C_{α} and C_{α}-H_{α} bond vectors form angles close to the magic angle with the helix axis (59.4° and 122.0° respectively) thus providing the maximum sensitivity for the interactions which are directed along these bonds. Alanine with a deuterated methyl group is therefore a natural choice for determining the topology of IMPs. Initial proof of concept has been carried out by labeling only a few residues at a time [149, 150] and the first systematic study was performed utilizing model Ala-rich peptides in a variety of lipid bilayers [151]. Since then deuterium NMR of methyl groups has been extensively applied for the investigation of antimicrobial peptides [152], IMPs [153], numerous model systems [154] and peptaibols [155].

Since deuterium NMR is recorded in a one-dimensional fashion typically employing a quadrupolar echo experiment [156] or quadrupolar CPMG [157], the spectral resolution precludes labeling of multiple sites, typically limiting the IMP to one or two labeled alanines. Unlike ^{1}H-^{15}N dipolar couplings, which retain a constant sign for transmembrane segments of IMPs, quadrupolar couplings oscillate between positive and negative values, but the sign typically cannot be determined experimentally, unless it exceeds ¾ of the quadrupolar coupling constant (i.e. >37 kHz for the methyl groups) in which case the sign must be positive. Such sign ambiguity necessitates employing multiple labels, or combining the methyl restraints with other O-SSNMR labels. Limited resolution that can be achieved in 1D experiment along with the complexity of the metabolic pathways limits deuterium NMR to the synthetic sequences.

The deuteron at an α-carbon presents an appealing supplement to the alanine methyl groups, since it is present in each of the canonical amino acids and its quadrupolar coupling undergoes major changes upon the transmembrane domain tilt or rotation. The attempts to employ $^2H_\alpha$ O-SSNMR has had limited success so far. In multiple single-span IMPs the backbone deuteron either could not be detected, or observed with extremely low sensitivity [151, 158]. Interestingly in several cases a significant increase in $^2H_\alpha$ signal intensity has been observed, which potentially relates to the peptide plane and/or whole body dynamics of IMPs [159].

These examples by no means cover all the uses of deuterium in oriented solid-state NMR of IMPs (for its use in solution and MAS NMR see above). Other applications include probing the aliphatic side chain dynamics [160–162], orientation of the Trp indoles [163], IMPs oligomerization [164], mobility of the lipidated IMPs [165, 166] as well as a multitude of studies of lipid bilayer membranes – IMPs hosts.

3.4.2.3   Fluorine Labeling in OSS NMR

For detailed considerations of ^{19}F O-SSNMR the reader is referred to the excellent recent review by Ulrich and co-workers [167], while we present a brief overview below. Fluorine is a highly appealing nucleus in biological NMR. High gyromagnetic ratio, 100% natural abundance of the NMR-active ^{19}F isotope and the lack of natural background leads to high sensitivity [168]. Care must be taken to exclude fluorinated solvents (e.g. trifluoroacetic acid, a frequent ion pairing additive) as well as fluorinated polymers from the probe assembly. Close Larmor frequencies of fluorine and hydrogen exert stringent requirements on the NMR hardware. Since biomolecules do not contain fluorine, unnatural amino acids, usually based on Phe, Pro or Aib, need to be introduced in the sequence synthetically, although promising results have been achieved with the genetic incorporation [169].

3.5   Isotopic Labeling for Protein-Protein Interaction Studies

A very useful application of methyl labeling (see Sect. 3.3.3) and uniform isotopic labeling (see Sect. 3.4.1.1) is found in the study of homo-oligomeric membrane proteins by NMR. Because of the symmetry of such molecules, the NMR signals are chemically equivalent; therefore only one set of resonances is observed. In order to obtain structural information about symmetric oligomers, asymmetric labeling strategies have been developed [91, 170, 171]. The objective of these strategies is to introduce "isotopic asymmetry" in the complexes. This can be done by labeling one of the protomers with a certain isotopic scheme and the other with a different scheme. Upon formation of the complex or oligomer, the intermolecular contacts can be unambiguously assigned. Pulse sequences can be designed to detect the dipolar contacts between the protomers [90, 170, 172].

We recently proposed two asymmetric labeling schemes to measure inter-protomer contacts in the pentameric phospholamban (PLN) for solution and solid-state NMR [90, 170]. PLN is homo-pentamer composed of five identical protomers (52 residues each). The transmembrane portion of each protomer consists of mainly hydrophobic amino acids Ile, Leu and Val, which are involved in keeping the oligomer together thorough hydrophobic interactions. The first labeling scheme was devised in order to probe inter-protomer contacts in detergent micelles by solution NMR. In this scheme, half of the protomers were labeled [U-^{2}H, ^{12}C, ^{14}N] and ^{13}CH$_3$ at the Ile$^{\delta 1}$ (using 2-ketobutyric acid-4-^{13}C,3,3-d$_2$ as precursor), whereas the other half was labeled [U-^{2}H, ^{12}C, ^{14}N] and ^{13}CH$_3$ at the Leu$^{\delta 1/2}$/Val$^{\gamma 1/}$ (using 2-keto-3-(methyl-d$_3$)-butyric acid-4-^{13}C as precursor). Using a methyl-methyl NOESY pulse sequence, it was possible to identify and unambiguously assign inter-protomer contacts, which were used for structure calculations (Fig. 3.7b). This I-LV methyl labeling scheme is very powerful since Ile$^{\delta 1}$ resonates at significantly different frequencies compared to Leu$^{\delta 1/2}$/Val$^{\gamma 1/2}$. Therefore the presence of inter-protomer contacts is straightforward to identify and correctly assign. This scheme can easily be extended to measure inter-protomer contacts between methyls and backbone amides, where half of the protein is uniformly (or selectively) labeled with ^{15}N at the amide groups in a deuterated background and half of the protein is methyl labeled at either Ile, or Leu/Val (Fig. 3.7a) [90]. A similar approach was used to identify inter-protomer contacts in lipid vesicles using MAS-NMR. In this case half of the protein was selectively labeled with ^{13}C using [^{13}C-Ile] amino acid and the other half was labeled with ^{13}C using [^{13}C-Leu] amino acid. The inter-protomer contacts were detected using a dipolar assisted rotational resonance (DARR) pulse sequence (Fig. 3.7c).

3.6 Post-expression Labeling

3.6.1 Post-expression Isotopic Labeling

There are several chemical methods to modify reactive amino acid side-chain groups after protein expression and purification [173]. By using isotopically labeled reagents, it is possible to selectively enrich amino acids with molecules containing NMR active isotopes. The most common residues whose side-chains can be chemically modified for NMR studies are cysteines, tyrosines and lysines.

The sulfhydryl group (−SH) of free cysteine in a protein can easily react in mild conditions with different chemical groups. Two applications that make use of the high nucleophilicity of free thiol groups in cysteines are the introduction of fluorine atoms and site directed methyl group substitution. In the first case, the NMR active ^{19}F is attached to cysteine by reaction of the free thiol with trifluoromethyl derivatives such as: 3-bromo-1,1,1-trifluoroacetone (BTFA) [174], trifluoroethylthio group (TET) [174], S-ethyl-trifluorothioacetate (SETFA) [176] and trifluoroacetamidosuccinican-hydride (TFASAN) [177]. This labeling approach has been successfully applied to the study of several proteins such as: citrate synthase [178], G-actin [179, 180], Myosin S-1/F-actin complex [181], SH3 domain [182], rhodopsin [175] and β2-Adrenergic Receptor [183]. Recently, Kay and co-workers introduced isotopically labeled methyl groups in cysteine side chains using methyl methanethiosulfonate to form ^{13}C-S-methylthiocysteine [184]. This labeling is very promising considering the advantages of observing methyl resonances by NMR and the fact that S-methylthiocysteine is very similar to a methionine residue, therefore it should not substantially alter the secondary structure of the protein. We have recently applied this approach to the selective methyl labeling of accessible cysteines in the 110 kDa integral membrane protein SERCA (sarcoplasmic reticulum Ca^{2+} ATPase) and obtained high-resolution solid and solution state NMR spectra (Fig. 3.8).

Another residue whose side chain can be modified is tyrosine. Richards et al. have proposed an electrochemical method for the nitration of the tyrosine ring at positions 3 in different proteins [185]. Tyrosine can also be mono-fluorinated by electrophilic substitution using acetyl hypofluorite in mild conditions and high yields (50–65%) [186].

Reductive methylation of lysine side chain has been used in many solution NMR studies to detect protein-protein interactions and ligand binding. The reaction occurs by addition of ^{13}C labeled form-aldehyde to the protein solution in reducing condition [187]. If sufficient formaldehyde is present, the

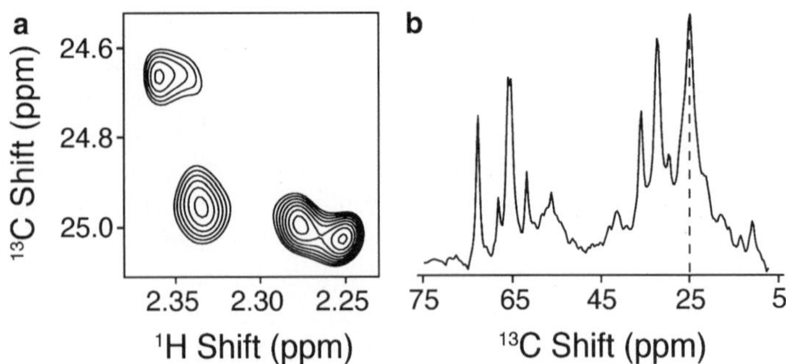

Fig. 3.8 Cysteine methylation of SERCA1a by methyl methanethiosulfonate (MMTS) reaction. (**a**) ^{1}H–^{13}C HSQC spectrum of ^{13}C methylthiocysteine in 100 mM ^{2}H dodecylphosphocholine acquired at 14.1 T field strength. (**b**) MAS one-dimensional cross-polarization of ^{13}C methylthiocysteine labeled SERCA1a in ^{2}H DMPC lipid vesicles run at −20 °C and spinning rate of 8,000 Hz acquired at 14.1 T field strength. Dashed lines indicate the peak corresponding to the labeled cysteines

side-chain of lysine residues will form a tertiary amine with two methyl-group substitutions [188]. This approach has been successfully applied by Kobilka and coworkers for the solution NMR study of the β2-Adrenergic Receptor [189].

3.6.2 Spin Labeling in NMR

Spin labeling refers to the covalent attachment of molecules with one or more unpaired electrons to proteins. Traditionally spin labeling has been used to study polypeptides by electron spin resonance; however, the effects of unpaired electron on the relaxation of nuclei is becoming routine in protein NMR studies [93, 190, 191]. Paramagnetic-based distance restraints have been used for the refinement of membrane protein structures [148] and for the positioning of membrane proteins in the lipid bilayers or detergent micelle [93].

Spin labeling is usually achieved post-translationally by *in vitro* chemical reactions involving cysteines through disulfide formation [192, 193] or lysines [173].

All these chemical methods must be used with caution to ensure that the reaction does not jeopardize the structural integrity or function of the protein. Furthermore, if the residues to be labeled are found buried in the core of soluble proteins or in transmembrane segments of membrane proteins, they might not be accessible to the labeling reagent.

3.7 Conclusions

The investigation of membrane proteins by NMR is a complex endeavor, but thanks to the development of improved instrumentation and production methods it is becoming increasingly feasible. New pulse sequences are continuously being devised that require specific labeling schemes, such as those described in this chapter. At the same time methods for the production of larger and more complex membrane proteins are also being actively developed.

Taken together, these accomplishments will permit an increasing number of medically relevant membrane proteins and protein complexes to be studied.

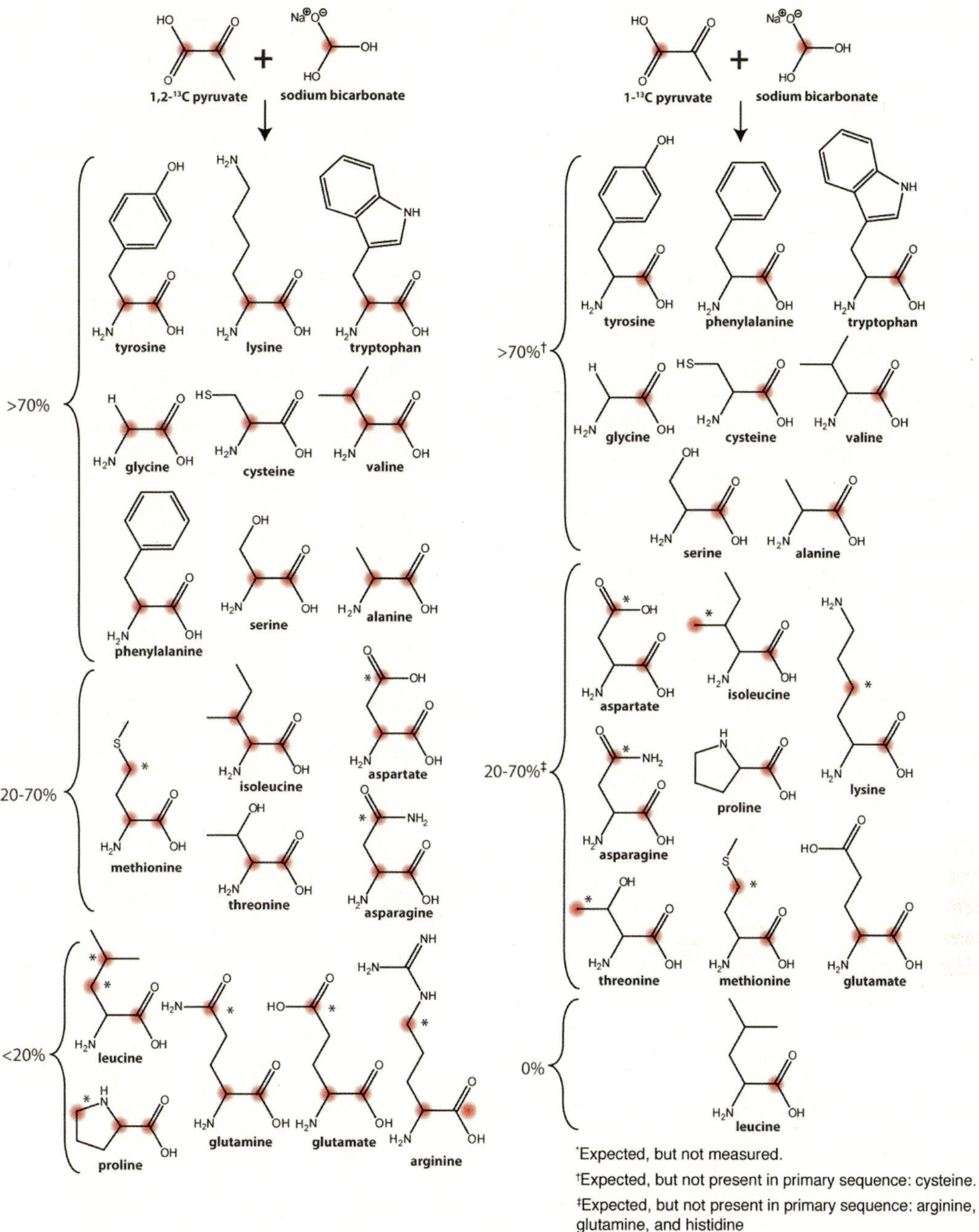

Fig. 3.9 Expected ¹³C distribution using pyruvate and sodium bicarbonate as the sole carbon sources in E. *coli* BL21(DE3). (a) 1,2-¹³C-pyruvate and NaH¹³CO₃ and (b) 1-¹³C-pyruvate and NaH¹³CO₃

Finally, we should point out that this chapter is not exhaustive of this field, which is in continuous evolution. Most of the examples reported are based on our own experience with membrane protein structural biology. The inevitable gaps present in this Chapter are filled in the other chapters of this book by outstanding scientist in the field of structural biology.

References

1. Crespi HL, Katz JJ (1969) High resolution proton magnetic resonance studies of fully deuterated and isotope hybrid proteins. Nature 224:560–562
2. Crespi HL, Rosenberg RM, Katz JJ (1968) Proton magnetic resonance of proteins fully deuterated except for 1H-leucine side chains. Science 161:795–796
3. Putter I, Barreto A, Markley JL, Jardetzky O (1969) Nuclear magnetic resonance studies of the structure and binding sites of enzymes. X. Preparation of selectively deuterated analogs of staphylococcal nuclease. Proc Natl Acad Sci USA 64:1396–1403
4. Markley JL, Putter I, Jardetzky O (1968) High-resolution nuclear magnetic resonance spectra of selectively deuterated staphylococcal nuclease. Science 161:1249–1251
5. Ohki S, Kainosho M (2008) Stable isotope labeling methods for protein NMR spectroscopy. Prog Nucl Magn Reson Spectrosc 53:208–226
6. Kim HJ, Howell SC, Van Horn WD, Jeon YH, Sanders CR (2009) Recent advances in the application of solution NMR spectroscopy to multi-span integral membrane proteins. Prog Nucl Magn Reson Spectrosc 55:335–360
7. Ruschak AM, Kay LE (2010) Methyl groups as probes of supra-molecular structure, dynamics and function. J Biomol NMR 46:75–87
8. Wallin E, von Heijne G (1998) Genome-wide analysis of integral membrane proteins from eubacterial, archaean, and eukaryotic organisms. Protein Sci 7:1029–1038
9. Ahram M, Litou ZI, Fang R, Al-Tawallbeh G (2006) Estimation of membrane proteins in the human proteome. In Silico Biol 6:379–386
10. Gilman AG (1987) G proteins: transducers of receptor-generated signals. Annu Rev Biochem 56:615–649
11. Wettschureck N, Offermanns S (2005) Mammalian G proteins and their cell type specific functions. Physiol Rev 85:1159–1204
12. Hille B (2001) Ion channels of excitable membranes. Sinauer Associates, Sunderland, pp 814, [8]
13. Brini M, Carafoli E (2009) Calcium pumps in health and disease. Physiol Rev 89:1341–1378
14. Traaseth NJ et al (2008) Structural and dynamic basis of phospholamban and sarcolipin inhibition of ca(2+)-ATPase. Biochemistry 47:3–13
15. Page RC et al (2006) Comprehensive evaluation of solution nuclear magnetic resonance spectroscopy sample preparation for helical integral membrane proteins. J Struct Funct Genomics 7:51–64
16. Eshaghi S et al (2005) An efficient strategy for high-throughput expression screening of recombinant integral membrane proteins. Protein Sci 14:676–683
17. Tate CG (2001) Overexpression of mammalian integral membrane proteins for structural studies. FEBS Lett 504:94–98
18. Sambrook J, Fritsch EF, Maniatis T (1989) Molecular cloning: a laboratory manual. Cold Spring Harbor Laboratory, Cold Spring Harbor
19. Ross A et al (2004) Optimised fermentation strategy for 13C/15N recombinant protein labelling in Escherichia coli for NMR-structure analysis. J Biotechnol 108:31–39
20. Cai M et al (1998) An efficient and cost-effective isotope labeling protocol for proteins expressed in Escherichia coli. J Biomol NMR 11:97–102
21. Marley J, Lu M, Bracken C (2001) A method for efficient isotopic labeling of recombinant proteins. J Biomol NMR 20:71–75
22. Studier FW (2005) Protein production by auto-induction in high density shaking cultures. Protein Expr Purif 41:207–234
23. Suzuki M, Mao L, Inouye M (2007) Single protein production (SPP) system in Escherichia coli. Nat Protoc 2:1802–1810
24. Schneider WM et al (2010) Efficient condensed-phase production of perdeuterated soluble and membrane proteins. J Struct Funct Genomics 11:143–154
25. Baneyx F, Mujacic M (2004) Recombinant protein folding and misfolding in Escherichia coli. Nat Biotechnol 22:1399–1408
26. Goldbourt A, Day LA, McDermott AE (2007) Assignment of congested NMR spectra: carbonyl backbone enrichment via the entner-doudoroff pathway. J Magn Reson 189:157–165
27. Kunji ER, Slotboom DJ, Poolman B (2003) Lactococcus lactis as host for overproduction of functional membrane proteins. Biochim Biophys Acta 1610:97–108
28. Janvilisri T, Shahi S, Venter H, Balakrishnan L, van Veen HW (2005) Arginine-482 is not essential for transport of antibiotics, primary bile acids and unconjugated sterols by the human breast cancer resistance protein (ABCG2). Biochem J 385:419–426
29. Koth CM, Payandeh J (2009) Strategies for the cloning and expression of membrane proteins. Adv Protein Chem Struct Biol 76:43–86

30. Lin-Cereghino J, Lin-Cereghino GP (2007) Vectors and strains for expression. Methods Mol Biol 389:11–26
31. Gossert AD et al (2011) A simple protocol for amino acid type selective isotope labeling in insect cells with improved yields and high reproducibility. J Biomol NMR 51(4):449–456
32. Werner K, Richter C, Klein-Seetharaman J, Schwalbe H (2008) Isotope labeling of mammalian GPCRs in HEK293 cells and characterization of the C-terminus of bovine rhodopsin by high resolution liquid NMR spectroscopy. J Biomol NMR 40:49–53
33. Stewart JM, Young JD (1984) Solid phase peptide synthesis. Pierce Chemical Co, Rockford, p 176
34. Klammt C et al (2007) Cell-free production of G protein-coupled receptors for functional and structural studies. J Struct Biol 158:482–493
35. Klammt C et al (2006) Cell-free expression as an emerging technique for the large scale production of integral membrane protein. FEBS J 273:4141–4153
36. Klammt C et al (2004) High level cell-free expression and specific labeling of integral membrane proteins. Eur J Biochem 271:568–580
37. Wu PS et al (2006) Amino-acid type identification in 15N-HSQC spectra by combinatorial selective 15N-labelling. J Biomol NMR 34:13–21
38. Ozawa K, Wu PS, Dixon NE, Otting G (2006) N-labelled proteins by cell-free protein synthesis. strategies for high-throughput NMR studies of proteins and protein-ligand complexes. FEBS J 273:4154–4159
39. Jeremy Craven C, Al-Owais M, Parker MJ (2007) A systematic analysis of backbone amide assignments achieved via combinatorial selective labelling of amino acids. J Biomol NMR 38:151–159
40. Parker MJ, Aulton-Jones M, Hounslow AM, Craven CJ (2004) A combinatorial selective labeling method for the assignment of backbone amide NMR resonances. J Am Chem Soc 126:5020–5021
41. Kainosho M, Guntert P (2009) SAIL – stereo-array isotope labeling. Q Rev Biophys 42:247–300
42. Xie H, Guo XM, Chen H (2009) Making the most of fusion tags technology in structural characterization of membrane proteins. Mol Biotechnol 42:135–145
43. Waugh DS (2005) Making the most of affinity tags. Trends Biotechnol 23:316–320
44. Arnau J, Lauritzen C, Petersen GE, Pedersen J (2006) Current strategies for the use of affinity tags and tag removal for the purification of recombinant proteins. Protein Expr Purif 48:1–13
45. Kapust RB, Waugh DS (2000) Controlled intracellular processing of fusion proteins by TEV protease. Protein Expr Purif 19:312–318
46. Abdullah N, Chase HA (2005) Removal of poly-histidine fusion tags from recombinant proteins purified by expanded bed adsorption. Biotechnol Bioeng 92:501–513
47. Jenny RJ, Mann KG, Lundblad RL (2003) A critical review of the methods for cleavage of fusion proteins with thrombin and factor xa. Protein Expr Purif 31:1–11
48. Kapust RB, Tozser J, Copeland TD, Waugh DS (2002) The P1′ specificity of tobacco etch virus protease. Biochem Biophys Res Commun 294:949–955
49. Buck B et al (2003) Overexpression, purification, and characterization of recombinant ca-ATPase regulators for high-resolution solution and solid-state NMR studies. Protein Expr Purif 30:253–261
50. Hu J et al (2007) Structural biology of transmembrane domains: efficient production and characterization of transmembrane peptides by NMR. Protein Sci 16:2153–2165
51. McIntosh LP, Dahlquist FW (1990) Biosynthetic incorporation of 15N and 13C for assignment and interpretation of nuclear magnetic resonance spectra of proteins. Q Rev Biophys 23:1–38
52. Hoogstraten CG, Johnson JE (2008) Metabolic labeling: taking advantage of bacterial pathways to prepare spectroscopically useful isotope patterns in proteins and nucleic acids. Concepts Magn Reson A 32A:34–55
53. Fiaux J, Bertelsen EB, Horwich AL, Wuthrich K (2004) Uniform and residue-specific 15N-labeling of proteins on a highly deuterated background. J Biomol NMR 29:289–297
54. Suzuki H et al (2005) Isotopic labeling of proteins by utilizing photosynthetic bacteria. Anal Biochem 347:324–326
55. LeMaster DM, LaIuppa JC, Kushlan DM (1994) Differential deuterium isotope shifts and one-bond 1H-13C scalar couplings in the conformational analysis of protein glycine residues. J Biomol NMR 4:863–870
56. Grzesiek S, Anglister J, Ren H, Bax A (1993) Carbon-13 line narrowing by deuterium decoupling in deuterium/carbon-13/nitrogen-15 enriched proteins. Application to triple resonance 4D J connectivity of sequential amides. J Am Chem Soc 115:4369–4370
57. Gardner KH, Kay LE (1998) The use of 2H, 13C, 15N multidimensional NMR to study the structure and dynamics of proteins. Annu Rev Biophys Biomol Struct 27:357–406
58. White SH, Wimley WC (1999) Membrane protein folding and stability: physical principles. Annu Rev Biophys Biomol Struct 28:319–365
59. Veglia G, Zeri AC, Ma C, Opella SJ (2002) Deuterium/hydrogen exchange factors measured by solution nuclear magnetic resonance spectroscopy as indicators of the structure and topology of membrane proteins. Biophys J 82:2176–2183
60. Oxenoid K, Kim HJ, Jacob J, Sonnichsen FD, Sanders CR (2004) NMR assignments for a helical 40 kDa membrane protein. J Am Chem Soc 126:5048–5049

61. Katz JJ, Crespi HL (1966) Deuterated organisms: cultivation and uses. Science 151:1187–1194
62. Meilleur F, Contzen J, Myles DA, Jung C (2004) Structural stability and dynamics of hydrogenated and perdeuterated cytochrome P450cam (CYP101). Biochemistry 43:8744–8753
63. Brockwell D et al (2001) Physicochemical consequences of the perdeuteriation of glutathione S-transferase from S. japonicum. Protein Sci 10:572–580
64. Schubert M, Smalla M, Schmieder P, Oschkinat H (1999) MUSIC in triple-resonance experiments: amino acid type-selective (1)H-(15)N correlations. J Magn Reson 141:34–43
65. Muchmore DC, McIntosh LP, Russell CB, Anderson DE, Dahlquist FW (1989) Expression and nitrogen-15 labeling of proteins for proton and nitrogen-15 nuclear magnetic resonance. Methods Enzymol 177:44–73
66. Waugh DS (1996) Genetic tools for selective labeling of proteins with alpha-15N-amino acids. J Biomol NMR 8:184–192
67. LeMaster DM, Kushlan DM (1996) Dynamical mapping of *E. coli* thioredoxin via 13C NMR relaxation analysis. J Am Chem Soc 118:9255–9264
68. Lin MT et al (2011) A rapid and robust method for selective isotope labeling of proteins. Methods 55:370–378
69. Vance CK, Kang YM, Miller AF (1997) Selective 15N labeling and direct observation by NMR of the active-site glutamine of fe-containing superoxide dismutase. J Biomol NMR 9:201–206
70. Maslennikov I et al (2010) Membrane domain structures of three classes of histidine kinase receptors by cell-free expression and rapid NMR analysis. Proc Natl Acad Sci USA 107:10902–10907
71. Sobhanifar S et al (2010) Cell-free expression and stable isotope labelling strategies for membrane proteins. J Biomol NMR 46:33–43
72. Makino S, Goren MA, Fox BG, Markley JL (2010) Cell-free protein synthesis technology in NMR high-throughput structure determination. Methods Mol Biol 607:127–147
73. Reckel S et al (2008) Transmembrane segment enhanced labeling as a tool for the backbone assignment of alpha-helical membrane proteins. Proc Natl Acad Sci USA 105:8262–8267
74. Cellitti SE et al (2008) In vivo incorporation of unnatural amino acids to probe structure, dynamics, and ligand binding in a large protein by nuclear magnetic resonance spectroscopy. J Am Chem Soc 130:9268–9281
75. Jones DH et al (2010) Site-specific labeling of proteins with NMR-active unnatural amino acids. J Biomol NMR 46:89–100
76. Jackson JC, Hammill JT, Mehl RA (2007) Site-specific incorporation of a (19)F-amino acid into proteins as an NMR probe for characterizing protein structure and reactivity. J Am Chem Soc 129:1160–1166
77. Xie J, Schultz PG (2005) Adding amino acids to the genetic repertoire. Curr Opin Chem Biol 9:548–554
78. Xie J, Schultz PG (2005) An expanding genetic code. Methods 36:227–238
79. Gerig JT (1994) Fluorine NMR of proteins. Prog Nucl Magn Reson Spectrosc 26(Part 4):293–370
80. Danielson MA, Falke JJ (1996) Use of 19F NMR to probe protein structure and conformational changes. Annu Rev Biophys Biomol Struct 25:163–195
81. Prosser RS, Luchette PA, Westerman PW (2000) Using O2 to probe membrane immersion depth by 19F NMR. Proc Natl Acad Sci USA 97:9967–9971
82. Kitevski-LeBlanc JL, Evanics F, Prosser RS (2009) Approaches for the measurement of solvent exposure in proteins by 19F NMR. J Biomol NMR 45:255–264
83. Skrisovska L, Schubert M, Allain FH (2010) Recent advances in segmental isotope labeling of proteins: NMR applications to large proteins and glycoproteins. J Biomol NMR 46:51–65
84. Goto NK, Gardner KH, Mueller GA, Willis RC, Kay LE (1999) A robust and cost-effective method for the production of val, leu, ile (delta 1) methyl-protonated 15N-, 13C-, 2H-labeled proteins. J Biomol NMR 13:369–374
85. Janin J, Miller S, Chothia C (1988) Surface, subunit interfaces and interior of oligomeric proteins. J Mol Biol 204:155–164
86. Miller S, Janin J, Lesk AM, Chothia C (1987) Interior and surface of monomeric proteins. J Mol Biol 196:641–656
87. Miller S, Lesk AM, Janin J, Chothia C (1987) The accessible surface area and stability of oligomeric proteins. Nature 328:834–836
88. Imai S, Osawa M, Takeuchi K, Shimada I (2010) Structural basis underlying the dual gate properties of KcsA. Proc Natl Acad Sci USA 107:6216–6221
89. Hiller S et al (2008) Solution structure of the integral human membrane protein VDAC-1 in detergent micelles. Science 321:1206–1210
90. Verardi R, Shi L, Traaseth NJ, Walsh N, Veglia G (2011) Structural topology of phospholamban pentamer in lipid bilayers by a hybrid solution and solid-state NMR method. Proc Natl Acad Sci USA 108:9101–9106
91. Oxenoid K, Chou JJ (2005) The structure of phospholamban pentamer reveals a channel-like architecture in membranes. Proc Natl Acad Sci USA 102:10870–10875
92. Zhou Y et al (2008) NMR solution structure of the integral membrane enzyme DsbB: functional insights into DsbB-catalyzed disulfide bond formation. Mol Cell 31:896–908

93. Zhou DH et al (2007) Solid-state protein-structure determination with proton-detected triple-resonance 3D magic-angle-spinning NMR spectroscopy. Angew Chem Int Ed Engl 46:8380–8383

94. Hologne M, Faelber K, Diehl A, Reif B (2005) Characterization of dynamics of perdeuterated proteins by MAS solid-state NMR. J Am Chem Soc 127:11208–11209

95. Paulson EK et al (2003) Sensitive high resolution inverse detection NMR spectroscopy of proteins in the solid state. J Am Chem Soc 125:15831–15836

96. Huang KY, Siemer AB, McDermott AE (2011) Homonuclear mixing sequences for perdeuterated proteins. J Magn Reson 208:122–127

97. Reif B et al (2012) Ultra-high resolution in MAS solid-state NMR of perdeuterated proteins: implications for structure and dynamics. J Magn Reson 216:1–12

98. Wickramasinghe NP, Kotecha M, Samoson A, Past J, Ishii Y (2007) Sensitivity enhancement in (13)C solid-state NMR of protein microcrystals by use of paramagnetic metal ions for optimizing (1)H T(1) relaxation. J Magn Reson 184:350–356

99. Akbey U et al (2010) Dynamic nuclear polarization of deuterated proteins. Angew Chem Int Ed Engl 49: 7803–7806

100. Lalli D et al (2011) Three-dimensional deuterium-carbon correlation experiments for high-resolution solid-state MAS NMR spectroscopy of large proteins. J Biomol NMR 51:477–485

101. Gopinath T, Veglia G (2012) Dual acquisition magic-angle spinning solid-state NMR-spectroscopy: simultaneous acquisition of multidimensional spectra of biomacromolecules. Angew Chem Int Ed Engl 51:2731–2735

102. Chekmenev EY et al (2006) Ion-binding study by 17O solid-state NMR spectroscopy in the model peptide gly-gly-gly at 19.6 T. J Am Chem Soc 128:9849–9855

103. Strandberg E et al (2004) Tilt angles of transmembrane model peptides in oriented and non-oriented lipid bilayers as determined by 2H solid-state NMR. Biophys J 86:3709–3721

104. Cady SD, Goodman C, Tatko CD, DeGrado WF, Hong M (2007) Determining the orientation of uniaxially rotating membrane proteins using unoriented samples: a 2H, 13C, AND 15N solid-state NMR investigation of the dynamics and orientation of a transmembrane helical bundle. J Am Chem Soc 129:5719–5729

105. Mani R et al (2006) Membrane-bound dimer structure of a beta-hairpin antimicrobial peptide from rotational-echo double-resonance solid-state NMR. Biochemistry 45:8341–8349

106. Buffy JJ, Waring AJ, Hong M (2005) Determination of peptide oligomerization in lipid bilayers using 19F spin diffusion NMR. J Am Chem Soc 127:4477–4483

107. Kandasamy SK et al (2009) Solid-state NMR and molecular dynamics simulations reveal the oligomeric ion-channels of TM2-GABA(A) stabilized by intermolecular hydrogen bonding. Biochim Biophys Acta 1788: 686–695

108. Liu W, Fei JZ, Kawakami T, Smith SO (2007) Structural constraints on the transmembrane and juxtamembrane regions of the phospholamban pentamer in membrane bilayers: Gln29 and Leu52. Biochim Biophys Acta 1768: 2971–2978

109. Jaroniec CP, MacPhee CE, Astrof NS, Dobson CM, Griffin RG (2002) Molecular conformation of a peptide fragment of transthyretin in an amyloid fibril. Proc Natl Acad Sci USA 99:16748–16753

110. Jaroniec CP et al (2004) High-resolution molecular structure of a peptide in an amyloid fibril determined by magic angle spinning NMR spectroscopy. Proc Natl Acad Sci USA 101:711–716

111. Tycko R (2006) Molecular structure of amyloid fibrils: insights from solid-state NMR. Q Rev Biophys 39:1–55

112. Doherty T, Su Y, Hong M (2010) High-resolution orientation and depth of insertion of the voltage-sensing S4 helix of a potassium channel in lipid bilayers. J Mol Biol 401:642–652

113. Gustavsson M, Traaseth NJ, Veglia G (2011) Probing ground and excited states of phospholamban in model and native lipid membranes by magic angle spinning NMR spectroscopy. Biochim Biophys Acta 1818:146–153

114. Vuister GW, Yamazaki T, Torchia DA, Bax A (1993) Measurement of two- and three-bond 13C-1H J couplings to the C delta carbons of leucine residues in staphylococcal nuclease. J Biomol NMR 3:297–306

115. Traaseth NJ, Veglia G (2011) Frequency-selective heteronuclear dephasing and selective carbonyl labeling to deconvolute crowded spectra of membrane proteins by magic angle spinning NMR. J Magn Reson 211:18–24

116. Baldus M, Petkova AT, Herzfeld J, Griffin RG (1998) Cross polarization in the tilted frame: assignment and spectral simplification in heteronuclear spin systems. Mol Phys 95:1197–1207

117. Banigan and Traaseth (2012), J Phys Chem B, 116(24):7138-44

118. McDowell LM, Lee M, McKay RA, Anderson KS, Schaefer J (1996) Intersubunit communication in tryptophan synthase by carbon-13 and fluorine-19 REDOR NMR. Biochemistry 35:3328–3334

119. Krishnarjuna B, Jaipuria G, Thakur A, D'Silva P, Atreya HS (2011) Amino acid selective unlabeling for sequence specific resonance assignments in proteins. J Biomol NMR 49:39–51

120. Vuister GW, Kim S, Wu C, Bax A (1994) 2D and 3D NMR study of phenylalanine residues in proteins by reverse isotopic labeling. J Am Chem Soc 116:9206–9210

121. Bystrov VF (1976) Spin–spin coupling and the conformational states of peptide systems. Prog Nucl Magn Reson Spectrosc 10:41–82

122. Hong M (1999) Resonance assignment of 13C/15N labeled solid proteins by two- and three-dimensional magic-angle-spinning NMR. J Biomol NMR 15:1–14
123. Takeuchi K, Gal M, Takahashi H, Shimada I, Wagner G (2011) HNCA-TOCSY-CANH experiments with alternate (13)C- (12)C labeling: a set of 3D experiment with unique supra-sequential information for mainchain resonance assignment. J Biomol NMR 49:17–26
124. Wand AJ, Bieber RJ, Urbauer JL, McEvoy RP, Gan Z (1995) Carbon relaxation in randomly fractionally 13C-enriched proteins. J Magn Reson B 108:173–175
125. Hong M, Jakes K (1999) Selective and extensive 13C labeling of a membrane protein for solid-state NMR investigations. J Biomol NMR 14:71–74
126. Higman VA et al (2009) Assigning large proteins in the solid state: a MAS NMR resonance assignment strategy using selectively and extensively 13C-labelled proteins. J Biomol NMR 44:245–260
127. Castellani F et al (2002) Structure of a protein determined by solid-state magic-angle-spinning NMR spectroscopy. Nature 420:98–102
128. Lundstrom P et al (2007) Fractional 13C enrichment of isolated carbons using [1–13C]- or [2–13C]-glucose facilitates the accurate measurement of dynamics at backbone calpha and side-chain methyl positions in proteins. J Biomol NMR 38:199–212
129. McDermott A (2009) Structure and dynamics of membrane proteins by magic angle spinning solid-state NMR. Annu Rev Biophys 38:385–403
130. Naito A (2009) Structure elucidation of membrane-associated peptides and proteins in oriented bilayers by solid-state NMR spectroscopy. Solid State Nucl Magn Reson 36:67–76
131. Marassi FM et al (2011) Structure determination of membrane proteins in five easy pieces. Methods 55:363–369
132. Bowie JU (2011) Membrane protein folding: how important are hydrogen bonds? Curr Opin Struct Biol 21:42–49
133. Bertram R et al (2003) Atomic refinement with correlated solid-state NMR restraints. J Magn Reson 163:300–309
134. Traaseth NJ et al (2009) Structure and topology of monomeric phospholamban in lipid membranes determined by a hybrid solution and solid-state NMR approach. Proc Natl Acad Sci USA 106:10165–10170
135. Shi L et al (2009) A refinement protocol to determine structure, topology, and depth of insertion of membrane proteins using hybrid solution and solid-state NMR restraints. J Biomol NMR 44:195–205
136. Straus SK, Scott WR, Schwieters CD, Marvin DA (2011) Consensus structure of Pf1 filamentous bacteriophage from X-ray fibre diffraction and solid-state NMR. Eur Biophys J 40:221–234
137. Vostrikov VV, Grant CV, Opella SJ, Koeppe 2nd RE (2011) On the combined analysis of ^{2}H and ^{15}N/^{1}H solid-state NMR data for determination of transmembrane peptide orientation and dynamics, Biophys J 101:2939–2947
138. Wu CH, Ramamoorthy A, Opella SJ (1994) High-resolution heteronuclear dipolar solid-state NMR spectroscopy. J Magn Reson Ser A 109:270–272
139. Sinha N et al (2007) Tailoring 13C labeling for triple-resonance solid-state NMR experiments on aligned samples of proteins. Magn Reson Chem 45(Suppl 1):S107–S115
140. Marassi FM, Opella SJ (2000) A solid-state NMR index of helical membrane protein structure and topology. J Magn Reson 144:150–155
141. Wang J et al (2000) Imaging membrane protein helical wheels. J Magn Reson 144:162–167
142. Page RC, Kim S, Cross TA (2008) Transmembrane helix uniformity examined by spectral mapping of torsion angles. Structure 16:787–797
143. Mote KR et al (2011) Multidimensional oriented solid-state NMR experiments enable the sequential assignment of uniformly (15)N labeled integral membrane proteins in magnetically aligned lipid bilayers. J Biomol NMR 51:339–346
144. Knox RW, Lu GJ, Opella SJ, Nevzorov AA (2010) A resonance assignment method for oriented-sample solid-state NMR of proteins. J Am Chem Soc 132:8255–8257
145. Lu GJ, Son WS, Opella SJ (2011) A general assignment method for oriented sample (OS) solid-state NMR of proteins based on the correlation of resonances through heteronuclear dipolar couplings in samples aligned parallel and perpendicular to the magnetic field. J Magn Reson 209:195–206
146. Nevzorov AA, Opella SJ (2003) Structural fitting of PISEMA spectra of aligned proteins. J Magn Reson 160:33–39
147. Asbury T et al (2006) PIPATH: an optimized algorithm for generating alpha-helical structures from PISEMA data. J Magn Reson 183:87–95
148. Shi L et al (2011) Paramagnetic-based NMR restraints lift residual dipolar coupling degeneracy in multidomain detergent-solubilized membrane proteins. J Am Chem Soc 133:2232–2241
149. Jones DH, Barber KR, VanDerLoo EW, Grant CW (1998) Epidermal growth factor receptor transmembrane domain: 2H NMR implications for orientation and motion in a bilayer environment. Biochemistry 37:16780–16787
150. Whiles JA et al (2001) Orientation and effects of mastoparan X on phospholipid bicelles. Biophys J 80:280–293

151. van der Wel PC, Strandberg E, Killian JA, Koeppe 2nd RE (2002) Geometry and intrinsic tilt of a tryptophan-anchored transmembrane alpha-helix determined by (2)H NMR. Biophys J 83:1479–1488
152. Strandberg E, Wadhwani P, Tremouilhac P, Durr UH, Ulrich AS (2006) Solid-state NMR analysis of the PGLa peptide orientation in DMPC bilayers: structural fidelity of 2H-labels versus high sensitivity of 19F-NMR. Biophys J 90:1676–1686
153. Resende JM et al (2009) Membrane structure and conformational changes of the antibiotic heterodimeric peptide distinctin by solid-state NMR spectroscopy. Proc Natl Acad Sci USA 106:16639–16644
154. Vostrikov VV, Hall BA, Greathouse DV, Koeppe 2nd RE, Sansom MS (2010) Changes in transmembrane helix alignment by arginine residues revealed by solid-state NMR experiments and coarse-grained MD simulations. J Am Chem Soc 132:5803–5811
155. Bertelsen K et al (2011) Long-term-stable ether-lipid vs conventional ester-lipid bicelles in oriented solid-state NMR: altered structural information in studies of antimicrobial peptides. J Phys Chem B 115:1767–1774
156. Davis JH, Maraviglia B, Weeks G, Godin DV (1979) Bilayer rigidity of the erythrocyte membrane2H-NMR of a perdeuterated palmitic acid probe. Biochim Biophys Acta 550:362–366
157. Larsen FH, Jakobsen HJ, Ellis PD, Nielsen NC (1998) QCPMG-MAS NMR of half-integer quadrupolar nuclei. J Magn Reson 131:144–147
158. Killian JA, Taylor MJ, Koeppe 2nd RE (1992) Orientation of the valine-1 side chain of the gramicidin transmembrane channel and implications for channel functioning. A ^{2}H NMR study. Biochemistry 31:11283–11290
159. Thomas R, Vostrikov VV, Greathouse DV, Koeppe 2nd RE (2009) Influence of proline upon the folding and geometry of the WALP19 transmembrane peptide. Biochemistry 48:11883–11891
160. Abu-Baker S et al (2007) Side chain and backbone dynamics of phospholamban in phospholipid bilayers utilizing 2H and 15N solid-state NMR spectroscopy. Biochemistry 46:11695–11706
161. Vold RL, Hoatson GL (2009) Effects of jump dynamics on solid state nuclear magnetic resonance line shapes and spin relaxation times. J Magn Reson 198:57–72
162. Vugmeyster L et al (2011) Slow motions in the hydrophobic core of chicken villin headpiece subdomain and their contributions to configurational entropy and heat capacity from solid-state deuteron NMR measurements. Biochemistry 50:10637–10646
163. van der Wel PC, Reed ND, Greathouse DV, Koeppe 2nd RE (2007) Orientation and motion of tryptophan interfacial anchors in membrane-spanning peptides. Biochemistry 46:7514–7524
164. Liu W, Crocker E, Siminovitch DJ, Smith SO (2003) Role of side-chain conformational entropy in transmembrane helix dimerization of glycophorin A. Biophys J 84:1263–1271
165. Struppe J, Komives EA, Taylor SS, Vold RR (1998) 2H NMR studies of a myristoylated peptide in neutral and acidic phospholipid bicelles. Biochemistry 37:15523–15527
166. Gaffarogullari EC et al (2011) A myristoyl/phosphoserine switch controls cAMP-dependent protein kinase association to membranes. J Mol Biol 411:823–836
167. Koch K, Afonin S, Ieronimo M, Berditsch M, Ulrich AS (2012) Solid-state (19)F-NMR of peptides in native membranes. Top Curr Chem 306:89–118
168. Luo W, Mani R, Hong M (2007) Side-chain conformation of the M2 transmembrane peptide proton channel of influenza a virus from 19F solid-state NMR. J Phys Chem B 111:10825–10832
169. Young TS, Schultz PG (2010) Beyond the canonical 20 amino acids: expanding the genetic lexicon. J Biol Chem 285:11039–11044
170. Traaseth NJ, Verardi R, Veglia G (2008) Asymmetric methyl group labeling as a probe of membrane protein homo-oligomers by NMR spectroscopy. J Am Chem Soc 130:2400–2401
171. Walters KJ et al (2001) Characterizing protein-protein complexes and oligomers by nuclear magnetic resonance spectroscopy. Methods Enzymol 339:238–258
172. Yang J, Tasayco ML, Polenova T (2008) Magic angle spinning NMR experiments for structural studies of differentially enriched protein interfaces and protein assemblies. J Am Chem Soc 130:5798–5807
173. Kosen PA (1989) Spin labeling of proteins. Methods Enzymol 177:86–121
174. Nelson DJ (1978) Fluorine-19 magnetic resonance of muscle calcium binding parvalbumin: PH dependency of resonance position and spin–lattice relaxation time. Inorg Chim Acta 27:L71–L74
175. Klein-Seetharaman J, Getmanova EV, Loewen MC, Reeves PJ, Khorana HG (1999) NMR spectroscopy in studies of light-induced structural changes in mammalian rhodopsin: applicability of solution (19)F NMR. Proc Natl Acad Sci USA 96:13744–13749
176. Adriaensens P et al (1988) Investigation of protein structure by means of 19F-NMR. A study of hen egg-white lysozyme. Eur J Biochem 177:383–394
177. Mehta VD, Kulkarni PV, Mason RP, Constantinescu A, Antich PP (1994) Fluorinated proteins as potential 19F magnetic resonance imaging and spectroscopy agents. Bioconjug Chem 5:257–261
178. Donald LJ, Crane BR, Anderson DH, Duckworth HW (1991) The role of cysteine 206 in allosteric inhibition of Escherichia coli citrate synthase. studies by chemical modification, site-directed mutagenesis, and 19F NMR. J Biol Chem 266:20709–20713

179. Phillips L, Separovic F, Cornell BA, Barden JA, dos Remedios CG (1991) Actin dynamics studied by solid-state NMR spectroscopy. Eur Biophys J 19:147–155
180. Brauer M, Sykes BD (1986) 19F nuclear magnetic resonance studies of selectively fluorinated derivatives of G- and F-actin. Biochemistry 25:2187–2191
181. Kay LE, Pascone JM, Sykes BD, Shriver JW (1987) 19F nuclear magnetic resonance as a probe of structural transitions and cooperative interactions in heavy meromyosin. J Biol Chem 262:1984–1988
182. Evanics F, Kitevski JL, Bezsonova I, Forman-Kay J, Prosser RS (2007) 19F NMR studies of solvent exposure and peptide binding to an SH3 domain. Biochim Biophys Acta 1770:221–230
183. Liu JJ, Horst R, Katritch V, Stevens RC, Wuthrich K (2012) Biased signaling pathways in beta2-adrenergic receptor characterized by 19F-NMR. Science 335:1106–1110
184. Religa TL, Ruschak AM, Rosenzweig R, Kay LE (2011) Site-directed methyl group labeling as an NMR probe of structure and dynamics in supramolecular protein systems: applications to the proteasome and to the ClpP protease. J Am Chem Soc 133:9063–9068
185. Richards PG, Coles B, Heptinstall J, Walton DJ (1994) Electrochemical modification of lysozyme: anodic reaction of tyrosine residues. Enzyme Microb Technol 16:795–801
186. Hebel D, Kirk KL, Cohen LA, Labroo VM (1990) First direct fluorination of tyrosine-containing biologically active peptides. Tetrahedron Lett 31:619–622
187. Abraham SJ, Hoheisel S, Gaponenko V (2008) Detection of protein-ligand interactions by NMR using reductive methylation of lysine residues. J Biomol NMR 42:143–148
188. Ivan R (1997) Macromolecular crystallography part A. In: Charles W, Carter J (eds) Methods in enzymology. Academic Press, New York, pp 171–179
189. Bokoch MP et al (2010) Ligand-specific regulation of the extracellular surface of a G-protein-coupled receptor. Nature 463:108–112
190. Su XC, Otting G (2010) Paramagnetic labelling of proteins and oligonucleotides for NMR. J Biomol NMR 46:101–112
191. Berardi MJ, Shih WM, Harrison SC, Chou JJ (2011) Mitochondrial uncoupling protein 2 structure determined by NMR molecular fragment searching. Nature 476:109–113
192. Trad CH, James W, Bhardwaj A, Butterfield DA (1995) Selective labeling of membrane protein sulfhydryl groups with methanethiosulfonate spin label. J Biochem Biophys Methods 30:287–299
193. Hubbell WL, Gross A, Langen R, Lietzow MA (1998) Recent advances in site-directed spin labeling of proteins. Curr Opin Struct Biol 8:649–656

Chapter 4
Isotope Labeling Methods for Relaxation Measurements

Patrik Lundström, Alexandra Ahlner, and Annica Theresia Blissing

Abstract Nuclear magnetic spin relaxation has emerged as a powerful technique for probing molecular dynamics. Not only is it possible to use it for determination of time constant(s) for molecular reorientation but it can also be used to characterize internal motions on time scales from picoseconds to seconds. Traditionally, uniformly ^{15}N labeled samples have been used for these experiments but it is clear that this limits the applications. For instance, sensitivity for large systems is dramatically increased if dynamics is probed at methyl groups and structural characterization of low-populated states requires measurements on $^{13}C\alpha$, $^{13}C\beta$ or ^{13}CO or $^{1}H\alpha$. Unfortunately, homonuclear scalar couplings may lead to artifacts in the latter types of experiments and selective isotopic labeling schemes that only label the desired position are necessary. Both selective and uniform labeling schemes for measurements of relaxation rates for a large number of positions in proteins are discussed in this chapter.

4.1 Introduction

Nuclear magnetic resonance (NMR) spin relaxation is a powerful tool for identifying and quantifying molecular dynamics on multiple time scales. For instance, measurements of the longitudinal (R_1) and transverse (R_2) auto-relaxation rates and the heteronuclear nuclear Overhauser enhancement (NOE) allows determination of the rotational diffusion tensor as well as characterization of the time scale and magnitude of bond vector motions through model-free analysis [1–3]. Slower motions on the μs-ms time-scale, often corresponding to larger structural rearrangements important for processes like protein folding, enzymatic catalysis and ligand binding can also be studied using Carr-Purcell-Meiboom-Gill (CPMG) [4, 5] and $R_{1\rho}$ [6, 7] relaxation dispersion. In these experiments the excess contribution to transverse relaxation is modeled in terms of exchange rates and chemical shift differences between exchanging states. In principle, all NMR active nuclei can be used as probes in relaxation experiments although some are more useful than others for reasons of simplicity, sensitivity and ease of interpretation. For instance, if the exchange contribution to transverse relaxation is quantified it is beneficial if this contribution is large compared to other contributions, which in practice excludes quadrupolar

P. Lundström (✉) • A. Ahlner • A.T. Blissing
Division of Molecular Biotechnology, Department of Physics,
Chemistry and Biology, Linköping University, SE-58183 Linköping, Sweden
e-mail: patlu@ifm.liu.se

H.S. Atreya (ed.), *Isotope Labeling in Biomolecular NMR*, Advances in Experimental Medicine and Biology 992,
DOI 10.1007/978-94-007-4954-2_4, © Springer Science+Business Media Dordrecht 2012

nuclei. On the other hand, for quadrupolar nuclei other relaxation mechanisms than the quadrupolar coupling can be disregarded, which simplifies the analysis.

In practice, all relaxation experiments involving proteins and other biomolecules require isotopic labeling. The most commonly used isotopic labeling schemes are uniform labeling with ^{15}N and/or ^{13}C. For reasons that are explained below, these methods are useful for the measurement of relaxation rates for certain positions in proteins while they fail for others. Different selective labeling techniques that can be used in these cases, advantages and disadvantages with these will also be discussed.

To be useful, a labeling scheme should optimally meet as many as possible of several, sometimes conflicting, criteria. For example, all desired sites should be labeled with high yield, sites that would interfere with experiments should not be labeled, the labeling protocol should be simple to implement, protein yields should be high, the isotopically labeled precursors should be cheap and as few different samples as possible should be required. The possibility to use certain samples for several different experiments should be exploited and the use of simple strategies such as uniformly ^{15}N, ^{13}C/^{15}N or ^{13}C/^{15}N/^{2}H labeled samples for certain relaxation experiments should not be overlooked.

4.2 Spin Relaxation and Chemical Exchange

4.2.1 Spin Relaxation

Relaxation is due to the stochastic modulation of spin couplings resulting from molecular motion. The quadrupolar coupling usually dominates relaxation for spins $I > \frac{1}{2}$ while for spins $I = \frac{1}{2}$ the most important relaxation mechanisms are due to dipole-dipole coupling and the chemical shift anisotropy. Relaxation can be described at different levels of rigor from the phenomenological description by the Bloch equations [8] to a fully quantum-mechanical treatment [9]. It turns out that a semi-classical approach, in which the spins are treated quantum-mechanically and the lattice is treated classically leads to correct answers for most relevant cases [10]. Application of this theory shows that relaxation for coupled spins is described by a set of coupled differential equations where cross-correlation and cross-relaxation lead to interconversion of operators and multi-exponential decays whereas absence of such mechanisms lead to mono-exponential decays. All relaxation rate constants are given by the coupling strength multiplied with a weighted sum of spectral densities. These are usually modeled according to the model-free formalism that takes global tumbling as well as internal motions on two time-scales faster than the global tumbling into account [1–3]. In the case of isotropic diffusion, the expression for the spectral density is then given by

$$J(\omega) = \frac{2}{5}\left[\frac{S^2 \tau_c}{1+(\omega\tau_c)^2} + \frac{\left(1-S_f^2\right)\tau_f'}{1+(\omega\tau_f')^2} + \frac{\left(S_f^2-S^2\right)\tau_s'}{1+(\omega\tau_s')^2}\right] \tag{4.1}$$

where $S^2 = S_f^2 S_s^2$; S_f^2 and S_s^2 are the generalized order parameters for fast and slow internal motions, respectively; $\tau_f' = \tau_f \tau_c / \left(\tau_f + \tau_c\right)$, $\tau_s' = \tau_s \tau_c / \left(\tau_s + \tau_c\right)$; τ_f and τ_s are the correlation times for the fast and slow internal motions respectively and τ_c is the correlation time for molecular tumbling.

The most common relaxation parameters measured to model protein dynamics are the longitudinal (R_1) and transverse (R_2) auto-relaxation rates and the heteronuclear NOE [11] that are given by Eqs. 4.2, 4.3, 4.4, and 4.5.

$$R_1 = \frac{d^2}{4}\left[3J(\omega_I) + 6J(\omega_I+\omega_S) + J(\omega_I-\omega_S)\right] + c^2 J(\omega_I) \tag{4.2}$$

$$R_2 = \frac{d^2}{8}\left[4J(0)+3J\left(\omega_I\right)+6J(\omega_S)+6J(\omega_I+\omega_S)+J(\omega_I-\omega_S)\right]+\frac{c^2}{6}\left[4J(0)+3J\left(\omega_I\right)\right] \qquad (4.3)$$

$$\sigma_{IS} = \frac{d^2}{4}\left[6J\left(\omega_I+\omega_I\right)-J(\omega_I-\omega_S)\right] \qquad (4.4)$$

$$NOE = 1+\frac{\sigma_{IS}}{R_1}\cdot\frac{\gamma_S}{\gamma_I} \qquad (4.5)$$

where $d = \mu_0\hbar\gamma_I\gamma_S\left\langle r_{IS}\right\rangle^{-3}/4\pi; c = \gamma_I B_0\Delta\sigma/\sqrt{3}; \mu_0$ is the permeability of vacuum; $\hbar$ is the reduced Planck constant; γ_I and γ_S are the magnetogyric ratios; r_{IS} is the internuclear distance; B_0 is the static magnetic field strength and $\Delta\sigma$ is the anisotropy of the (axially symmetric) chemical shift tensor.

For spins $I=\frac{1}{2}$ the auto-relaxation rates depend on dipolar and chemical shift anisotropy. It can be shown that if there are no cross-correlations between these mechanisms, the decay of transverse magnetization is mono-exponential as desired. However, for longitudinal magnetization the dipolar interaction with another spin will lead to cross-relaxation. Similarly, the NOE will not be quantitated accurately if the spin belongs to a dipolar coupled network. Cross-correlations between dipole-dipole and chemical shift anisotropy mechanisms or different dipole-dipole mechanisms also lead to multi-exponential decays that prevent the extraction of accurate parameters. These problems can however be corrected by inverting the dipolar field of the coupled nucleus by 180° pulses at appropriate time-points during the relaxation delay. For backbone ^{15}N positions this is an easy task and dedicated pulse sequences that lead to minimal errors have been described [11, 12]. The reason why this is feasible is because relaxation in this case is totally dominated by the interaction with the attached proton and the ^{15}N chemical shift anisotropy. Importantly, the ^{15}N spin is isolated from other ^{15}N spins in the sense that such dipolar and scalar couplings are vanishingly small. For the other two common spin $I=\frac{1}{2}$ nuclei in proteins, ^{13}C and ^{1}H, the situation is less favorable and the naive approach of using uniformly labeled samples for these experiments often fails. For measurements of transverse relaxation rates an additional concern is the homonuclear scalar coupling.

4.2.2   Chemical Exchange

Chemical exchange is the result of a stochastic modulation of the chemical shift on a time scale that is slower than the correlation time for molecular tumbling. This modulation may indeed be due to chemical reactions as the name suggests but more commonly it results from conformational exchange that for proteins include processes like folding, ligand binding and enzymatic catalysis [13–15]. Chemical exchange manifests as an excess contribution to transverse relaxation and thus line broadening and is classified as slow, intermediate or fast depending on the size of the exchange rate constant in relation to the difference in resonance frequencies between exchanging states. The maximal information that can be extracted from measurements of exchange broadening is the populations of all exchanging states as well as exchange rate constants and (the magnitude of) the difference in resonance frequencies between exchanging states as shown in Eq. 4.6. If chemical exchange is fast on the chemical shift time-scale, the exchange contribution to transverse relaxation, R_{ex}, for exchange between sites A and B is

$$R_{ex} = \frac{p_A p_B \Delta\omega_{AB}^2}{k_{ex}} \qquad (4.6)$$

where P_A and P_B are the populations of the exchanging states, $\Delta\omega_{AB}$ is the difference in resonance frequencies for states A and B and $k_{ex} = k_{AB} + k_{BA}$ is the exchange rate constant. For slow exchange, R_{ex} for states A and B equals k_{AB} and k_{BA}, respectively, whereas a considerably more complicated equation governs intermediate exchange.

Chemical exchange on the intermediate to fast time-scale is measured by Carr-Purcell-Meiboom-Gill (CPMG) [16, 17] or $R_{1\rho}$ [18, 19] relaxation dispersion experiments where the effective transverse relaxation rate $R_{2,eff}$ is a function of the effective field of either a CPMG pulse train or the amplitude and offset of the spinlock field. The obtained dispersion profiles are fitted to the Bloch-McConnell equations [20], either numerically or to expressions that cover special cases [21, 22]. To characterize slow exchange, ZZ-exchange experiments are commonly employed [23]. These experiments only report on the exchange rate between exchanging states but all chemical shifts are, of course, readily obtained from the peak positions in the spectra.

As for measurements of fast dynamics, most studies hitherto have focused on using ^{15}N as a probe for millisecond dynamics. However, since excited state chemical shifts can be extracted from the experiments and the chemical shifts can be used as restraints in structure calculations [24–26] it is highly desirable to measure relaxation dispersions also for other nuclei to characterize protein excited states structurally. Another reason for measurements on different nuclei is to increase the confidence in the extracted parameters when fitting the data to more complex models like three-site exchange [27].

In CPMG experiments, scalar coupling evolution may result in artifacts for at least two reasons. First, the relaxation rates of in-phase and anti-phase operators may be very different, especially if the coupled nucleus is a proton. If care is not taken, the magnetization will spend different amounts of time in-phase and anti-phase, even during a constant time delay, depending on the number of refocusing pulses that are applied and the effective relaxation rate is thus not only modulated by chemical exchange. Relaxation compensation elements that explicitly average in-phase and anti-phase relaxation rates for all choices of the number of CPMG pulses [4] or continuous wave decoupling of the coupled nucleus to keep the magnetization in-phase [5] minimize artifacts when these techniques can be applied. Both these methods are however hard to apply if the coupled nuclei are of the same species. A second problem in the homonuclear case is that the scalar coupling evolves differently in the slow and fast pulsing limits since the spins get increasingly more strongly coupled as the pulsing rate increases. In $R_{1\rho}$ experiments Hartmann-Hahn transfer for certain choices of offset and amplitude of the effective field present a similar problem. The reason for the widespread use of backbone ^{15}N as a probe for molecular dynamics largely is that one does not need to be concerned with such problems.

Another way to eliminate problems associated with homonuclear scalar couplings is to let the coupled nuclei evolve as multiple-quantum operators during the relaxation delay since this renders the scalar coupling between the nuclei inactive. Multiple-quantum relaxation rates typically report on cross-correlated couplings. These may be cross-correlations between two different dipole-dipole interactions, two chemical shift anisotropy interactions or one dipole-dipole and one chemical shift anisotropy interaction. Multiple-quantum relaxation rates thus report on the angle between the principal frames of the two couplings and can be used to obtain information on dihedral angles in proteins [28, 29]. In the case of fast exchange, the excess contribution to the difference between double-quantum and zero-quantum contributions, $\Delta R_{MQ,ex}$, is given by

$$\Delta R_{MQ,ex} = \frac{4 p_A p_B \Delta\omega_I \Delta\omega_S}{k_{ex}} \tag{4.7}$$

where P_A and P_B are the populations for states A and B and $\Delta\omega_I$ and $\Delta\omega_S$ are the associated differences in resonance frequencies for spins I and S [30]. It is noteworthy that $\Delta R_{MQ,ex}$ is four times as sensitive to chemical exchange as R_{ex} (single-quantum) if $\Delta\omega_I$ and $\Delta\omega_S$ are of similar magnitude.

An interesting feature of cross-correlated chemical exchange is that $\Delta R_{MQ,ex}$ reports on the relative signs of $\Delta\omega_I$ and $\Delta\omega_S$.

To measure the above mentioned parameters using NMR spectroscopy, the protein sample requires isotopic labeling. Some experiments require very precise positioning of labels, which will be reviewed later on. First, we turn to the metabolic pathways responsible for amino acid synthesis in *E. coli* to shine light on how these pathways can be utilized for labeling purposes.

4.3 Biosynthesis of Amino Acids

Since proteins used for NMR are samples almost exclusively produced by over expression in *E. coli*, the focus will be on over expression in this organism. Cell-free synthesis will be discussed briefly at the end of this chapter. To appreciate how different labels are incorporated into specific positions in proteins we must study how amino acids are synthesized. The main references for this section are the textbooks in biochemistry by Voet and Voet [31] and in bacterial metabolism by Gottschalk [32].

Working backwards from the amino acids, the last step of biosynthesis is typically transamination of the corresponding α-ketoacid where the amide group is derived from Gln or Glu and the α-proton from solvent. Specifically this means that the isotopic composition at Hα will be identical to the solvent composition, ignoring kinetic isotope effects. Nitrogen is usually supplied to the growth medium in the form of NH_4Cl so that enrichment with ^{15}N is achieved by using $^{15}NH_4Cl$ as the nitrogen source.

The precursors for the α-ketoacids of all amino acids are surprisingly few. Working from the more complex ones to the more simple ones we have phosphoenolpyruvate and erythose-4-phosphate (Phe, Trp, Tyr), ribose-5-phosphate (His, Trp), α-ketoglutarate (Arg, Gln, Glu, Pro), oxaloacetate (Asp, Asn, Lys, Met, Thr), pyruvate (Ala, Leu, Lys) and 3-phosphoglutarate (Gly, Ser, Cys). If we want to know from which molecules most positions in amino acid residues are derived it suffices to know the biosynthesis of these compounds. It turns out that 3-phosphoglutarate, phosphoenolpyruvate and pyruvate are intermediates in glycolysis, oxaloacetate is produced either from carboxylation of phosphoenolpyruvate or from the tricarboxylic acid (TCA) cycle. The precursor α-ketoglutarate is exclusively produced in the TCA cycle. Positions 4 and 5 of this molecule are derived from acetyl-S-CoA while positions 1–3 are derived from positions 4–2 of oxaloacetate (note the descending order in the latter case). The precursors for the aromatic side-chains are intermediates of the pentose phosphate pathway.

The three metabolic pathways will be reviewed briefly and are presented in Figs. 4.1, 4.2, and 4.3. To illustrate the concepts of selective labeling and scrambling, the carbon atom initially at position 2 of glucose is highlighted in red.

4.3.1 Glycolysis

Glycolysis starts with the conversion of glucose to 1,3-bisphosphofructose through two intermediates. This molecule is subsequently cleaved into dihydroxyacetonephosphate and glyceraldehydephosphate. While dihydroxyacetonephosphate is a dead end in the biosynthesis of amino acids, this molecule is readily interconverted into glyceraldehydephosphate when this pool is depleted. Glyceraldehyde-phosphate is oxidized into pyruvate through a number of intermediates two of which are 3-phosphoglutarate and phosphoenolpyruvate. The carbons at positions 1–3 of these molecules, which are identical to positions 1–3 or 6–4 (note the descending order in the latter case) of glucose, correspond to CO, Cα and Cβ, respectively, for the amino acids derived from glycolytic intermediates.

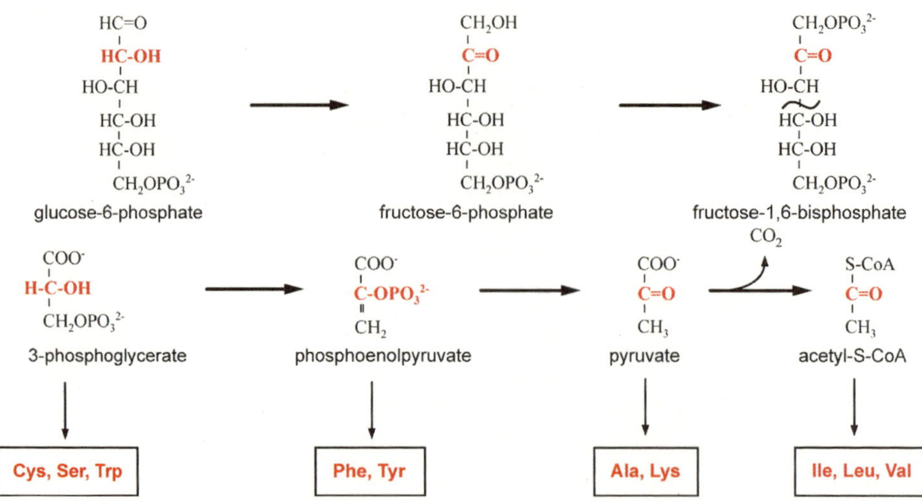

Fig. 4.1 Glycolysis. Carbon atoms originally at position 2 of glucose are highlighted in *red*. Steps that are of little relevance to the text are not shown. Amino acids derived from various intermediates are indicated

4.3.2 The TCA Cycle

The starting point for the TCA cycle (Fig. 4.2) is oxaloacetate that may be the product from the previous pass of the cycle or synthesized by carboxylation of pyruvate or phosphoenolpyruvate. The labeling might thus be different depending on the relative fluxes of these pathways and will also depend on the isotopic composition of bicarbonate in the solvent. Oxaloacetate is fused to acetyl to yield citrate which isomerizes and is subsequently decarboxylated at position 1 of oxaloacetate, to yield α-ketoglutarate, the precursor of Glu, Gln, Arg and Pro. α-ketoglutarate is decarboxylated, so that the position corresponding to position 4 of oxaloacetate leaves, to yield succinate. Through three intermediates succinate in turn is converted back to oxaloacetate. It is noteworthy that while succinate is symmetric in the sense that positions 1 and 4 as well as positions 2 and 3 are equivalent, oxaloacetate is not. This adds more complexity to the labeling patterns of amino acids.

4.3.3 The Pentose Phosphate Pathway

The pentose phosphate pathway serves multiple purposes, including being a source of reducing power (NADPH) and to provide the cell with building blocks for nucleotides and aromatic amino acids. As is seen in Fig. 4.3, it comprises multiple steps and except for the oxidative steps all are readily reversible. How the pentose phosphate pathway is run is thus dependent on the requirements of the cell. For instance, if NADPH is needed the pentose phosphate pathway is run forward as indicated in the figure whereas if building blocks are needed it may be run in reverse by the use of fructose-6-phosphate and glyceraldehyde-3-phosphate as substrates. The resulting labeling patterns for applicable amino acids thus depend on the relative balance of these modes.

The oxidative branch of the pentose phosphate pathway involves decarboxylation of position 1 from 6-phosphoglucose to yield ribulose-5-phosphate that is readily isomerized to ribose-5-phosphate and xylulose-5-phosphate. In the non-oxidative branch, these pentoses combine to form first seduheptalose-7-phosphate and glyceraldehyde-3-phosphate and then erythrose-4-phosphate and fructose-6-phosphate. Erythrose-4-phosphate is used in the biosynthesis of the amino acids Phe, Trp and Tyr but it can also combine with xylulose-5-phoshate to give fructose-6-phosphate and glyceraldehyde-3-phosphate.

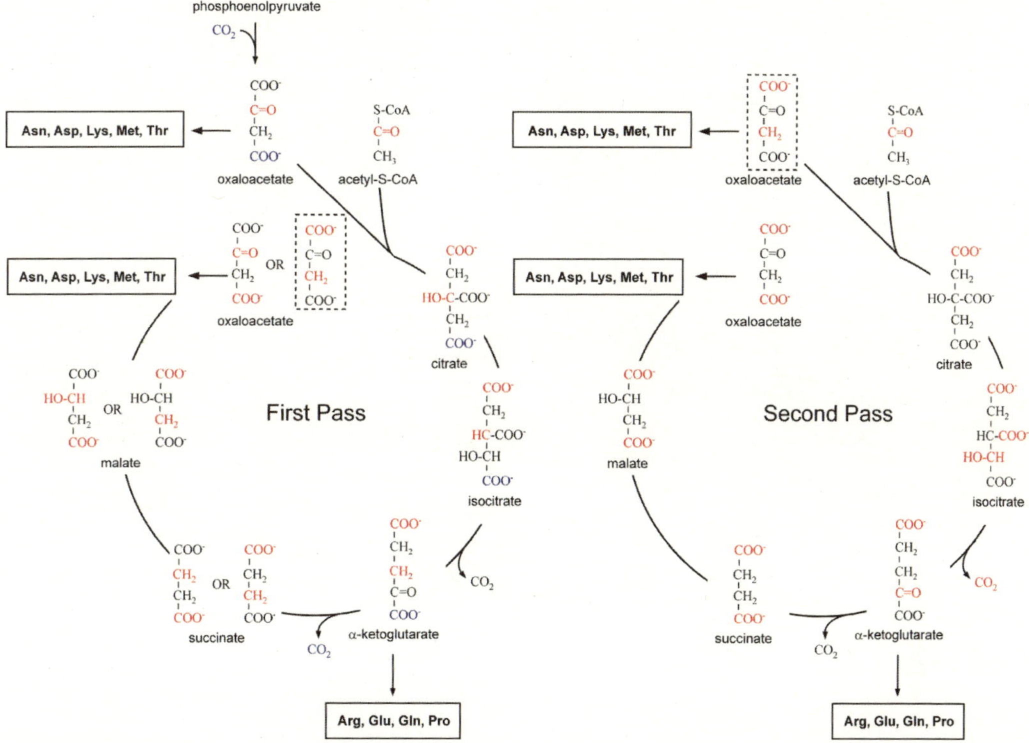

Fig. 4.2 The TCA cycle. Two passes are shown and the amino acids derived from the intermediates oxaloacetate and α-ketoglutarate are indicated. Some steps of little relevance to this text have been omitted. Carbon atoms originally at position 2 of glucose are highlighted in *red* and the carbon atom originating from carbon dioxide is shown in *blue*. Residues that are derived from the intermediates oxaloacetate and α-ketoglutarate are indicated. It is noteworthy that Cα for the residues derived from oxaloacetate and Cβ for the residues derived from α-ketoglutarate are isotopically enriched in the first pass of the cycle whereas Cα for the α-ketoglutarate group of residues is enriched in the second pass

4.3.4 Scrambling

Scrambling of label may result from additional pathways than those considered but also from the complexity of the considered pathways themselves. For example, in the TCA cycle oxaloacetate can be produced by carboxylation of phosphoenolpyruvate but also as the last step in the cycle. By simply following the fate of various carbon atoms in the two cases it is clear that the labeling pattern may be different (Fig. 4.2). Furthermore, the oxaloacetate product in one pass of the cycle can be used as substrate in the next pass, yielding yet another labeling pattern. The isotopomer composition of the amino acids derived from oxaloacetate may thus be extremely complex. The same considerations hold for the amino acids derived from α-ketoglutarate.

The pentose phosphate pathway is even more prone to scrambling since several of the reactions are rapid and readily reversible. This applies to the isomerization of the pentoses and also by the transaldolase and transketolase reactions. Hence, the pathway can, and is, run in different modes depending on the cellular demands. The isotopomer compositions of amino acids made from building blocks of the pentose phosphate pathway are thus notoriously hard to predict. It should also be noted that the products of the pentose phosphate pathway, fructose-6-phosphate and glyceraldehyde-3-phosphate, are glycolytic intermediates that might be labeled differently than if produced in glycolysis, as illustrated in Figs. 4.1 and 4.3.

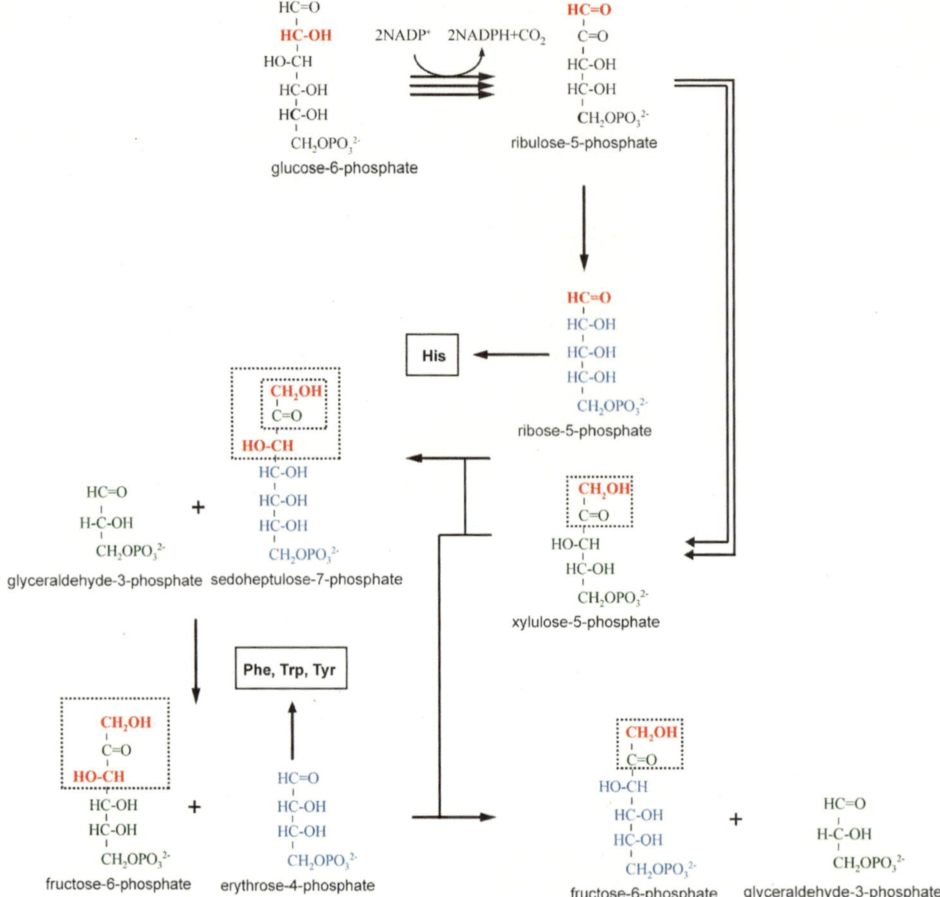

Fig. 4.3 The pentose phosphate pathway. Carbon atoms originally at position 2 of glucose are highlighted in *red*. Other *color coding* and *boxes* are used to keep track of which fragments that are used for the different positions of the various compounds. Amino acids that are derived from pentose phosphate pathway intermediates are indicated

4.4 Protein Expression and Purification

Isotopically enriched proteins are usually expressed in M9 minimal medium [33], referred to as M9 medium in the following. M9 medium is primarily composed of 6 g/L Na_2HPO_4, 3 g/L KH_2PO_4 and 0.5 g/L NaCl. These salts are dissolved in H_2O or D_2O or a mixture of the two depending on labeling scheme. This medium is supplemented with 1 mM $MgSO_4$, 0.1 mM $CaCl_2$, 10 mg/L biotin, 10 mg/L thiamine and antibiotics. Depending on labeling scheme, stock solutions for these components should be dissolved in H_2O or D_2O. A carbon source, most commonly 2–3 g/L glucose, and a nitrogen source, almost always 0.5–1 g/L NH_4Cl are also added to the medium. An expression protocol that has frequently been used to produce uniformly and selectively labeled samples is as follows [34]:

1. Transfer one or more freshly transformed *E. coli* colonies of BL21(DE3) strain to 30 mL LB (in H_2O) supplemented with the appropriate antibiotic(s) and grow cells at 37°C in a shaking incubator until $OD_{600} = 1.0$ is reached.
2. Spin down the cells at $1{,}200 \times g$, 15 min. at room temperature (25°C).

3. Resuspend a fraction of the cells in 10% of the isotopically labeled M9 medium to achieve OD_{600} of 0.1–0.2. Grow the cells at 37°C until $OD_{600} = 1.0$. Pour the starter culture directly into the remaining 90% of the isotopically enriched M9 medium.
4. Grow the cells at 37°C until $OD_{600} = 0.6$–1.0.
5. Induce over expression with 0.5–1 mM IPTG. Perform over expression either at 37°C for 2–5 h or at room temperature or 16°C overnight. The final OD_{600} will depend on the growth medium and on the duration of over expression. This step should be modified if selectively labeled precursors or amino acids are used. In this case these compounds are added to the growth medium 1 h before over expression is induced.

Purification is usually performed by lysing the cells and subjecting the lysate to different methods of chromatography. These include affinity chromatography, ion-exchange chromatography and size-exclusion chromatography. A particular concern is when the protein has to be purified from inclusion bodies. In this case the inclusion bodies must be solubilized in 6 M GdnCl or 8 M urea after lysis. After a preliminary purification step, the protein is refolded and purification is then continued. Refolding is done by exchange to a buffer that favors the folded state and can be performed by several different methods including dialysis, on column or by rapid dilution [35]. The refolding protocol typically has to be optimized differently for each protein. Denaturation and refolding is also usually necessary if complete exchange of protons to deuterons, or vice versa, at amide positions is required.

4.5 Labeling at Specific Positions in Proteins

4.5.1 Backbone and Side-Chain Nitrogen Positions

For completeness the discussion will start by ^{15}N since it is the most common probe for molecular dynamics for a number of reasons. For HSQC-type experiments involving ^{15}N relaxation it suffices to produce a uniformly ^{15}N labeled sample by expression in M9 medium where 0.5–1 g/L $^{15}NH_4Cl$ is the sole nitrogen source using the protocol outlined above. Experiments measuring ^{15}N R_1, R_2 and the $[^{1}H]$-^{15}N NOE [11, 12] and interpreted according to the model-free formalism [1–3] have been the standard method for estimating the diffusion tensor and the backbone flexibility of proteins since the early 1990s. Backbone ^{15}N positions have also been used extensively to study microsecond to millisecond dynamics in proteins. Relaxation compensated CPMG pulse sequences that average relaxation of in-phase and anti-phase operators equally regardless of the repetition rate of the refocusing pulses [4] during a constant time relaxation delay [36] means that artifact-free ^{15}N CPMG dispersions can be recorded. For exchange rates in excess of a few thousand per second, rotating-frame dispersion experiments are better suited [6]. The experiments can be designed to measure $R_{1\rho}$-R_1 as suggested originally or the pure $R_{1\rho}$ rate [37].

Since nitrogen also is present in some side-chains these positions can also be studied and they can be used to probe formation and disruption of salt-bridges (Arg, Lys) and to monitor protonation/deprotonation (His).

Although not required, the experiments for quantifying dynamics on all time-scales work equally well for samples that are also labeled with ^{13}C. This is useful since this means that the $^{13}C/^{15}N$ labeled sample usually used for resonance assignments and NOESY experiments can be used to measure relaxation at ^{15}N sites. As will be described below, such samples can also be used to probe dynamics at carbonyl and certain other positions. Furthermore, perdeuteration, achieved by using deuterated glucose as the carbon source and 100% D_2O for protein expression, dramatically enhances sensitivity for large proteins. A very versatile sample is thus a uniformly $^{13}C/^{15}N/^2H$ labeled sample with deuterium at amide positions back exchanged to protons.

Table 4.1 Chemical shift ranges and scalar coupling constant for selected ^{13}C positions in proteins

Position	Chemical shift range (ppm)	Homonuclear one-bond couplings
^{13}CO	170–180	$J_{COCA} = 55$ Hz
$^{13}C\alpha$	40–70	$J_{CACO} = 55$ Hz, $J_{CACB} = 35$ Hz
$^{13}C\beta$	15–75	$J_{CBCA} = 35$ Hz, $J_{CBCG} = 35$ Hz
$^{13}C^{methyl}$	5–30	$J_{CC} = 35$ Hz
$^{13}C^{aromatic}$	120–150	$J_{CC} = 60$ Hz

4.5.2 Carbon Positions

For the other important heteronucleus, ^{13}C, the situation is more complicated. All uniformly ^{13}C-labeled amino acid residues except Gly comprise spin systems of at least three, Cα, Cβ and CO, scalar coupled ^{13}C nuclei. The chemical shift range and important scalar couplings for different positions are given in Table 4.1. The one bond scalar coupling constant between $^{13}C\alpha$ and ^{13}CO is 55 Hz, between aliphatic ^{13}C positions it is 35 Hz and between aromatic ^{13}C positions it is around 60 Hz. In addition there may be significant three-bond couplings, notably between $^{13}C^{methyl}$ and $^{13}C\alpha$ for Leu residues and between $^{13}CO^{backbone}$ and $^{13}CO^{side-chain}$. Different strategies need to be employed depending on which nucleus that is probed and this section will be a survey of different labeling schemes and tricks in pulse sequences needed to probe dynamics of many of these positions in relaxation experiments. Although simultaneous labeling by ^{15}N is not necessary except for in experiments where magnetization is transferred through this nucleus it is useful to routinely include it in any labeling scheme since it facilitates a convenient ^{15}N-HSQC check of sample integrity.

4.5.2.1 ^{13}CO at Backbone and Side-Chains Positions

The backbone ^{13}CO chemical shift is sensitive to the backbone dihedral angles and can be used as a probe for secondary structure whereas the side-chain ^{13}CO chemical shift reports on electrostatic interactions. In addition, the relaxation properties of this nucleus are quite favorable compared to other ^{13}C positions in protonated samples. Another simplifying feature compared to other ^{13}C positions is that for $^{13}CO^{backbone}$ the only covalently bound carbon is Cα. From Table 4.1 it follows that they resonate approximately 15 kHz apart already at a static magnetic field of 11.7 T. It is thus feasible to manipulate ^{13}CO and $^{13}C^{\alpha}$ separately using band-selective RF-pulses. This was utilized by Ishima et al. who designed a HNCO-type CPMG experiment for the measurement of ^{13}CO relaxation dispersions in a uniformly $^{13}C/^{15}N$ labeled sample and found good correlation between extracted rate constants from ^{13}CO and ^{15}N CPMG experiments [38]. One scalar coupling that was not refocused in the pulse sequence was the three bond coupling between backbone ^{13}CO and side-chain ^{13}CO in Asx residues. This manifested as artifacts for residues of these types since the coupling evolves differently in the slow and fast pulsing limits. This issue was addressed in a later communication describing a similar CPMG experiment for backbone ^{13}CO sites in proteins [39]. By including a refocusing element (termed J-refocusing element) in the middle of the CPMG period it was possible to selectively invert the side-chain but not backbone ^{13}CO so that the coupling was refocused by the end of the CPMG period. This effectively removed the problem with artifacts resulting from these couplings. Another method of measuring ^{13}CO dispersions on uniformly labeled samples was reported by Mulder and Akke who developed a rotating-frame relaxation experiment and measured R_2 rates for the proteins calbindin D$_{9k}$ and the E140Q mutant of the C-terminal domain of calmodulin [40].

If desired, selective labeling of CO backbone positions in proteins can be achieved by using 3 g/L [3-^{13}C]-pyruvate and 3 mM NaH$^{13}CO_3$ as the carbon sources in the growth medium [41]. The use of labeled bicarbonate increases the fractional incorporation of label for residue types derived from α-ketoglutarate (Fig. 4.2). This strategy leads to a fractional incorporation of label of 70–90% for residues derived from glycolytic intermediates and about 25% for residues derived from TCA cycle intermediates. The only residue type that is not significantly labeled is Leu. It has however been shown that the pairwise root-mean-square-deviation of extracted excited state chemical shifts from measurements on selectively and uniformly labeled samples is essentially zero [39]. Because of this there are few reasons to use the selective labeling strategy since a uniformly labeled sample provides superior sensitivity, is more versatile and that the complication due to $^{13}CO^{backbone}$-$^{13}CO^{side-chain}$ scalar couplings can be circumvented by including the J-refocusing element. The presence of the three bond $^{13}CO^{backbone}$-$^{13}CO^{backbone}$ coupling for adjacent residues is a concern for both labeling strategies although it is scaled down, especially for some pairs of residue types, using the selective labeling.

Uniform or partial deuteration of aliphatic and aromatic positions is not required for these experiments but is still useful since the sensitivity increases somewhat for small proteins and more dramatically so for larger ones, especially if the strategy is combined with TROSY type experiments. A $^{15}N/^{13}C/^2H$ labeled sample can additionally be used for measurements of amide proton relaxation rates (described below). One sample can thus be used to probe dynamics for three different nuclei for all amino acids except Pro (or residues preceding Pro in the case of ^{13}CO).

Uniformly ^{13}C labeled samples can also be used to measure millisecond dynamics at ^{13}CO side-chain positions in proteins. In a recent application Hansen and Kay measured CPMG dispersions for ^{13}CO side-chains in uniformly labeled proteins [42]. Similar concerns as for backbone positions apply and the J-refocusing element described above was included to refocus couplings to backbone ^{13}CO.

4.5.2.2 $^{13}C\alpha$ Positions

For $^{13}C\alpha$, the situation is significantly more complicated. In addition to the coupling to ^{13}CO, which can be refocused, one must consider the coupling to $^{13}C\beta$, which in general cannot. This necessitates the development of selective labeling schemes that label $^{13}C\alpha$ but not $^{13}C\beta$. This can be done in different ways. Perhaps the simplest approach is to use randomly ^{13}C labeled glucose as the carbon source. This will result in some isolated $^{13}C\alpha$ positions (or as $^{13}C\alpha$-^{13}CO spin-pairs which can be handled using band-selective pulse) and some problematic $^{13}C\alpha$-$^{13}C\beta$ spin pairs. A related method based on using a mixture of differently labeled acetate molecules has also been described by Wand et al. who used 15% [2-^{13}C]-acetate, 15% [2-^{13}C]-acetate and 70% unlabeled acetate and measured relaxation rates at various carbon positions in ubiquitin [43]. The main drawback of these methods is that the pulse sequence must contain an element that edits out ^{13}C-^{13}C spin pairs. This introduces a fixed delay in the pulse sequences, leading to reduced sensitivity. Furthermore, the approach is not effective in yielding a large fraction of isolated $^{13}C\alpha$ moieties.

A significantly better approach was proposed by LeMaster and Kushlan [44]. They used a bacterial strain deficient in the enzymes sdh-1 and mdh-1 as to disrupt the TCA cycle to reduce scrambling and used [2-^{13}C]-glycerol as the carbon source for expression. They reported high levels of isotopic enrichment at Cα without simultaneous enrichment at Cβ for 13 residue types. Ile and Val were highly enriched at Cα but unfortunately also at Cβ. The only residues that were not enriched were Arg, Gln, Glu, Leu and Pro using this strategy. However, also these can get highly enriched at Cα if a second sample is prepared by expression in the same strain with [1,3-$^{13}C_2$]-glycerol as the carbon source. Thus, if two different samples are produced, $^{13}C\alpha$ relaxation rates can be measured for all residue types except Ile and Val. This labeling scheme has since been the major method for obtaining samples with alternate ^{13}C-^{12}C labeling for solid state NMR applications [45]. It has not yet been established whether the method can be used for recording artifact-free $^{13}C\alpha$ CPMG dispersions. That has however been established using a similar strategy based on 3 g/L [2-^{13}C]-glucose as the carbon source and ordinary BL21(DE3) cells [41, 46]. Using this labeling scheme and a pulse sequence optimized for

^{13}Cα dispersions, accurate excited state chemical shifts could be extracted for an SH3 domain from Abp1p. Also, clean mono-exponential decays were observed in R_1 and $R_{1\rho}$ experiments for ubiquitin showing that cross-correlation and cross-relaxation artifacts are absent. The obvious drawback with this method compared to the one of LeMaster and Kushlan is that incorporation of label is only half as effective. There are however a few things that speak in its favor. The first is that one sample suffices for relaxation measurements involving 17 residue types. Additionally, faster growth rates and higher protein yields are expected using glucose as the carbon source in cells with an intact TCA cycle.

In $R_{1\rho}$ experiments the weak form of the homonuclear scalar coupling is inactive and by clever positioning of the RF-carrier, magnetization transfer due to the strong scalar coupling can be scaled down sufficiently to not constitute a problem even for uniformly ^{13}C labeled samples. Yamazaki et al. showed that accurate measurements of $R_{1\rho}$ could indeed be obtained at ^{13}Cα positions for all amino acid residues except Ser and Thr on uniformly ^{13}C labeled samples whereas measurements of R_1 and the heteronuclear [^{1}Hα]-^{13}C NOE were more troublesome because of cross-correlation and cross-relaxation effects [47]. The $R_{1\rho}$ measurements involving uniformly ^{13}C labeled samples have subsequently been extended to probe microsecond-millisecond dynamics for ^{13}Cα positions in proteins [48].

4.5.2.3 ^{13}Cβ Positions

For Cβ an analysis of glycolysis and the TCA cycle shows that the inverse of the schemes for labeling Cα should be effective at obtaining high level enrichment for many residue types. Thus using [1,3-^{13}C$_2$]-glycerol or [1-^{13}C]-glucose as the carbon source should be workable strategies. In the case of relaxation experiments involving Cβ it is in general not sufficient that Cα positions stay unlabeled but the same must also apply for Cγ. It turns out that even when only considering the TCA cycle it is not feasible to get isolated ^{13}Cβ moieties for certain residue types as shown in Fig. 4.4.

LeMaster and Kushlan used the cell-line described above with two lesions in the TCA cycle to prevent scrambling and measured ^{13}Cβ relaxation rates for a sample of thioredoxin produced by over expression with [1,3-^{13}C$_2$]-glycerol as the carbon source [44]. Kay and co-workers implemented a similar strategy with 3 g/L [1-^{13}C]-glucose as the main carbon source in a cell-line in which sdh-1 was knocked out [49]. To reduce the extent of carboxylation of phosphoenolpyruvate with labeled carbon dioxide, ^{12}CO$_2$ was added to the medium in the form of 20 mM natural abundance sodium bicarbonate. Figure 4.4 shows there will be no ^{13}Cβ-^{13}Cα or ^{13}Cβ-^{13}Cγ spin-pairs as a result of scrambling in the TCA cycle if it is disrupted this way. Unfortunately only 11 residue types will be labeled to more than 30% using this strategy. However, four additional residue types are available if an additional sample using 3 g/L [2-^{13}C]-glucose as the carbon source is produced (Fig. 4.2). In this case there is no need for using bacterial strains with lesions in the TCA cycle although the use of such strains leads to increased fractional incorporation of label, perhaps at the expense of slower growth rates and overall lower protein yields.

The current methodology for specific labeling at Cβ positions is far from optimal which is evident from the limited number of ^{13}Cβ relaxation experiments that have been reported. A further complicating issue with this position is that there are one, two or three attached protons depending on residue type, necessitating three different versions of pulse sequences for the relaxation experiments in order to probe all residue types [49].

4.5.2.4 Methyl Side-Chains

Methyl side-chains are attractive probes for molecular dynamics since they provide high sensitivity and are ubiquitous in well-folded proteins. In many relaxation experiments it is necessary that covalently linked carbon positions are not isotopically labeled and several methods for achieving this have

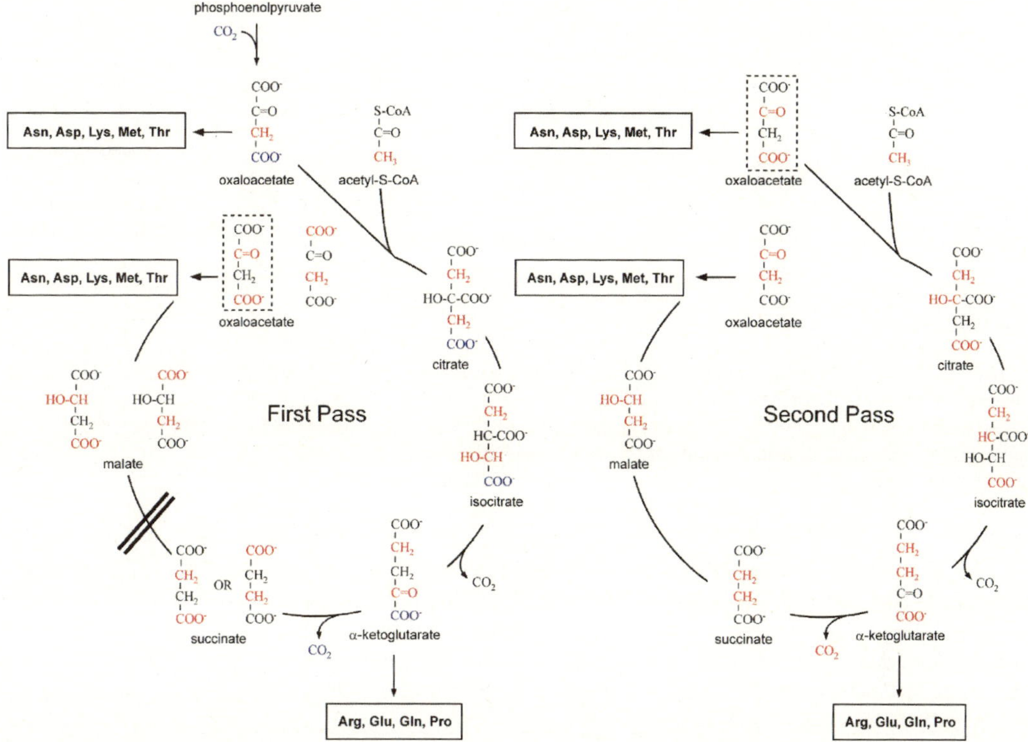

Fig. 4.4 Illustration of scrambling of label in the TCA cycle if [1-^{13}C]-glucose is used as the carbon source. Carbon positions originally at position 1 of glucose are highlighted in *red* and carbon dioxide that is used to phosphorylate phosphoenolpyruvate is colored *blue*. The *double slash* indicates the end point of the pathway if the enzyme succinate dehydrogenase is knocked out

been proposed. Because the methyl group of Met is isolated from other carbon nuclei, selective ^{13}C labeling at the methyl groups is not necessary and uniformly ^{13}C labeled samples can be used for relaxation experiments [50]. A simple and cheap strategy to obtain isolated ^{13}C methyl groups for most other cases is to use 3 g/L [1-^{13}C]-glucose as the carbon source. A straightforward analysis of the biosynthetic pathways shows that this leads to enrichments of almost 50% at the methyl side-chains of Ala, Leu, Val, and Ileγ2 without enrichment at neighboring carbon nuclei [46]. This strategy has been used to record methyl CPMG dispersions for proteins of different sizes such as the FF domain from human FBP11, 71 residues, 8.6 kDa, and a complex between *E. coli* NAD(P)H:FRE, 232 residues, 27 kDa and FAD [51, 52].

The same methyl side-chains can also be labeled using other strategies. Mulder et al. used [3-^{13}C]-pyruvate as carbon source which results in close to 100% incorporation of label at the above mentioned methyl side-chains [53]. The degree of incorporation of label is thus doubled compared to if [1-^{13}C]-glucose is used. However, it should be noted that [3-^{13}C]-pyruvate is significantly more expensive and that bacterial growth is slow using this carbon source, perhaps with reduced protein yields as a result. Contrary to when [1-^{13}C]-glucose was used the authors noted scrambling for Ala residues resulting in a mixture of ^{12}Cα-^{13}Cβ and ^{13}Cα-^{13}Cβ spin-pairs. Contributions to the NMR signal from the latter can however be removed by incorporation of an editing element into the pulse sequence, causing decreased sensitivity [53]. The scheme suggested by LeMaster and Kushlan, using [1,3-^{13}C$_2$]-glycerol as the carbon source [44] should produce similar results.

A cleaner way of selectively labeling certain methyl side-chains is to add the commercially available compounds α-ketobutyrate (precursor of Ile) and/or α-ketoisovalerate (precursor of Leu, Val) that are

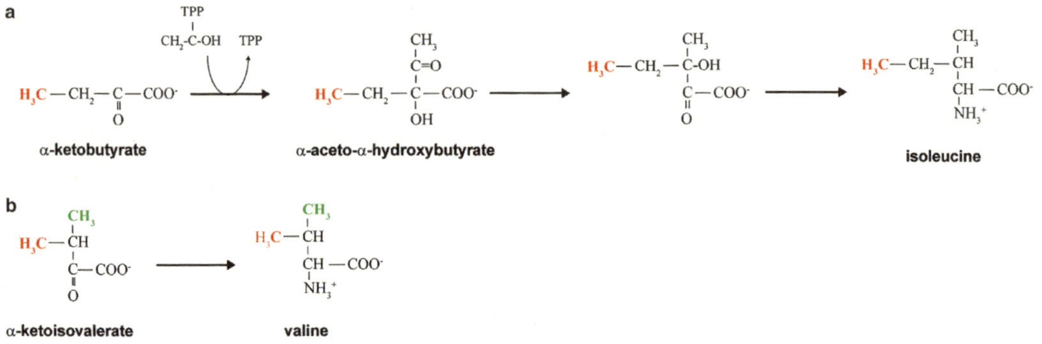

Fig. 4.5 Synthesis of the amino acids (**a**) Ile and (**b**) Val from the precursors α-ketobutyrate and α-ketoisovalerate, respectively. The methyl group of α-ketobutyrate is highlighted in *red* and the two methyl groups of α-ketoisovalerate are shown in *red* and *green*. Both or either can be labeled with ^{13}C. The methyl groups of Leu will be labeled in the same manner as the ones in Val if α-ketoisovalerate is added to the growth medium

specifically labeled with ^{13}C at the methyl groups to the growth medium [54]. This is usually referred to as ILV labeling and is shown in Fig. 4.5. The methyl groups of Ileδ1, Leu and Val are labeled to 90% if the precursors are supplied in concentrations of 50 mg/L for α-ketobutyrate and 100 mg/L for α-ketoisovalerate 1 h prior to induction of protein expression. Although these compounds can potentially be degraded into precursors for other amino acids the authors noticed essentially no labeling at other positions. A very useful feature of the method is that the labeling of the methyl groups can be customized. In addition of having all methyl groups ^{13}CH$_3$ it is also possible to have them ^{13}CH$_2$D or ^{13}CHD$_2$. One can also label different methyl groups differently. By using α-ketoisovalerate labeled with ^{13}CH$_3$ at one methyl group and with ^{12}CD$_3$ at the other, so that only the *proR* or *proS* methyl groups of Leu and Val are detectable, resulting in less crowded spectra. Finally, for applications involving high molecular weight proteins non-methyl positions of α-ketoisovalerate can be deuterated in-house by incubation at elevated pH [54]. Using this approach it is possible to record relaxation experiments for systems as large as the proteasome 20S core particle of 670 kDa [55].

Recently Ruschak et al. suggested a method to instead label Ileγ2 to enable measurements at that position in large proteins [56]. The scheme is based on the precursor α-aceto-α-hydroxybutyrate that is ^{13}C/^{1}H labeled only at the relevant methyl group and ^{12}C/^{2}H labeled elsewhere. For reasons of stability the compound is purchased in its ethyl ester form and de-esterified by incubation with esterase. An amount corresponding to 100 mg/L of the acid form is added to the growth medium. Contrary to what is observed for the ILV labeling scheme, scrambling leads to the presence of weak correlations of *proR* Valγ and Leuδ [56]. However, these do not complicate the interpretation of the spectra considerably.

Certain relaxation experiments involving methyl side-chains are feasible for uniformly ^{13}C or ^{13}C/^{15}N labeled samples. Brath et al. produced a uniformly ^{13}C/^{15}N-labeled, partially ^{2}H labeled sample of FKBP12 by over expression in M9 medium supplemented with 50% D$_2$O. They used this sample to measure $R_{1\rho}$ in methyl groups of the ^{13}CHD$_2$ variety [57]. In this case evolution of the strong three-bond ^{13}C^{methyl}-^{13}Cα couplings in Leu residues is a concern for certain combinations of offsets and spin-lock field strengths. It should be added that other ratios of H$_2$O/D$_2$O ratios during over expression, for instance 100% D$_2$O, also lead to a large fraction of ^{13}CHD$_2$ isotopomers.

4.5.2.5 Carbon Positions in Aromatic Side-Chains

Besides methyl groups the other most common residues in the protein interior are the side-chains of the aromatic residues. Additionally the side-chain of His is for instance frequently located at the active site of enzymes and is involved in catalysis. Despite their important role for protein structure and function, NMR relaxation experiments involving aromatic side-chains are rare and only a few applications and one selective labeling scheme will be reported here.

Hass et al. have exploited the fact that the $C\epsilon 1$ position of His residues is isolated from scalar couplings to other carbon positions and thus used a uniformly $^{13}C/^{15}N$ labeled sample to measure CPMG dispersions on the protein plastocyanin. The extracted chemical shift differences agreed well with those obtained from chemical shift titrations [58].

[1-^{13}C]-glucose was used to label aromatic side-chains of Phe and Tyr at $C\delta$, His at $C\delta 2$ and $C\epsilon 1$ and Trp $C\delta 1$ and $C\epsilon 3$ (and $C\epsilon 2$ although it cannot be seen in HSQC type experiments). The labeling efficiency is about 50%. This labeling scheme allowed relaxation dispersions to be measured for the aromatic side-chains of the E140Q mutant of the C-terminal domain of calmodulin [59]. Boyer and Lee used this labeling scheme to also probe fast dynamics at 13 aromatic positions for the protein eglin C to compare differences in dynamics between the wild type protein and the V54A mutant [60].

4.5.3 Proton Positions

The main benefit of measuring relaxation dispersions for protons is that their high magnetogyric ratio facilitates high repetition rates and high effective fields in CPMG and $R_{1\rho}$ experiments, respectively. Motions that are one order of magnitude faster than can be measured in ^{15}N relaxation dispersion experiments may thus be captured. The main complications when measuring relaxation rates for protons are their sizable dipolar and scalar couplings with remote protons. Both these difficulties can be reduced significantly by perdeuteration or partial deuteration and by including refocusing elements for homonuclear couplings in the pulse sequences. Labeling schemes and applications for relaxation experiments involving ^{1}HN, $^{1}H\alpha$ and $^{1}H^{methyl}$ will now follow.

4.5.3.1 Amide Proton Positions

The amide proton is coupled to the alpha proton with a coupling constant that ranges from 3 to 12 Hz depending on secondary structure [61]. Since prohibitively long pulses are needed to refocus this coupling during a CPMG pulse train used to measure transverse relaxation rates, it is necessary to remove it by perdeuteration [62, 63]. The labeling scheme for these experiments is thus M9 medium supplemented with 1 g/L $^{15}NH_4Cl$ and 3 g/L [$^{13}C_6,^2H_7$]-glucose in 100% D_2O. It has been shown that accurate protein excited state chemical shifts can be extracted from CPMG experiments performed on such samples [41]. An excellent alternative to CPMG experiments for the amide proton that can be used for both protonated and perdeuterated samples is rotating-frame relaxation experiments [64–66].

4.5.3.2 Alpha Proton Positions

For the alpha protons similar problems as for the amide protons exist. However, when running CPMG dispersion experiments one must also consider couplings to beta protons. Because of the small separation in chemical shifts between $^{1}H\alpha$ and $^{1}H\beta$ it is not possible to use pulses that are selective to $^{1}H\alpha$ in the CPMG pulse train. This means that if the beta positions are protonated artifacts similar to the ones that were present in the carbonyl CPMG experiment result. However in this case the problem exists for all residue types and is worse because the magnetization can be further transferred to the gamma position. Even if a refocusing element, akin to the one used in the ^{13}CO dispersion experiment, is included, artifact-free CPMG dispersions are not possible for fully protonated samples. The solution was to combine this strategy with a selective labeling scheme where the protein was produced by over expression in a medium containing 3 g/L [$^{13}C_6$, $^2H_7$]-glucose as the carbon source and 50% D_2O/50% H_2O as solvent [67]. This leads to deuteration at beta positions of 50–88% depending on residue type. Unfortunately, it also leads to an overall sensitivity loss of 50% since protonation at alpha positions is decreased by 50% for all residue types. Using this labeling scheme in combination with a refocusing element in the middle of the relaxation delay, it was possible to extract accurate chemical shifts of protein excited states [67].

4.5.3.3 Methyl Proton Positions

Mulder and coworkers have developed a CPMG relaxation dispersion experiment that measures millisecond dynamics on $^{13}CHD_2$ groups [68]. They achieve the labeling by expressing calbindin D_{9k} in a medium supplemented with protonated 2 g/L [$^{13}C_6$]-glucose in D_2O as described [69]. The experiment was also applied to the transcriptional activator NtrC^r. In this case [1-^{13}C]-glucose was the carbon source. The benefit of this is that non-constant time evolution NMR experiments can be employed, which may increase overall sensitivity for large proteins despite the fact that the methyl groups are only labeled to 50%. Kay and coworkers instead used an ILV labeled sample to measure these dynamics on the 20S proteasome core particle [70]. It should be noted that $^{13}CHD_2$ groups, especially in a highly deuterated background and with other carbon positions being ^{12}C, are sensitive probes for measurements of fast dynamics because of very favorable relaxation properties and because that no cross-correlation effects complicate relaxation behavior.

4.5.4 Deuterium Positions

Since deuterium is a spin $I = 1$ nucleus, its dominating relaxation mechanism is the quadrupolar interaction. This mechanism dominates to the extent that other relaxation mechanisms, including chemical exchange, can be safely neglected, which simplifies the analysis. Furthermore, in this case auto-relaxation rates can be measured for five different operators leading to high levels of confidence in the extracted parameters. Because of the large intrinsic relaxation rates, relaxation measurements involving deuterium spins are in practice limited to methyl groups. Kay and coworkers used a uniformly ^{13}C labeled, fractionally deuterated sample and pulse sequences that select for the CH_2D isotopomer to measure the five relaxation rates in order to calculate order parameters for side-chains of the B1 domain of peptostreptococcal protein L [71, 72]. Wand and coworkers have used the same approach to characterize side-chain dynamics in calmodulin [73]. Of course, other labeling schemes, like ILV labeling with $^{13}CH_2D$ labeled methyl groups, can also be employed for measurements of these relaxation rates. This has been used to probe dynamics of the 20S core particle of the proteasome [55].

4.6 Labeling by Cell-Free Synthesis

Rather than expressing the protein in a cell, the proteins for NMR applications can also be expressed *in vitro* [74]. The DNA or mRNA for the target protein is added to a cell extract containing the transcription and translation machinery of the cell, along with a variety of other compounds including amino acids, nucleoside triphosphates (NTPs) and several enzymes. A way to regenerate energy is also required. The reaction mixture is typically only a few μL-mL large, and yields of several mg of protein per mL of reaction mixture can be achieved. Chaperones, detergents and other compounds that facilitate folding can also be added.

Cell-free protein expression has several advantages both for creating specific isotopic labeling schemes, but also for efficient expression, generally. One is speed since the entire over expression protocol is normally conducted in a few hours, compared to one to several days when using over expression in *E. coli*. Additionally the purification protocol is usually simpler. Another important aspect is that because of the short time of expression and more well-defined conditions, there is less scrambling so that more of the label ends up at the desired place. It is very easy to label individual amino acids simply by adding them to the reaction mixture in labeled form and adding the other amino acids unlabeled. Scrambling due to *E. coli* metabolism is minimized and can often be further reduced by adding specific inhibitors to the reaction mixture. Although labeled amino acids may be quite expensive, the small amounts required for cell-free protein expression makes this method competitive also from a financial point of view. If only one or a few of the amino acids are supplied in labeled form, costs are reduced further. Cell-free protein expression can be very useful for proteins which do not express well in cells, for example because of toxicity. Cell-free expression is also proving to be successful for membrane proteins which can be very difficult to express in large amounts in cells [75].

A powerful application of cell-free synthesis is stereo-array isotope labeling (SAIL) [76]. This protocol produces alternate labeling in the following way. First, stereo-selective replacement of one ^{1}H in methylene groups by ^{2}H; second, replacement of two ^{1}H in each methyl group by ^{2}H; third, stereo-selective modification of the prochiral methyl groups of Leu and Val such that one methyl is ^{12}CD$_3$ and the other is ^{13}CHD$_2$; and last, labeling of six-membered aromatic rings by alternating ^{12}CD and ^{13}CH moieties. Although the method was developed for structure determination, it is easy to imagine applications involving relaxation as well. For instance, the labeling scheme should be optimal for probing dynamics at aromatic side-chains. Unfortunately the method is still too expensive to be an option for most laboratories.

4.7 Concluding Remarks

As should be evident from this chapter, relaxation experiments are feasible for many different nuclear species and positions in proteins. For many positions, uniform labeling with ^{15}N, ^{15}N/^{13}C or ^{15}N/^{13}C/^{2}H is adequate for obtaining robust results. These include ^{15}N and ^{13}CO at backbone and side-chain positions as well as ^{1}HN at the protein backbone. In other cases, it is necessary to use selective strategies to remove scalar and dipolar interactions. The selective labeling protocols range from being very simple, like substituting selectively labeled glucose for unlabeled or uniformly labeled glucose, to being more intricate, like using selectively labeled glycerol as the carbon source in genetically engineered bacterial strains or supplementing the growth medium with customized precursors for a subset of the amino acids.

The original methods of measuring R_1, R_2 and NOE in uniformly ^{15}N labeled samples have been, and still are, extremely useful for determining the diffusion tensor and characterizing sub-nanosecond motions of proteins and CPMG experiments recorded for these protein samples have been instrumental in increasing our understanding of processes like protein folding, ligand binding and enzymatic catalysis. It is however increasingly clear that relaxation experiments probing dynamics at other sites provide complementary information and in some cases has opened an avenue to understanding processes and characterizing intermediate protein states that have previously eluded us. For instance, by using excited state chemical shifts extracted from CPMG experiments for ^{15}N, ^{1}HN, ^{1}Hα, ^{13}Cα and ^{13}CO as the restraints in structure calculations it is now possible to determine structures of transiently populated folding intermediates [77] and by using ILV labeled samples it is possible to characterize dynamics of high molecular weight systems and to correlate these dynamics with function [55]. Although the development of labeling schemes of the last decade has enabled all this it will be exciting to continue to improve these methods to be able to tackle new biological questions during the next decade.

References

1. Lipari G, Szabo A (1982) Model-free approach to the interpretation of nuclear magnetic-resonance relaxation in macromolecules.1. Theory and range of validity. J Am Chem Soc 104:4546–4559
2. Lipari G, Szabo A (1982) Model-free approach to the interpretation of nuclear magnetic-resonance relaxation in macromolecules. 2. Analysis of experimental results. J Am Chem Soc 104:4559–4570
3. Clore GM, Szabo A, Bax A, Kay LE, Driscoll PC, Gronenborn AM (1990) Deviations from the simple two-parameter model-free approach to the interpretation of ^{15}N nuclear magnetic relaxation of proteins. J Am Chem Soc 112:4989–4991
4. Loria JP, Rance M, Palmer AG 3rd (1999) A relaxation-compensated Carr-Purcell-Meiboom-Gill sequence for characterizing chemical exchange by NMR spectroscopy. J Am Chem Soc 121:2331–2332
5. Hansen DF, Vallurupalli P, Kay LE (2008) An improved ^{15}N relaxation dispersion experiment for the measurement of millisecond time-scale dynamics in proteins. J Phys Chem B 112:5898–5904
6. Akke M, Palmer AG 3rd (1996) Monitoring macromolecular motions on microsecond to millisecond time scales by $R_{1\rho}$-R_1 constant relaxation time NMR spectroscopy. J Am Chem Soc 118:911–912
7. Korzhnev DM, Orekhov VY, Dahlquist FW, Kay LE (2003) Off-resonance $R_{1\rho}$ relaxation outside of the fast exchange limit: an experimental study of a cavity mutant of T4 lysozyme. J Biomol NMR 26:39–48
8. Bloch F (1946) Nuclear induction. Phys Rev 70:460–474
9. Abragam A (1961) Principles of nuclear magnetism. Oxford University Press, Oxford
10. Redfield AG (1957) On the theory of relaxation processes. IBM J Res Dev 1:19–31
11. Kay LE, Torchia DA, Bax A (1989) Backbone dynamics of proteins as studied by ^{15}N inverse detected heteronuclear NMR-spectroscopy – application to staphylococcal nuclease. Biochemistry 28:8972–8979
12. Farrow NA, Muhandiram R, Singer AU, Pascal SM, Kay CM, Gish G, Shoelson SE, Pawson T, Formankay JD, Kay LE (1994) Backbone dynamics of a free and a phosphopeptide-complexed Src homology-2 domain studied by ^{15}N NMR relaxation. Biochemistry 33:5984–6003
13. Boehr DD, McElheny D, Dyson HJ, Wright PE (2006) The dynamic energy landscape of dihydrofolate reductase catalysis. Science 313:1638–1642
14. Sugase K, Dyson HJ, Wright PE (2007) Mechanism of coupled folding and binding of an intrinsically disordered protein. Nature 447:1021–1025
15. Korzhnev DM, Salvatella X, Vendruscolo M, Di Nardo AA, Davidson AR, Dobson CM, Kay LE (2004) Low-populated folding intermediates of Fyn SH3 characterized by relaxation dispersion NMR. Nature 430:586–590
16. Carr HY, Purcell EM (1954) Effects of diffusion on free precession in nuclear magnetic resonance experiments. Phys Rev 94:630–638
17. Meiboom S, Gill D (1958) Modified spin-echo method for measuring nuclear relaxation times. Rev Sci Instrum 29:688–691
18. Jones GP (1966) Spin–lattice relaxation in the rotating frame: weak collision case. Phys Rev 148:332–335
19. Davis DG, Perlman ME, London RE (1994) Direct measurements of the dissociation-rate constant for inhibitor-enzyme complexes via the $T_{1\rho}$ and T_2 (CPMG) methods. J Magn Reson Ser B 104:266–275
20. McConnell HM (1958) Reaction rates by nuclear magnetic resonance. J Chem Phys 28:430–431
21. Palmer AG 3rd, Kroenke CD, Loria JP (2001) Nuclear magnetic resonance methods for quantifying microsecond-to-millisecond motions in biological macromolecules. Methods Enzymol 339:204–238
22. Palmer AG 3rd, Massi F (2006) Characterization of the dynamics of biomacromolecules using rotating-frame spin relaxation NMR spectroscopy. Chem Rev 106:1700–1719
23. Jeener J, Meier MH, Bachmann P, Ernst RR (1979) Investigation of exchange processes by 2-dimensional NMR spectroscopy. J Chem Phys 71:4546–4553
24. Cornilescu G, Delaglio F, Bax A (1999) Protein backbone angle restraints from searching a database for chemical shift and sequence homology. J Biomol NMR 13:289–302
25. Shen Y, Lange O, Delaglio F, Rossi P, Aramini JM, Liu GH, Eletsky A, Wu YB, Singarapu KK, Lemak A, Ignatchenko A, Arrowsmith CH, Szyperski T, Montelione GT, Baker D, Bax A (2008) Consistent blind protein structure generation from NMR chemical shift data. Proc Natl Acad Sci USA 105:4685–4690
26. Cavalli A, Salvatella X, Dobson CM, Vendruscolo M (2007) Protein structure determination from NMR chemical shifts. Proc Natl Acad Sci USA 104:9615–9620
27. Korzhnev DM, Neudecker P, Mittermaier A, Orekhov VY, Kay LE (2005) Multiple-site exchange in proteins studied with a suite of six NMR relaxation dispersion experiments: an application to the folding of a Fyn SH3 domain mutant. J Am Chem Soc 127:15602–15611
28. Skrynnikov NR, Konrat R, Muhandiram DR, Kay LE (2000) Relative orientation of peptide planes in proteins is reflected in carbonyl-carbonyl chemical shift anisotropy cross-correlated spin relaxation. J Am Chem Soc 122:7059–7071

29. Kloiber K, Konrat R (2000) Measurement of the protein backbone dihedral angle phi based on quantification of remote CSA/DD interference in inter-residue 13C'(i - 1)-13Calpha(i) multiple-quantum coherences. J Biomol NMR 17:265–268

30. Kloiber K, Konrat R (2000) Differential multiple-quantum relaxation arising from cross-correlated time-modulation of isotropic chemical shifts. J Biomol NMR 18:33–42

31. Voet D, Voet JG (1995) Biochemistry. Wiley, Hoboken

32. Gottschalk G (1986) Bacterial metabolism. Springer, New York

33. Maniatis T, Sambrook J, Fritsch EF (1982) Molecular cloning: a laboratory manual. Cold Spring Harbor Laboratory, Cold Spring Harbor, pp 68–69

34. Lundström P, Vallurupalli P, Hansen DF, Kay LE (2009) Isotope labeling methods for studies of excited protein states by relaxation dispersion NMR spectroscopy. Nat Protoc 4:1641–1648

35. Middelberg APJ (2002) Preparative protein refolding. Trends Biotechnol 20:437–443

36. Mulder FAA, Skrynnikov NR, Hon B, Dahlquist FW, Kay LE (2001) Measurement of slow (μs-ms) time scale dynamics in protein side chains by ^{15}N relaxation dispersion NMR spectroscopy: application to Asn and Gln residues in a cavity mutant of T4 lysozyme. J Am Chem Soc 123:967–975

37. Korzhnev DM, Skrynnikov NR, Millet O, Torchia DA, Kay LE (2002) An NMR experiment for the accurate measurement of heteronuclear spin-lock relaxation rates. J Am Chem Soc 124:10743–10753

38. Ishima R, Baber J, Louis JM, Torchia DA (2004) Carbonyl carbon transverse relaxation dispersion measurements and ms-μs timescale motion in a protein hydrogen bond network. J Biomol NMR 29:187–198

39. Lundström P, Hansen DF, Kay LE (2008) Measurement of carbonyl chemical shifts of excited protein states by relaxation dispersion NMR spectroscopy: comparison between uniformly and selectively (13)C labeled samples. J Biomol NMR 42:35–47

40. Mulder FAA, Akke M (2003) Carbonyl ^{13}C transverse relaxation measurements to sample protein backbone dynamics. Magn Reson Chem 41:853–865

41. Hansen DF, Vallurupalli P, Lundström P, Neudecker P, Kay LE (2008) Probing chemical shifts of invisible states of proteins with relaxation dispersion NMR spectroscopy: how well can we do? J Am Chem Soc 130:2667–2675

42. Hansen AL, Kay LE (2011) Quantifying millisecond time-scale exchange in proteins by CPMG relaxation dispersion NMR spectroscopy of side-chain carbonyl groups. J Biomol NMR 50:347–355

43. Wand AJ, Bieber RJ, Urbauer JL, McEvoy RP, Gan ZH (1995) Carbon relaxation in randomly fractionally ^{13}C-enriched proteins. J Magn Reson Ser B 108:173–175

44. LeMaster DM, Kushlan DM (1996) Dynamical mapping of E. coli thioredoxin via ^{13}C NMR relaxation analysis. J Am Chem Soc 118:9255–9264

45. Castellani F, van Rossum B, Diehl A, Schubert M, Rehbein K, Oschkinat H (2002) Structure of a protein determined by solid-state magic-angle-spinning NMR spectroscopy. Nature 420:98–102

46. Lundström P, Teilum K, Carstensen T, Bezsonova I, Wiesner S, Hansen DF, Religa TL, Akke M, Kay LE (2007) Fractional ^{13}C enrichment of isolated carbons using [1-^{13}C]- or [2-^{13}C]-glucose facilitates the accurate measurement of dynamics at backbone C$^{\alpha}$ and side-chain methyl positions in proteins. J Biomol NMR 38:199–212

47. Yamazaki T, Muhandiram R, Kay LE (1994) NMR experiments for the measurement of carbon relaxation properties in highly enriched, uniformly ^{13}C, ^{15}N-labeled proteins – application to ^{13}C$^{\alpha}$ carbons. J Am Chem Soc 116:8266–8278

48. Lundström P, Akke M (2005) Microsecond protein dynamics measured by ^{13}C$^{\alpha}$ rotating-frame spin relaxation. Chembiochem 6:1685–1692

49. Lundström P, Lin H, Kay LE (2009) Measuring ^{13}C$^{\beta}$ chemical shifts of invisible excited states in proteins by relaxation dispersion NMR spectroscopy. J Biomol NMR 44:139–155

50. Skrynnikov NR, Mulder FAA, Hon B, Dahlquist FW, Kay LE (2001) Probing slow time scale dynamics at methyl-containing side chains in proteins by relaxation dispersion NMR measurements: application to methionine residues in a cavity mutant of T4 lysozyme. J Am Chem Soc 123:4556–4566

51. Korzhnev DM, Religa TL, Lundström P, Fersht AR, Kay LE (2007) The folding pathway of an FF domain: characterization of an on-pathway intermediate state under folding conditions by ^{15}N, ^{13}C$^{\alpha}$ and ^{13}C-methyl relaxation dispersion and ^{1}H/^{2}H-exchange NMR spectroscopy. J Mol Biol 372:497–512

52. Lundström P, Vallurupalli P, Religa TL, Dahlquist FW, Kay LE (2007) A single-quantum methyl ^{13}C-relaxation dispersion experiment with improved sensitivity. J Biomol NMR 38:79–88

53. Mulder FAA, Hon B, Mittermaier A, Dahlquist FW, Kay LE (2002) Slow internal dynamics in proteins: application of NMR relaxation dispersion spectroscopy to methyl groups in a cavity mutant of T4 lysozyme. J Am Chem Soc 124:1443–1451

54. Goto NK, Gardner KH, Mueller GA, Willis RC, Kay LE (1999) A robust and cost-effective method for the production of Val, Leu, Ile (δ1) methyl-protonated ^{15}N-, ^{13}C-, ^{2}H-labeled proteins. J Biomol NMR 13:369–374

55. Sprangers R, Kay LE (2007) Quantitative dynamics and binding studies of the 20S proteasome by NMR. Nature 445:618–622

56. Ruschak AM, Velyvis A, Kay LE (2010) A simple strategy for ^{13}C,^{1}H labeling at the Ile-gamma 2 methyl position in highly deuterated proteins. J Biomol NMR 48:129–135

57. Brath U, Akke M, Yang DW, Kay LE, Mulder FAA (2006) Functional dynamics of human FKBP12 revealed by methyl ^{13}C rotating frame relaxation dispersion NMR spectroscopy. J Am Chem Soc 128:5718–5727

58. Hass MA, Hansen DF, Christensen HE, Led JJ, Kay LE (2008) Characterization of conformational exchange of a histidine side chain: protonation, rotamerization, and tautomerization of His61 in plastocyanin from Anabaena variabilis. J Am Chem Soc 130:8460–8470

59. Teilum K, Brath U, Lundström P, Akke M (2006) Biosynthetic ^{13}C labeling of aromatic side chains in proteins for NMR relaxation measurements. J Am Chem Soc 128:2506–2507

60. Boyer JA, Lee AL (2008) Monitoring aromatic picosecond to nanosecond dynamics in proteins via (13)C relaxation: expanding perturbation mapping of the rigidifying core mutation, V54A, in Eglin C. Biochemistry 47:4876–4886

61. Schmidt JM, Blumel M, Löhr F, Rüterjans H (1999) Self-consistent (3)J coupling analysis for the joint calibration of Karplus coefficients and evaluation of torsion angles. J Biomol NMR 14:1–12

62. Ishima R, Torchia DA (2003) Extending the range of amide proton relaxation dispersion experiments in proteins using a constant-time relaxation-compensated CPMG approach. J Biomol NMR 25:243–248

63. Orekhov VY, Korzhnev DM, Kay LE (2004) Double- and zero-quantum NMR relaxation dispersion experiments sampling millisecond time scale dynamics in proteins. J Am Chem Soc 126:1886–1891

64. Ishima R, Wingfield PT, Stahl SJ, Kaufman JD, Torchia DA (1998) Using amide ^{1}H and ^{15}N transverse relaxation to detect millisecond time-scale motions in perdeuterated proteins: application to HIV-1 protease. J Am Chem Soc 120:10534–10542

65. Lundström P, Akke M (2005) Off-resonance rotating-frame amide proton spin relaxation experiments measuring microsecond chemical exchange in proteins. J Biomol NMR 32:163–173

66. Eichmuller C, Skrynnikov NR (2005) A new amide proton $R_{1\rho}$ experiment permits accurate characterization of microsecond time-scale conformational exchange. J Biomol NMR 32:281–293

67. Lundström P, Hansen DF, Vallurupalli P, Kay LE (2009) Accurate measurement of alpha proton chemical shifts of excited protein states by relaxation dispersion NMR spectroscopy. J Am Chem Soc 131:1915–1926

68. Otten R, Villali J, Kern D, Mulder FAA (2010) Probing microsecond time scale dynamics in proteins by methyl ^{1}H Carr-Purcell-Meiboom-Gill relaxation dispersion NMR measurements. Application to activation of the signaling protein NtrC(r). J Am Chem Soc 132:17004–17014

69. LeMaster DM (1990) Deuterium labeling in NMR structural analysis of larger proteins. Q Rev Biophys 23:133–174

70. Baldwin AJ, Religa TL, Hansen DF, Bouvignies G, Kay LE (2010) ^{13}CHD$_2$ methyl group probes of millisecond time scale exchange in proteins by ^{1}H relaxation dispersion: an application to proteasome gating residue dynamics. J Am Chem Soc 132:10992–10995

71. Millet O, Muhandiram DR, Skrynnikov NR, Kay LE (2002) Deuterium spin probes of side-chain dynamics in proteins. 1. Measurement of five relaxation rates per deuteron in ^{13}C-labeled and fractionally ^{2}H-enriched proteins in solution. J Am Chem Soc 124:6439–6448

72. Skrynnikov NR, Millet O, Kay LE (2002) Deuterium spin probes of side-chain dynamics in proteins. 2. Spectral density mapping and identification of nanosecond time-scale side-chain motions. J Am Chem Soc 124:6449–6460

73. Frederick KK, Marlow MS, Valentine KG, Wand AJ (2007) Conformational entropy in molecular recognition by proteins. Nature 448:325–329

74. Kigawa T, Muto Y, Yokoyama S (1995) Cell-free synthesis and amino acid-selective stable isotope labeling of proteins for NMR analysis. J Biomol NMR 6:129–134

75. Schwarz D, Daley D, Beckhaus T, Dotsch V, Bernhard F (2010) Cell-free expression profiling of E. coli inner membrane proteins. Proteomics 10:1762–1779

76. Kainosho M, Torizawa T, Iwashita Y, Terauchi T, Ono AM, Guntert P (2006) Optimal isotope labelling for NMR protein structure determinations. Nature 440:52–57

77. Korzhnev DM, Religa TL, Banachewicz W, Fersht AR, Kay LE (2010) A transient and low-populated protein-folding intermediate at atomic resolution. Science 329:1312–1316

Chapter 5
Stereo-Array Isotope Labeling Method for Studying Protein Structure and Dynamics

Yohei Miyanoiri, Mitsuhiro Takeda, and Masatsune Kainosho

Abstract The stereo-array isotope labeling (SAIL) method utilizes proteins with isotope labeling patterns optimized with regard to an intended NMR study. The SAIL proteins are prepared by incorporating chemically synthesized amino acids into target proteins, using a cell-free protein synthesis system or a cellular expression system. Over the past decade, the SAIL method has been facilitating a wide variety of new investigations, including high-resolution structure determinations of large proteins and investigations of protein dynamics. In this chapter, the applications of SAIL-related approaches are introduced.

5.1 Introduction

Advances in protein NMR studies are based on the continuous development of isotope labeling techniques [1–4]. The enrichment of carbon and nitrogen atoms in proteins by ^{13}C and ^{15}N, respectively, enables the application of a wide variety of multi-dimensional experiments [5–7]. On the other hand, the replacement of ^{1}H by ^{2}H in a protein simplifies the NMR spectra and mitigates proton spin diffusion [8–10]. When combined with the transverse relaxation optimized spectroscopy (TROSY) method [11–13], NMR studies of large proteins become possible. Therefore, the successful use of stable isotope labeling is a key to achieve NMR studies of proteins.

However, conventional isotope labeling schemes, such as uniform $^{13}C/^{15}N$ labeling (UL) and random fractional deuteration, are not necessarily optimal with regard to NMR studies. For example, in NMR analyses of large proteins, the use of a UL protein generates the problem of signal overlapping and severe line-broadening, while a protein with random fractional deuteration suffers from complicated spectra, due to the presence of numerous different isotopomers. In addition, analyses of the side-chain

Y. Miyanoiri • M. Takeda
Structural Biology Research Center, Graduate School of Science,
Nagoya University, Furo-cho, Chikusa-ku, Nagoya 464-8602, Japan

M. Kainosho (✉)
Structural Biology Research Center, Graduate School of Science,
Nagoya University, Furo-cho, Chikusa-ku, Nagoya 464-8602, Japan

Center for Priority Areas, Graduate School of Science and Technology, Tokyo Metropolitan University,
1-1 Minami-ohsawa, Hachioji 192-0397, Japan
e-mail: kainosho@tmu.ac.jp

H.S. Atreya (ed.), *Isotope Labeling in Biomolecular NMR*, Advances in Experimental Medicine and Biology 992,
DOI 10.1007/978-94-007-4954-2_5, © Springer Science+Business Media Dordrecht 2012

dynamics are not possible with conventionally labeled proteins. Therefore, further refinement of the isotope labeling schemes is an important key to advance NMR studies of proteins.

The SAIL method was developed to address the problems that cannot be overcome by conventional isotope labeling schemes. In this approach, amino acids with designed isotope labeling patterns (SAIL amino acids) are chemically synthesized and incorporated into the protein to be studied [14, 15]. This strategy allows the production of proteins with a wide variety of isotope labeling patterns, which cannot be generated by conventional protein production methods.

This chapter presents an overview of the current status of the SAIL method. In the first part, the full SAIL approach is introduced. The full SAIL approach is directed toward determining the three-dimensional structures of proteins larger than 25 kDa. In the second part, the residue-selective SAIL approach is described. This approach is focused on obtaining valuable local information about proteins.

5.2 Full SAIL Protein

5.2.1 Design of SAIL Amino Acids

Expanding the molecular range of proteins amenable to NMR structure determination is an important subject. The main obstacles for protein NMR studies of large proteins are the extensive signal overlapping and severe line-broadening of the peaks. To address this problem, the successful use of protein deuteration has long been recognized as a key method. However, the perdeuteration of a protein removes all side-chain non-exchangeable protons, and thus the side-chain relevant NOEs are no longer available [16–18]. The random fractional deuteration method [19, 20] also suffers from the presence of numerous isotopomers and the dilution of ^{1}H. Methyl protonation on a deuterated background is a powerful approach to probe specific methyl groups [13]. In this labeling pattern, however, the proton atoms are localized and unevenly distributed in the protein molecule, and thus it is insufficient to determine the hydrophobic core formed by non-methyl groups. In the SAIL method, a site-specific isotope labeling scheme was designed, as follows (Fig. 5.1):

1. Stereo-selective replacement of one ^{1}H in methylene groups by ^{2}H.
2. Replacement of two ^{1}H in each methyl group by ^{2}H.
3. Stereo-selective modification of the prochiral methyl groups of Leu and Val, such that one methyl is $-^{12}C(^2H)_3$ and the other is $-^{13}C^1H(^2H)_2$.
4. Labeling of six-membered aromatic rings by alternating $^{12}C-^2H$ and $^{13}C-^1H$ moieties.

Twenty SAIL amino acids were chemically synthesized, based on these design concepts [14, 21, 22]. These SAIL amino acids are now commercially available from SAIL Technologies (http://www.sail-technologies.com/). Note that these isotope labeling schemes are designed for proteins around 25–50 kDa. Further reduction of the proton density is needed to determine the structures of proteins larger than 50 kDa [23].

5.2.2 Cell-Free Synthesis of SAIL Proteins

For the production of full-SAIL proteins, a prerequisite is the efficient incorporation of the 20 SAIL amino acids in their intact forms into a target protein. In the case of conventional cellular expression systems, however, the incorporation of certain amino acids (e.g., Gly, Ser, Asp and Glu) into target proteins is hampered by metabolic conversion in the cells. To overcome this problem, a cell-free protein

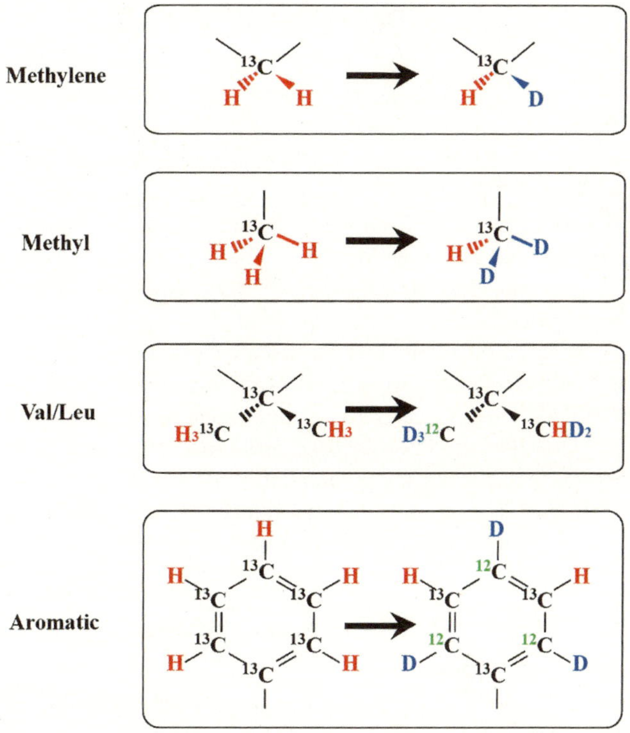

Fig. 5.1 Isotope labeling
schemes of SAIL amino acids

production system is utilized for the production of full-SAIL proteins [24, 25]. In the cell-free protein
synthesis, the problematic amino-acid metabolism is highly suppressed, and thus the isotopes of the
labeled amino acids are not scrambled or lost. The details of the *E. coli* cell-free expression system
are described in another chapter in this book.

5.2.3   NMR Experiments on SAIL Proteins

A comparison between the NMR spectra of a SAIL protein and those of a UL protein reveals that
the SAIL method drastically improves the spectral quality. As compared to those of the UL protein,
the NMR spectra of the SAIL protein are simplified and the peak sensitivity is increased, due to the
reduced transverse relaxation. Furthermore, the fewer long-range couplings result in further signal
sharpening. The signal intensities for methylene groups are 3–7 times higher with SAIL than with
uniform labeling under the same conditions [14]. In the case of an aromatic ring, the absence of one-
bond ^{13}C-^{13}C scalar coupling eliminates the need for the constant time data collection method [26].

The SAIL isotope labeling pattern is designed to facilitate the assignment strategy based on
through-bond connectivity, since the scalar-coupling connectivity required for backbone and side
chain assignments is preserved in all amino acids. However, the magnetization transfer is facilitated
by the extensive deuteration, along with the side-chains [22]. In addition, the stereo- and regio-specific
deuteration of the methylene and pro-chiral methyl groups eliminates the need for their stereo-specific
assignments. Thus, the almost complete assignment of the side-chain peaks can readily be achieved
in the SAIL proteins, as compared to the UL proteins. Most heteronuclear multidimensional NMR
experiments could benefit from the SAIL scheme. A minor modification is that deuterium decoupling
should be applied during chemical shift encoding on aliphatic ^{13}C carbons, to avoid splitting by the
^{13}C-^{2}H coupling [27].

5.2.4   Structure Determination of SAIL Proteins

The isotope labeling pattern of the SAIL proteins is also favorable for collecting NOE distance restraints. As compared to UL proteins, the number of ^{1}H atoms is decreased to less than half in SAIL proteins, which simplifies NOESY spectra and mitigates proton spin diffusion. While the expected number of NOEs in a SAIL protein is smaller than that in the corresponding UL protein, the distance restraints obtainable from the SAIL protein are more quantitative and contribute to the determination of the protein conformation, due to the stereo-specific assignments involving protons. In addition, the reduced spin diffusion in the SAIL proteins enables the recording of NOESY spectra with longer NOE mixing times and more qualitative NOEs, as compared to the corresponding UL proteins.

With the quantitative NOE distance restraints, a high-resolution protein structure can be determined. The CYANA software is used for automated NOE assignment and structure calculations [28]. The SAIL method has been applied to determine the three-dimensional structures of several proteins [14, 26, 29, 30]. The spectra from SAIL proteins are simplified as compared to those from UL proteins, and thus fully automated NMR structure determination, including peak resonances, NOE peaks and structure calculations, can be achieved without human intervention [31, 32].

5.3   Residue-Selective SAIL Method

Residue-selective labeling of a protein by a SAIL amino acid is an important method along with the full-SAIL approach. In this method, target proteins are residue-selectively labeled by one or more SAIL amino acid(s) to obtain detailed local information on protein structure and dynamics. An important advantage of the residue-selective SAIL method over the full-SAIL approach is the applicability of cellular expression for the protein production. In the following, the cellular expression of a SAIL protein is first described, and then several applications based on the residue-selective SAIL method are introduced.

5.3.1   Cellular Expression of Residue-Selective SAIL Proteins

The production of a full-SAIL protein necessitates the use of a cell-free expression system, due to the incorporation of all 20 amino acids into the target protein. In the case of the residue-selective SAIL method, however, only certain kinds of SAIL amino acids must be incorporated into the target protein. Therefore, as long as the selected amino acids are efficiently incorporated into the protein, cellular expression can be used for the protein production. Especially, the availability of *E. coli* cells with robust protein synthesis expands the kinds of proteins amenable to SAIL-based NMR studies.

A concern about cellular expression lies in the incorporation rate of the added SAIL amino acid and the labeling efficiency in the produced protein. In terms of the labeling efficiency, the use of an auxotrophic *E. coli* strain ensures almost complete labeling efficiency in the produced protein. As auxotrophic *E. coli* cells are unable to synthesize the specific amino acid, the cells exclusively utilize the supplemented amino acid for protein synthesis. However, the protein yields from such auxotrophic strains are frequently lower than those from conventionally utilized autotrophic strains. Especially, when the amount of the added amino acid is decreased to reduce the cost, the growth of the auxotrophic strain is likely to suffer. Therefore, the optimal amount of supplemented amino acid should be determined by a pilot experiment with a small-scale culture, to find the right compromise between the yield of the protein and the amount of the labeled amino acid. Note that a target protein with an acceptable level of labeling efficiency can also be achieved by using conventional *E. coli*

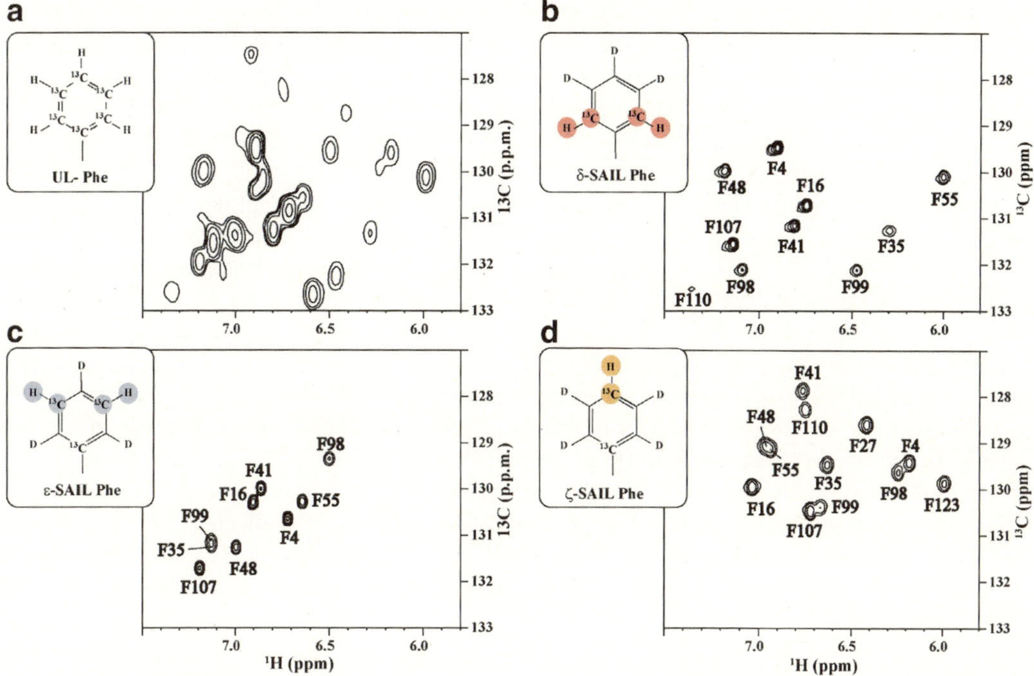

Fig. 5.2 ^{1}H-^{13}C HSQC spectra of *E. coli* peptidyl prolyl *cis-trans* isomerase b (18.2 kDa) selectively labeled by UL Phe (**a**), δ-SAIL Phe (**b**), ε-SAIL Phe (**c**) and ζ-SAIL Phe (**d**). The isotope labeling pattern of each Phe is displayed along with the corresponding spectra. Observed peaks are labeled with their assignments in (**b**), (**c**) and (**d**) (Reproduced from Takeda et al. [34]. With permission)

autotrophic strains as an expression system. While the dilution of the added amino acid occurs in the autotrophic strains, the growth and the protein production level of the autotrophic strain are much higher than those of the auxotrophic strain, which offsets the dilution of the isotope.

5.3.2 Phe and Tyr with Alternate Isotope Labeling Patterns

The aromatic atoms of the side-chain phenyl rings of Phe and Tyr residues in proteins are important subjects of NMR study, from the viewpoints of both structure determination and dynamics analysis. In an NMR structure determination, the phenyl rings frequently constitute the hydrophobic core of a folded protein, and thus the rings serve as sources of valuable NOE restraints. In a dynamics analysis, the rotational movement of the rings about their C_{β}-C_{γ} axis provides information about the large amplitude breathing motions of proteins [33].

However, the quality of the NMR spectra obtained from a conventional uniformly UL protein is often poor, due to spectral crowding and severe line-broadening. To address this issue, δ-SAIL Phe/Tyr, ε-SAIL Phe/Tyr and ζ-SAIL Phe were synthesized [34]. The isotope labeling scheme of their aromatic rings is common, in that selected proton/carbon moieties are kept as ^{1}H and ^{13}C, respectively, and the other aromatic proton and carbon atoms are ^{12}C and ^{2}H labeled, except for the ^{13}C$_{\gamma}$ atoms of ε- SAIL Phe/Tyr and ζ-SAIL Phe. However, they differ in the positions of the ^{13}C/^{1}H moieties, and the data obtainable from proteins labeled with these Phe/Tyr are different and complementary to each other.

A comparison of the ^{1}H-^{13}C HSQC spectra in the aromatic region, recorded on 18.2 kDa *E. coli* peptidyl prolyl *cis-trans* isomerase (EPPIb) proteins selectively labeled with UL-Phe, δ-SAIL Phe, ε-SAIL Phe, or ζ-SAIL Phe, is shown in Fig. 5.2. It is obvious that the signal overlapping and line-broadening observed for UL Phe-labeled EPPIb was mitigated in the three SAIL Phe-labeled proteins. This improvement of the spectral quality arose from the systematic elimination of the relevant scalar and dipolar interactions within the ring. Unambiguous assignments of the aromatic ring resonances of SAIL Tyr/Phe can readily be achieved by the NOE connectivity between H_δ and H_β for δ-SAIL Tyr/Phe, or by the through-bond ^{13}C-^{13}C scalar coupling connectivities through $^{13}C_\gamma$ for ε-SAIL Tyr/Phe and ζ-SAIL Phe [34, 35].

Note that 12 correlation peaks were observed in ζ-SAIL Phe-labeled EPPIb, while only 9 peaks were observed for the δ- and ε-SAIL Phe-labeled samples. This distinct number of observable peaks arises from the fact that three peaks at the δ- and ε-positions are exchange-broadened due to the ring flipping motion, on the order of milliseconds. As the proton-carbon moieties at the ζ-positions are located on the rotation axis of the ring flipping motion, their line-shapes are not affected by the ring flipping rate. On the other hand, those at the δ- and ε-positions inter-convert between equivalent positions, and the exchange rate is in the intermediate range on the NMR chemical shift time scale.

A comparison of the NMR structures of full SAIL EPPIb proteins in which Phe/Tyr are δ-SAIL Phe/δ-SAIL Tyr, ε-SAIL Phe/ ε-SAIL Tyr or ζ-SAIL Phe/δ-SAIL Tyr reveals that the structure with ζ-SAIL Phe/δ-SAIL Tyr is the best converged. The reason for this is that the ζ-position is likely to give rise to long-range restraints. In addition, the line-shape at the ζ-position is unaffected by the ring flipping motion of the aromatic ring about its C_β-C_γ axis. The NOE restraints involving H_δ, H_ε and H_ζ are highly complementary for protein structure determination. Therefore, when the three different types of NOE restraints are combined, the structural quality of EPPIb becomes fairly well defined [34].

5.3.3 SAIL Trp

Trp residues are also frequently embedded in the hydrophobic cores of proteins, and therefore the determination of the precise orientation of the indole ring is crucial for protein NMR structure analyses. However, the observation and assignment of the indole ring signals are difficult when using UL proteins, due to the tight ^{13}C-^{13}C couplings and ^1H-^1H dipolar interactions in the indole ring, as in the cases of the aromatic rings in Phe and Tyr. As a result, the indole ring atoms of Trp produce broadened signals with severe overlapping in the ^1H-^{13}C HSQC spectrum, even in medium-sized proteins. To overcome these problems, we synthesized Trp with a systematically optimized isotope labeling pattern [36].

The synthesized Trp, referred to as $[^{12}C_\gamma, {}^{12}C_{\varepsilon2}]$ SAIL Trp, has three ^{13}C-^1H pairs at the $\delta1$, $\eta2$ and $\varepsilon3$ positions in its indole ring (Fig. 5.3a). As the spin systems of the three ^{13}C-^1H pairs are isolated from each other, the scalar couplings and dipolar interactions in the indole ring are significantly reduced. These optimizations of the relaxation pathway can improve the observation of indole ring signals. Actually, when we studied the UL-Trp labeled Myb-R2R3 protein, almost all of the aromatic CH signals were broadened and could not be detected in the ^1H-^{13}C HSQC spectrum. On the other hand, the $[^{12}C_\gamma, {}^{12}C_{\varepsilon2}]$ SAIL Trp labeled Myb-R2R3 provided a drastically improved ^1H-^{13}C HSQC spectrum. In the SAIL protein, all 18 expected CH signals ($\delta1$, $\eta2$ and $\varepsilon3$ for six Trp residues) are observed with extremely high sensitivity (Fig. 5.3b)

The $[^{12}C_\gamma, {}^{12}C_{\varepsilon2}]$-SAIL Trp also facilitates the sequence specific assignment of the indole ring signals. We can easily assign the $\delta1$ and $\varepsilon3$ signals by using the intra-residue NOEs from aliphatic/amide protons (i.e., $^1H_{\beta3}$, $^1H_\alpha$ and ^1H_N). The $\eta2$ signal is assigned by correlating the $^1H_{\varepsilon3}/^{13}C_{\varepsilon3}$ resonances to $^1H_{\eta2}/^{13}C_{\eta2}$ via the $^1H_{\varepsilon3}$-$^{13}C_{\eta2}$ or $^1H_{\eta2}$-$^{13}C_{\varepsilon3}$ three-bond scalar coupling (~8 Hz). In the aromatic SAIL amino acids, irrelevant scalar and dipolar interactions are systematically eliminated, enabling the correlation experiments based on long range coupling [34, 35, 37]. Based on the intra-residue NOE and scalar coupling connectivity, the complete assignment of the Trp indole ring signals was achieved for the Myb-R2R3 protein [36].

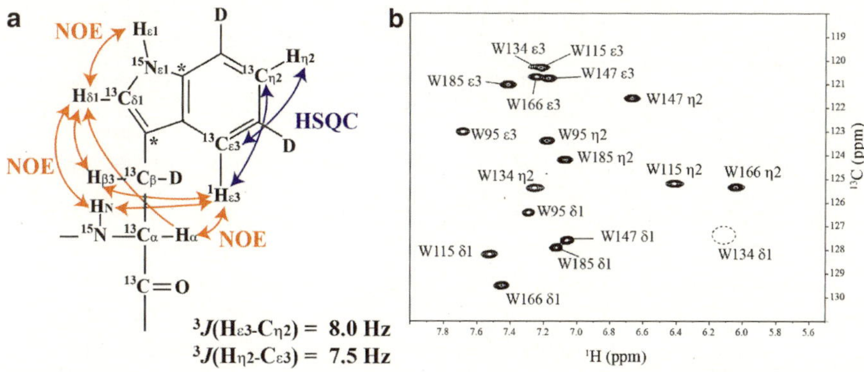

Fig. 5.3 Isotope labeling pattern of [^{12}C$_\gamma$, ^{12}C$_{\varepsilon2}$] SAIL Trp and the magnetization transfer pathway for the assignment. The ^{12}C atoms are not shown in the figure (**a**). The magnetization transfer pathways for assignment are depicted as *arrows*, with the names of the experiments. ^3J(C–H) scalar couplings are depicted. When this Trp is incorporated into the Myb-R2R3 protein, well resolved peaks at the δ1, ε3 and η2 positions were observed (**b**) (Reproduced from Miyanoiri et al. [36]. With permission)

Another key feature of the [^{12}C$_\gamma$, ^{12}C$_{\varepsilon2}$] SAIL Trp is that the intra-residue NOE pattern provides information about the χ_2 angle. The χ_2 angle of Trp has preferences for three different regions: ~+90°, ~0° and ~−90° [38]. When the χ_2 angles are +90° and −90°, strong NOEs between ^{1}H$_{\beta3}$ and ^{1}H$_{\varepsilon3}$ and between ^{1}H$_{\beta3}$ and ^{1}H$_{\delta1}$ are observed, respectively. When the χ_2 angle is ~0°, a weak NOE between ^{1}H$_{\beta3}$ and ^{1}H$_{\varepsilon3}$ and a strong NOE between ^{1}H$_{\delta1}$ and ^{1}H$_\alpha$ are observed. In the case of the Myb-R2R3 protein, strong NOEs between ^{1}H$_{\beta3}$-^{1}H$_{\varepsilon3}$ (i.e., the χ_2 angle is ~+90°) were observed for four of the six Trp residues, and those between ^{1}H$_{\beta3}$ and ^{1}H$_{\delta1}$ (i.e., the χ_2 angle is ~−90°) were observed for the remaining two Trp residues. These results were coincident with the χ_2 angle values of the Trps in the crystal structure of Myb-R2R3.

5.3.4   NMR Hydrogen Exchange Study of Polar Side-Chain Groups

Hydrogen bonds involving the side-chain hydroxyl (OH) or sulfhydryl (SH) groups of Ser, Thr, Tyr and Cys residues are structurally and functionally important interactions in proteins. To characterize the individual polar side-chain groups responsible for the side-chain hydrogen bonds, the evaluation of their hydrogen exchange rates with solvent water provides valuable information about them. However, the chemical exchange rates of such polar side-chain groups are frequently faster than the NMR chemical shift time scale, and thus it has been assumed that the exchanging protons are rarely observable by NMR.

To exclusively detect the slowly exchanging OH/SH groups and to evaluate their hydrogen exchange rates, we developed a robust and simple approach based on the SAIL method [30, 37, 39]. In this approach, the carbon atoms attached to the OH/SH groups (i.e., C$_\zeta$ atom of Tyr and C$_\beta$ atoms of Ser, Thr and Cys) are directly observed by NMR in an equimolar H$_2$O/D$_2$O mixture. The key principle of this method is that the chemical shifts of the carbons attached to OH and OD are different, due to the deuterium isotope shift effect. When carbon atoms are attached to slowly exchanging OH/SH groups, they give rise to isotopomer-resolved split peaks in the H$_2$O/D$_2$O solution, while the rapidly exchanging OH/SH groups generate averaged peaks. Therefore, based on the line-shape of the carbon peaks in an H$_2$O/D$_2$O solution, the slowly exchanging OH/SH groups can readily be identified. A prerequisite for

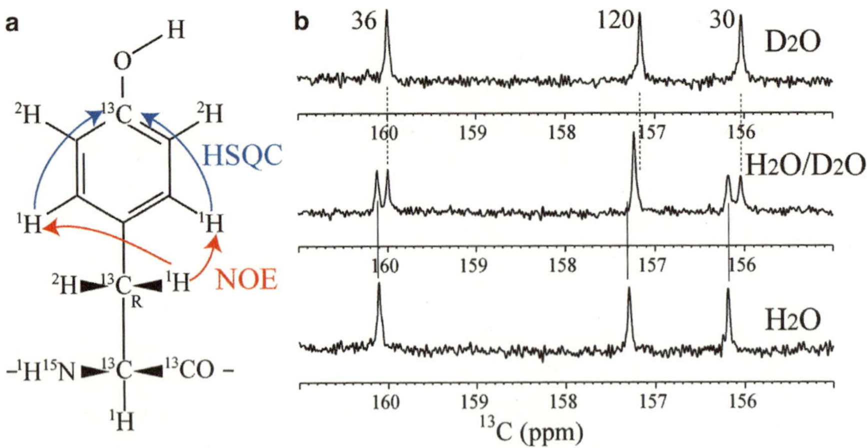

Fig. 5.4 The $^{13}C_\zeta$ peaks in ζ-SAIL Tyr are assigned by the NOE connectivities between the β3- and δ-protons, and then connected to the ζ-carbon though ^3J coupling by an HSQC experiment (**a**). When the $^{13}C_\zeta$ carbons are observed in an equimolar H$_2$O/D$_2$O (1:1) solution, the isotopomers of ^{13}C(–OH) and ^{13}C(–OD) are resolved when the exchange rate is slower than the size of the isotope shift effect (**b**) (Reproduced from Takeda et al. [37]. With permission)

this method is that the line-width of the carbon peaks must be sharp enough to resolve the isotope shift effect. To this end, the isotope labeling patterns of the OH/SH-harboring amino acids are designed such that irrelevant scalar and dipolar interactions are extensively eliminated. In the case of Tyr residues, for example, only the ζ-carbon is enriched by ^{13}C in the phenyl ring. For the assignment of the ζ-carbon, the ε-protons are deuterated and the $^{13}C_\zeta$ atoms are assigned by correlating ^{1}H$_\beta$ to ^{1}H$_\delta$ by the NOE and then to $^{13}C_\zeta$ via three-bond scalar coupling (Fig. 5.4a) [39]. When the Tyr $^{13}C_\zeta$ peaks in the 18.2 kDa *E. coli* peptidyl prolyl *cis-trans* isomerase (EPPIb) were observed in the H$_2$O/D$_2$O (1:1) solution, two out of the three $^{13}C_\zeta$ peaks gave rise to isotopomer-resolved peaks (Fig. 5.4b). Furthermore, the ^{13}C exchange experiment for the resolved peaks allows quantitative evaluation of the H/D exchange rates. The results for the EPPIb revealed that about 30% of the OH/SH protons in this globular protein were slowly exchanging. We successfully elucidated their hydrogen bonding patterns, based on the NOEs involving their OH/SH protons, and determined the precise side-chain conformations [30].

5.3.5 Conformational Analysis of a Disulfide Bond by Quantitative NOEs Across the Bond

Disulfide bonds within proteins play important roles in stabilizing their specific conformations. Therefore, their bonding patterns and conformations are an important subject of study. The NOEs between the β-protons of cysteine residues across disulfide bonds in proteins provide direct information about the connectivities and conformations of these important cross-links. With conventional UL-proteins, however, fast spin diffusion processes mediated by strong dipolar interactions between the *geminal* β-protons prohibit the quantitative measurements, and thus the analyses, of long-range NOEs across disulfide bonds (Fig. 5.5a). To overcome this problem, we proposed a method using proteins selectively labeled with an equimolar mixture of L-[β-^{13}C;α,β_2-^{2}H$_2$] Cys ((2R,3S)-[3-^{13}C;2,3-^{2}H$_2$] Cys) and L-[β-^{13}C;α,β_3-^{2}H$_2$] Cys ((2R,3R)-[3-^{13}C;2,3-^{2}H$_2$] Cys), but otherwise fully deuterated. Since either one of the *prochiral* methylene protons, namely β2 (*proS*) or β3 (*proR*), is always replaced with a deuteron and no other protons remain in proteins prepared by this labeling scheme, all four of the expected NOEs for the β-protons across disulfide bonds could be measured without any spin diffusion interference (Fig. 5.5b). Therefore, the NOEs for the β2 and β3 pairs across each of the disulfide bonds

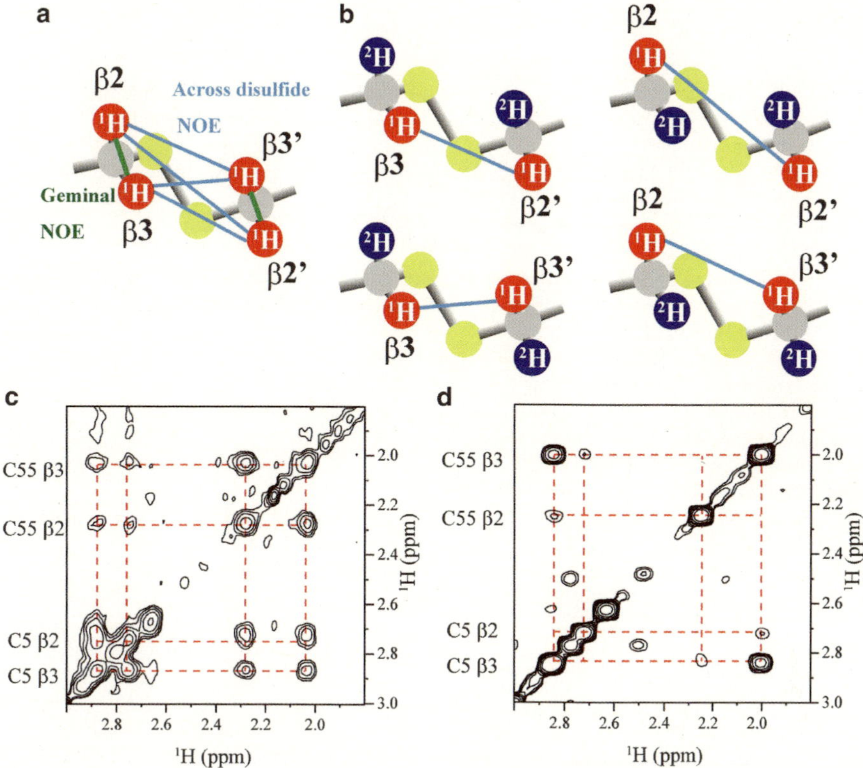

Fig. 5.5 The observation of across-disulfide NOEs. In a disulfide bond, dipolar interactions between the geminal protons cause problematic spin diffusion and hamper the quantitative analysis of the four across-disulfide NOEs (**a**). By incorporating equimolar amounts of (2R,3S)- and (2R,3R)-[β-^{13}C;α,β-^{2}H$_2$] Cys into a target protein, quantitative across-disulfide NOEs are observed in the complete absence of the geminal dipolar interactions (**b**). In conventional UL proteins, the geminal NOEs are observed and spin diffusion hampers the quantitative analysis of the across-disulfide NOEs (**c**). In the ((2R, 3RS)-[3-^{13}C;2,3-^{2}H$_2$] Cys, [U-^{2}H])-labeled protein, the geminal NOEs are eliminated and quantitative NOEs are observed (**d**) (Reproduced from Takeda et al. [40]. With permission)

could be observed with high sensitivity, even though they are 25% of the theoretical maximum for each pair. When compared with conventional UL proteins, proton spin diffusion is highly suppressed in the (L-[β-^{13}C;α,β$_2$/β$_3$-^{2}H$_2$] Cys, [U-^{2}H])-labeled protein, enabling the observation of quantitative across-disulfide NOEs (Fig. 5.5c, d). With the NOE information, the disulfide bond connectivities can be unambiguously established for proteins with multiple disulfide bonds. Another advantage of this isotope labeling scheme over the conventional UL scheme is that the cross-relaxation rate can be evaluated accurately using long mixing times, by virtue of the elimination of the spin diffusion (Fig. 5.5c, d). Based on the precise proton distances of the four β-proton pairs derived from the quantitative NOEs, the conformation of a disulfide bond, namely χ2 and χ3, can be determined [40].

5.4 Conclusions

Overviews of the full-SAIL and residue-selective SAIL methods have been provided in this chapter. A key feature of the SAIL method lies in the extensive optimization of the isotope labeling pattern with regard to an intended purpose. With the increasing demand for investigating the structures of large proteins and the functionally relevant protein dynamics, the optimization of stable labeling schemes for proteins will become even more important.

References

1. Kainosho M (1997) Isotope labelling of macromolecules for structure determinations. Nat Struct Biol 4:854–857
2. Goto NK, Kay LE (2000) New developments in isotope labeling strategies for protein solution NMR spectroscopy. Curr Opin Struct Biol 10:585–592
3. Lian LY, Middleton DA (2001) Labelling approaches for protein structural studies by solution-state and solid-state NMR. Prog Nucl Magn Reson Spectrosc 39:171–190
4. Ohki S, Kainosho M (2008) Stable isotope labeling methods for protein NMR spectroscopy. Prog Nucl Magn Reson Spectrosc 53:208–226
5. Ikura M, Kay LE, Bax A (1990) A novel approach for sequential assignment of proton, carbon-13, and nitrogen-15 spectra of larger proteins: heteronuclear triple-resonance three-dimensional NMR spectroscopy. Application to calmodulin. Biochemistry 29:4659–4667
6. Kay LE, Ikura M, Tschudin R, Bax A (1990) Three-dimensional triple-resonance NMR spectroscopy of isotopically enriched proteins. J Magn Reson 89:496–514
7. Clore GM, Gronenborn AM (1994) Multidimensional heteronuclear nuclear magnetic resonance of proteins. Methods Enzymol 239:349–363
8. Markley JL, Putter I, Jardetzky O (1968) High-resolution nuclear magnetic resonance spectra of selectively deuterated staphylococcal nuclease. Science 161:1249–1251
9. Crespi HL, Rosenberg RM, Katz JJ (1968) Proton magnetic resonance of proteins fully deuterated except for ^{1}H-leucine side chains. Science 161:795–796
10. Gardner KH, Kay LE (1998) The use of ^{2}H, ^{13}C, ^{15}N multidimensional NMR to study the structure and dynamics of proteins. Annu Rev Biophys Biomol Struct 27:357–406
11. Pervushin K, Riek R, Wider G, Wüthrich K (1997) Attenuated T_2 relaxation by mutual cancellation of dipole-dipole coupling and chemical shift anisotropy indicates an avenue to NMR structures of very large biological macromolecules in solution. Proc Natl Acad Sci USA 94:12366–12371
12. Pervushin K, Riek R, Wider G, Wüthrich K (1998) Transverse relaxation-optimized spectroscopy (TROSY) for NMR studies of aromatic spin systems in ^{13}C-labeled proteins. J Am Chem Soc 120:6394–6400
13. Tugarinov V, Hwang PM, Ollerenshaw JE, Kay LE (2003) Cross-correlated relaxation enhanced ^{1}H-^{13}C NMR spectroscopy of methyl groups in very high molecular weight proteins and protein complexes. J Am Chem Soc 125:10420–10428
14. Kainosho M, Torizawa T, Iwashita Y, Terauchi T, Ono AM, Güntert P (2006) Optimal isotope labelling for NMR protein structure determinations. Nature 440:52–57
15. Kainosho M, Güntert P (2009) SAIL-Stereo-array isotope labeling. Q Rev Biophys 7:1–54
16. Venters RA, Huang CC, Farmer BT 2nd, Trolard R, Spicer LD, Fierke CA (1995) High-level ^{2}H/^{13}C/^{15}N labeling of proteins for NMR studies. J Biomol NMR 5:339–344
17. Venters RA, Metzler WJ, Spicer LD, Mueller L, Farmer BT 2nd (1995) Use of ^{1}H$_N$-^{1}H$_N$ NOEs to determine protein global folds in perdeuterated proteins. J Am Chem Soc 117:9592–9593
18. Venters RA, Farmer BT 2nd, Fierke CA, Spicer LD (1996) Characterizing the use of perdeuteration in NMR studies of large proteins: ^{13}C, ^{15}N and ^{1}H assignments of human carbonic anhydrase II. J Mol Biol 264:1101–1116
19. LeMaster DM, Richards FM (1988) NMR sequential assignment of *Escherichia coli* thioredoxin utilizing random fractional deuteration. Biochemistry 27:142–150
20. Nietlispach D, Clowes RT, Broadhurst RW, Ito Y, Keeler J, Kelly M, Ashurst J, Oschkinat H, Domaille PJ, Laue ED (1996) An approach to the structure determination of larger proteins using triple resonance NMR experiments in conjunction with random fractional deuteration. J Am Chem Soc 118:407–415
21. Terauchi T, Kobayashi K, Okuma K, Oba M, Nishiyama K, Kainosho M (2008) Stereoselective synthesis of triply isotope-labeled Ser, Cys, and Ala: amino acids for stereoarray isotope labeling technology. Org Lett 10:2785–2787
22. Terauchi T, Kamikawai T, Vinogradov MG, Starodubtseva EV, Takeda M, Kainosho M (2011) Synthesis of stereoarray isotope labeled (SAIL) lysine via the "head-to-tail" conversion of SAIL glutamic acid. Org Lett 13:161–163
23. Ikeya T, Terauchi T, Güntert P, Kainosho M (2006) Evaluation of stereo-array isotope labeling (SAIL) patterns for automated structural analysis of proteins with CYANA. Magn Reson Chem 44:S152–S157
24. Torizawa T, Shimizu M, Taoka M, Miyano H, Kainosho M (2004) Efficient production of isotopically labeled proteins by cell-free synthesis: a practical protocol. J Biomol NMR 30:311–325
25. Takeda M, Ikeya T, Güntert P, Kainosho M (2007) Automated structure determination of proteins with the SAIL-FLYA NMR method. Nat Protoc 2:2896–2902
26. Takeda M, Chang CK, Ikeya T, Güntert P, Chang YH, Hsu YL, Huang TH, Kainosho M (2008) Solution structure of the C-terminal dimerization domain of SARS coronavirus nucleocapsid protein solved by the SAIL-NMR method. J Mol Biol 380:608–622

27. Grzesiek S, Anglister J, Ren H, Bax A (1993) Carbon-13 line narrowing by deuterium decoupling in deuterium/carbon-13/nitrogen-15 enriched proteins. Application to triple resonance 4D J connectivity of sequential amides. J Am Chem Soc 115:4369–4370
28. Güntert P (2003) Automated NMR structure calculation with CYANA. Prog NMR Spectrosc 43:105–125
29. Takeda M, Sugimori N, Torizawa T, Terauchi T, Ono AM, Yagi H, Yamaguchi Y, Kato K, Ikeya T, Jee J, Güntert P, Aceti DJ, Markley JL, Kainosho M (2008) Structure of the putative 32 kDa myrosinase binding protein from Arabidopsis (At3g16450.1) determined by SAIL-NMR. FEBS J 275:5873–5884
30. Takeda M, Jee J, Ono AM, Terauchi T, Kainosho M (2011) Hydrogen exchange study on the hydroxyl groups of serine and threonine residues in proteins and structure refinement using NOE restraints with polar side-chain groups. J Am Chem Soc 133:17420–17427
31. Ikeya T, Takeda M, Yoshida H, Terauchi T, Jee JG, Kainosho M, Güntert P (2009) Automated NMR structure determination of stereo-array isotope labeled ubiquitin from minimal sets of spectra using the SAIL-FLYA system. J Biomol NMR 44:261–272
32. Ikeya T, Jee JG, Shigemitsu Y, Hamatsu J, Mishima M, Ito Y, Kainosho M, Güntert P (2011) Exclusively NOESY-based automated NMR assignment and structure determination of proteins. J Biomol NMR 50:137–146
33. Wagner G, DeMarco A, Wüthrich K (1976) Dynamics of the aromatic amino acid residues in the globular conformation of the basic pancreatic trypsin inhibitor (BPTI). I. ^{1}H NMR studies. Biophys Struct Mech 2:139–158
34. Takeda M, Jee J, Ono AM, Terauchi T, Kainosho M (2010) Application of SAIL phenylalanine and tyrosine with alternative isotope-labeling patterns for protein structure determination. J Biomol NMR 46:45–49
35. Torizawa T, Ono AM, Terauchi T, Kainosho M (2005) NMR assignment methods for the aromatic ring resonances of phenylalanine and tyrosine residues in proteins. J Am Chem Soc 127:12620–12626
36. Miyanoiri Y, Takeda M, Jee J, Ono AM, Okuma K, Terauchi T, Kainosho M (2011) Alternative SAIL-Trp for robust aromatic signal assignment and determination of the $\chi(2)$ conformation by intra-residue NOEs. J Biomol NMR 51:425–435
37. Takeda M, Jee J, Ono AM, Terauchi T, Kainosho M (2009) Hydrogen exchange rate of tyrosine hydroxyl groups in proteins as studied by the deuterium isotope effect on Cζ chemical shifts. J Am Chem Soc 131:18556–18562
38. Dunbrack RL Jr, Cohen FE (1997) Bayesian statistical analysis of protein side-chain rotamer preferences. Protein Sci 6:1661–1681
39. Takeda M, Jee J, Terauchi T, Kainosho M (2010) Detection of the sulfhydryl groups in proteins with slow hydrogen exchange rates and determination of their proton/deuteron fractionation factors using the deuterium-induced effects on the ^{13}Cβ NMR signals. J Am Chem Soc 132:6254–6260
40. Takeda M, Terauchi T, Kainosho M (2012) Conformational analysis by quantitative NOE measurements of the β-proton pairs across individual disulfide bonds in proteins. J Biomol NMR 52:127–139

Chapter 6
Amino Acid Selective Labeling and Unlabeling for Protein Resonance Assignments

Garima Jaipuria, B. Krishnarjuna, Somnath Mondal, Abhinav Dubey, and Hanudatta S. Atreya

Abstract Structural characterization of proteins by NMR spectroscopy begins with the process of sequence specific resonance assignments in which the ^{1}H, ^{13}C and ^{15}N chemical shifts of all backbone and side-chain nuclei in the polypeptide are assigned. This process requires different isotope labeled forms of the protein together with specific experiments for establishing the sequential connectivity between the neighboring amino acid residues. In the case of spectral overlap, it is useful to identify spin systems corresponding to the different amino acid types selectively. With isotope labeling this can be achieved in two ways: (i) amino acid selective labeling or (ii) amino acid selective 'unlabeling'. This chapter describes both these methods with more emphasis on selective unlabeling describing the various practical aspects. The recent developments involving combinatorial selective labeling and unlabeling are also discussed.

6.1 Introduction

One of the first steps in elucidating the structure or dynamics of proteins is sequence specific resonance assignments. In this part of the structure determination process the identity of each resonance or peak observed in a given spectrum is established. In the case of small proteins (molecular mass <10 kDa) this can be accomplished using a set of two-dimensional (2D) homonuclear (^{1}H-^{1}H) correlation experiments [1, 2]. As the size of the protein increases beyond 10 kDa the number of observed resonances also increases. Consequently, given that chemical shift correlations of a given type are generally confined to the same region of the spectrum irrespective of the amino acid type or protein (e.g., amide ^{1}H signals almost always occur in the range of 6–10 ppm or ^{13}C$^\alpha$-^{1}H$^\alpha$ correlations occur in the range

G. Jaipuria • S. Mondal • H.S. Atreya (✉)
NMR Research Centre, Indian Institute of Science, Bangalore 560012, India
hsatreya@sif.iisc.ernet.in

Solid State and Structural Chemistry Unit, Indian Institute of Science,
Bangalore 560012, India

B. Krishnarjuna
NMR Research Centre, Indian Institute of Science, Bangalore 560012, India

A. Dubey
NMR Research Centre, Indian Institute of Science, Bangalore 560012, India

IISc Mathematics Initiative, Indian Institute of Science,
Bangalore 560012, India

H.S. Atreya (ed.), *Isotope Labeling in Biomolecular NMR*, Advances in Experimental Medicine and Biology 992,   95
DOI 10.1007/978-94-007-4954-2_6, © Springer Science+Business Media Dordrecht 2012

of 45–65 ppm etc.) the increase in the number of resonances results in their overlap with each other in the spectrum. Such spectral congestion can be alleviated by resorting to heteronuclear three or higher dimensional NMR experiments which provide resolution by correlating a larger number of different chemical shifts [2] using a protein sample enriched uniformly with $^{13}C/^{15}N/^2H$ isotopes. However, in systems which are plagued with severe spectral overlap and/or chemical shift degeneracy such as proteins in solid state [3] (also discussed in Chap. 3 of this book), membrane [4, 5] or structurally disordered proteins [6] high dimensional shift correlations fail at times to provide sufficient resolution to accomplish unambiguous resonance assignments. Hence reduction of spectral overlap or spectral 'simplification' becomes pivotal. One solution is to reduce the number of peaks observed in the spectrum by detecting peaks selectively from a given amino acid type in the protein. This also helps in sequence specific resonance assignments if the type of residues *around* (i.e., the N-terminal or the C-terminal neighbor of) the selected residue can be identified and placed in the primary sequence. This chapter focuses on the different isotope labeling schemes to achieve the objectives of spectral simplification and sequence specific resonance assignments.

Systematic study of isotope labeling in proteins started in 1960s and one of the first amino acid type selective labeling method involved incorporation of specific protonated amino acids against a deuterated background [7]. Subsequently, selective incorporation of $^{13}C/^{15}N$-labeled amino acids against an unlabeled ($^{12}C/^{14}N$) background was developed [8, 9]. In general, selective identification and assignment of different amino acid types can be accomplished in one of three ways: (i) using NMR experiments which detect a single or a set of amino acid-type(s) based on a specific magnetization transfer [10–18], (ii) selective $^{13}C/^{15}N$ labeling of amino acids using cell-based or cell-free methods [8, 19, 20] or/and (iii) selective unlabeling of amino acids using cell-based methods [21–25]. In the first approach, NMR experiments are implemented which exploit the topology of the side-chain of different amino acids. The magnetization transfer pathways in these experiments are designed to specifically detect a given amino acid type in the spectrum [10–18]. This approach suffers a loss in sensitivity with increase in size of proteins due to the long delay periods used in the pulse sequences for obtaining the desired magnetization transfer. Moreover, these experiments cannot be used for proteins which lose side-chain ^1H upon deuteration.

The second approach involves selective labeling ($^{13}C/^{15}N$) of specific amino acid types in the protein while keeping the rest of the amino acids in the unlabeled form (i.e., $^{12}C/^{14}N$). This is achieved either by cell-based [8, 9] or cell-free (*in vitro*) methods [20, 26, 27]. In the cell-based method the host organism is supplied with the desired isotopically labeled ($^{13}C/^{15}N$) amino acid, while supplying the rest of the amino acids in unlabeled form [8]. Isotope labeling of a specific site in the desired amino acid type can also be achieved using the appropriate site-selectively labeled amino acids or their precursors (e.g. methyl groups [28, 29]). In the cell-free based approach selective labelling is accomplished *in vitro* using the necessary components of the machinery used by the cells for protein biosynthesis [20, 26, 27, 30]. In recent years, this approach has emerged as a preferred alternative due to the following dis-advantages of the cell-based methods: (i) mis-incorporation of isotope labels at undesired sites (also known as 'isotope scrambling') and (ii) difficulty to produce certain types of proteins such as membrane proteins and/or those which are toxic to the cells expressing them. These are overcome to some extent in the cell-free methods. Cell free methods are also preferable in terms of the costs involved if site-selectively labeled amino acids need to be incorporated [31]. However, the cell-based methods have remained popular due to ease of expressing proteins with high yields especially in organisms such as *E. Coli*.

One of the drawbacks of the selective labeling approach described above is the requirement of expensive $^{13}C/^{15}N/^2H$ (site selectively) enriched amino acids resulting in high costs especially if the protein yields are low, An inexpensive alternative is the method of amino acid selective "unlabeling" or 'reverse'/'inverse' labeling [21–25]. This involves selective unlabeling of specific amino acid-types in the protein keeping the other amino acids uniformly $^{13}C/^{15}N$ labeled. This is accomplished by supplying the host microorganism with $^{15}NH_4Cl$ or/and ^{13}C ($^1H/^2H$) -D-glucose as the sole source of nitrogen and carbon, respectively, along with the $^{12}C/^{14}N$ form of the desired amino acid(s) to be selectively unlabeled. As a result, resonances from residues that are unlabeled are not observed in the NMR spectra. A comparison of the spectrum is then made with a spectrum acquired on a control sample involving uniform $^{13}C/^{15}N$ labeling. This enables the assignment of chemical shifts of nuclei belonging to the unlabeled residue. This method requires relatively inexpensive sources of isotope labels (only unlabeled amino acids are used along with $^{15}NH_4Cl$ and/or ^{13}C-D-glucose). It has been used in a large number of applications such as spectral simplification [22, 24, 32, 33], spin-system or amino acid-type identification [21, 23, 34], sequence specific resonance assignments [25], stereospecific assignment of the pro-chiral methyl groups of Val and Leu in large molecular weight proteins [35] and measurement of residual dipolar couplings (RDC) [36]. Its application has been proposed for structural studies of membrane and paramagnetic proteins [37, 38].

In addition to spectral simplification selective labeling or unlabeling is also useful for amino-acid type identification and sequence specific resonance assignments. This is achieved by using a specific combination of amino acid types for selective identification. Different samples of the protein are prepared in which different set of amino acid types are selectively labeled (sometime site-selectively) [39–53]. This approach is termed as "*combinatorial selective labeling*". In a similar vein combinatorial amino acid selective unlabeling can be carried out in a combinatorial approach [21, 25]. In combinatorial selective labeling/unlabeling the combination of amino acid types are decided based on different critera such as the primary sequence of the protein of interest and the ease with which the set of amino acid types selected can be distinguished in the spectrum. Different algorithms have also been proposed for choosing an optimal combination of amino acid for combinatorial labeling [44, 50, 53, 54].

This chapter will focus primarily on two aspects: (1) amino acid selective unlabeling strategies for protein resonance assignment and structure determination and (2) combinatorial schemes involving selective labelling or unlabeling. The method for producing proteins with specific amino acid labeled selectively has been described/reviewed extensively in the past especially for proteins expressed in *E. Coli* [8, 9, 19, 55]. In Chaps. 10 and 11 of this book, different strategies for selective labeling of proteins expressed in higher organisms such as Yeast, insect cells and mammalian cells have been described. In Chap. 9 of this book amino acid selective labeling using the cell-free method is described. The procedure for producing selectively unlabeled protein samples with cell-based methods is described below.

6.2 Amino Acid Selective Unlabeling

6.2.1 Sample Preparation

As of date, all selective unlabeling approaches have been cell-based. The procedure for preparing a selectively unlabeled protein sample using an *E. coli* expression system is similar to that employed in selective amino acid labeling [8, 9]. That is, the desired amino acid type is added in unlabeled form (i.e., $^{12}C/^{14}N$) against a background of ^{13}C and ^{15}N labeling. The procedure typically employed when using an *E. coli* expression system is given below:

1. Inoculate a 5 ml Luria Broth medium (LB/rich medium) with a single colony taken from a freshly transformed plate.
2. After 8–10 h of growth, transfer it to 1 l of LB medium and continue to grow till the optical density (O. D.) reaches 0.6.
3. Centrifuge the cells at 5,000 rpm for 10 min, re-suspend the cell pellet in 1 l of 1X minimal medium containing the desired amino acid to be selectively unlabeled at a concentration of 1.0 g/l. The minimal media consists of M9 salts containing 1 g/l $^{15}NH_4Cl$ supplemented with 0.250 g/l of $MgSO_4 \cdot 2H_2O$, 0.015 g/l of $CaCl_2$, 4.0 g/l [$^{13}C_6$] glucose. The minimal medium can be supplemented with vitamins/other salts (such as $ZnSO_4$) if required.

4. Allow the cells to grow for another 45 min and induce the desired protein expression suitably (e.g., using IPTG).

The procedure for harvesting/lysis of cells and protein purification remains the same as used for preparing a uniformly labeled sample. The amount of unlabeled amino acid to be used depends on the overall protein yield. For good protein expression up to 1 g/l of the desired amino acid to be selectively unlabeled can be used. In case of lower yields of the protein, the exogenously added amino acid can be reduced to 0.5 g/l. In case of amino acids which undergo isotope scrambling, larger amounts are required to be added, which is discussed below.

6.2.1.1 Identifying Peaks Corresponding to Unlabeled Amino Acids

There are two ways to identify peaks corresponding to the selectively unlabeled amino acid residue in the 2D [^{15}N-^{1}H] HSQC spectrum. The first method involves appropriate scaling and subtraction of the spectrum from that of the reference sample containing uniformly ^{15}N labeled protein (which will be referred to as the 'difference spectrum'). The peaks in the difference spectrum correspond to the specific amino acid residue being unlabeled. A drawback of this method is the imperfect cancellation of undesired peaks upon subtraction owing to the fact the two protein samples (reference and the selectively unlabeled) may have different concentrations and hence different signal-to-noise (S/N) ratios. In such cases a second method [25] can be used. In this method, first a reference or 'control' peak in the spectrum is chosen which is known to be un-affected by unlabeling (for instance a Gly residue). Next, the ratio of volume of each peak in the 2D HSQC spectrum of the uniformly labeled sample (also referred to as the reference sample) to the volume of the control peak in the same spectrum is taken. This ratio is called as IR_{ref}. In a similar manner, the ratio of volume of each peak in the 2D HSQC spectrum of the selectively unlabeled sample to the volume of the control residue in the same spectrum is taken. This ratio is called IR_{unlab}. The ratio, IR_{unlab}/IR_{ref}, is then independent of the differences in the signal-to-noise across the two spectra being compared (uniform and selectively unlabeled) and can be used in concert with the 2D difference spectrum. It also provides information on the extent to which the unlabeling of the particular amino acid has occurred. This is illustrated in Fig. 6.1b where IR_{unlab}/IR_{ref} is plotted for each residue in a Lys selectively unlabeled sample of ubiquitin. A peak which is absent in the spectrum acquired for the selectively unlabeled sample will have $IR_{unlab,i}/IR_{ref,i}$ ~0, whereas for any peak unaffected by selective unlabeling $IR_{unlab,i}/IR_{ref,i}$ with the ~1 for any peak unaffected by selective unlabeling. In practice, a range of values from 0 to 1 are observed for the unlabeled residues reflecting the extent to which a given amino-acid type has been unlabeled. We consider $IR_{unlab,i}/IR_{ref,i}$ <0.5 to indicate that the residue i has undergone unlabeling or has been effected by unlabeling and $IR_{unlab,i}/IR_{ref,i}$ >0.5 is taken as an indication of no effect of unlabeling.

6.2.1.2 Misincorporation of ^{14}N at Undesired Sites (Isotope Scrambling)

The extent of misincorporation of ^{14}N at the undesired sites (also referred to as isotope scrambling) is presented in Table 6.1. The overall nature of isotope scrambling is similar to that observed in the selective labeling approach [8], which is expected as the biosynthetic pathways involved are identical in the two methods. Figure 6.1c illustrates the istope scrambling for Gly. In a Gly selectively unlabeled spectrum of Ubiquitin (Fig. 6.1c), all Ser are unlabeled. Out of the 20 amino acid types, the best ones suited for selective unlabeling based on their abundance and cross labeling are: Ala, Arg, Asn, His, Lys, Pro. Unlabeling of Thr results in unlabeling of Gly and Ser but not vice-versa (i.e., Gly or Ser unlabel each other but not Thr; Fig. 6.1). Thr unlabeling also results in unlabeling of ^{13}CO and ^{13}C$^{\delta1}$ of Ile. However, due to the ease with which Gly, Ser, and Thr can be distinguished from each other based on their ^{13}C$^{\alpha/\beta}$, misincorporation of ^{14}N among these three amino acids is not harmful.

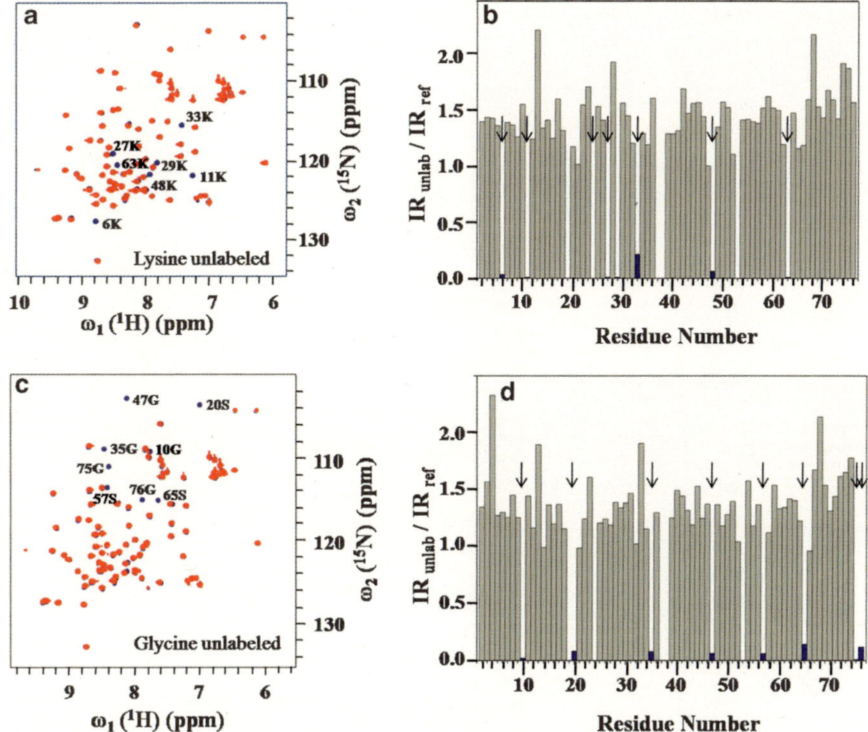

Fig. 6.1 (**a**) An overlay of selected region of 2D [^{15}N-^{1}H] HSQC spectrum of uniformly ^{15}N labeled (*blue*) and lysine selectively unlabeled (*red*) samples of ubiquitin. Assignments for lysine residues are indicated by the residue number. (**b**) A plot of IR$_{unlab,i}$/IR$_{ref,i}$: IR$_{unlab,i}$ = I$_i^{unlab}$/I$_{control}^{unlab}$ and IR$_{ref,i}$ = I$_i^{ref}$/I$_{control}^{ref}$ where *i* denotes the residue number, *I* denotes volume of the peak and 'control' denotes a residue which does not undergo any effect of unlabeling in both selectively unlabeled and the reference sample. In this case the control residue chosen was G47. All residues that undergo the desired unlabeling are indicated as *blue bars*, also indicated with an *arrow* on the *top*. In the case of (K) sample almost complete unlabeling of Lys is observed (i.e., IR$_{unlab,i}$/IR$_{ref}$ ~0). (**c**) Spectrum displaying scrambling of Glycine to Serine in the overlay of selected region of 2D [^{15}N-^{1}H] HSQC spectrum of uniformly ^{15}N labeled (*blue*) and Glycine selectively unlabeled (*red*) samples of ubiquitin. (**d**) A plot similar to (**b**) for Glycine unlabeled ubiquitin where K6 was chosen as the control residue. *Arrows* in (**d**) point towards the residues having IR$_{unlab,i}$/IR$_{ref}$ ~0. Note that E24 and G53 were not assigned and hence are absent along with Pro

The amino acids Ile, Leu, Val inter-convert strongly to each other. For such amino acids which undergo isotope scrambling, larger amounts of the amino acid has to be added to the growth medium. This is illustrated in Fig. 6.2a, where IR$_{unlab}$/IR$_{ref}$ is plotted as a function of residue number for five samples prepared with different amounts of Ile added. The extent of unlabeling and isotope scrambling of Leu and Val (as seen in the ratio IR$_{unlab}$/IR$_{ref}$) increases as the amount of Ile added is increased and reaches a stable value for 1.0 g/l of Ile. In contrast to this, for selective unlabeling of Lys which does not undergo isotope scrambling (Table 6.1), even 100 mg/l is sufficient for achieving complete unlabeling. This is evident from Fig. 6.2b, where the ratio IR$_{unlab}$/IR$_{ref}$ remains <0.5 even for small amounts of Lys added. Unlike in the selective labeling approach, unlabeling of glutamine is not detrimental. Its conversion to glutamic acid gets diluted by transamination of the latter to other amino acids. Only weak unlabeling of proline (via Glutamic acid) is observed in the samples prepared with Gln selectively unlabeled. On the other hand, glutamic acid or aspartic acid strongly unlabels many other amino acid types.

More than one amino acid type can be selectively unlabeled in a given sample. The particular combination of amino acids chosen depends on the relative ease with which the different amino acid types can be distinguished from one another in the spectrum. Such discrimination can be

Table 6.1 Extent of isotope scrambling in selective unlabeling

Selectively unlabeled amino acid type	Amino acid type unlabeled due to cross metabolism
Alanine	Tryptophan
Arginine	–
Asparagine	–
Aspartic acid	Phenylalanine, Tyrosine Threonine, Asparagine Lysine
Glutamine	Proline
Glycine	Serine
Histidine	–
Isoleucine	Leucine Valine
Leucine	Isoleucine Valine
Lysine	–
Methionine	
Phenylalanine	Tyrosine
Proline	–
Serine	Glycine
Threonine	Gly, Ser, Ile ($C^{\gamma 2}$ and $C^{\delta 1}$)
Valine	Leucine Isoleucine
Tyrosine	Phenylalanine

made on the basis of $^{13}C^{\alpha}$ and $^{13}C^{\beta}$ chemical shifts which are well-known indicators of amino acid-types [56, 57] and can be obtained from a 3D HNCACB spectrum acquired for the control sample. The 20 amino acids can be divided into nine groups (Table 6.2; labeled I–IX) with each group having a distinct range of $^{13}C^{\alpha}$ and $^{13}C^{\beta}$ chemical shifts. For instance, Arg and Ala can be chosen together for selective unlabeling as their $^{13}C^{\beta}$ shifts resonate in different spectral regions (Table 6.2). Similarly Ser and Gly which undergo intercoversion (Table 6.1) can be distinguished due to their distinct $^{13}C^{\alpha}$ and $^{13}C^{\beta}$ chemical shifts (Table 6.2). In the case of Ile/Leu/Val, their interconversion does not hamper analysis because these three amino acids belong to distinct categories (Table 6.2).

A method to reduce isotope scrambling was proposed recently by Rasia et al. for *E. Coli* based protein expression systems [58]. This is based on feeding the host organism with suitable metabolic precursors of specific amino acids such as Ile, Leu, Phe and Tyr. The metabolic precursors used are: α-ketobutyrate for Ile, α-ketoisovalerate for Leu and Val, phenylpyruvate for Phe and 4-hydroxy phenylpyruvate for Tyr. The precursors are directly converted to their respective amino acids without scrambling. An additional advantage of this method is that these amino acids are generated with a unique $^{13}C/^{12}C$ labeling pattern which can be exploited for their identification and assignment.

6.2.2 Applications of Selective Unlabeling

The application of selective unlabeling can be divided into two categories: (1) spectral simplification and (2) spin-system identification and resonance assignments.

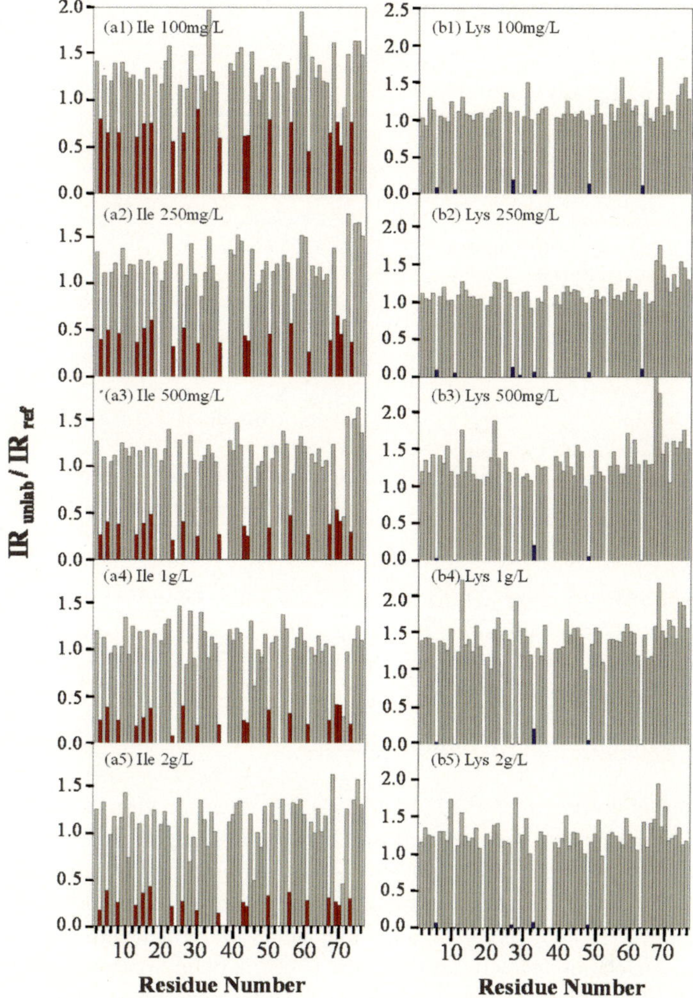

Fig. 6.2 The effect of amount of unlabeled amino acid added to the growth medium on mis-incorporation of ^{14}N label for isoleucine (*a1–a5*) and Lysine (*b1–b5*) in ubiquitin. Though Lysine residue gets completely unlabeled at almost all concentrations (*blue bars*), optimum unlabeling for Ile is achieved at an amount of 1 g (*a4*) of isoleucine (*red bars*) added per litre of culture. That is, beyond this amount any addition of Ile does not affect the extent of unlabeling. Plots of $IR_{unlab,i}/IR_{ref,i}$: $IR_{unlab,i} = I_i^{unlab}/I_{control}^{unlab}$ and $IR_{ref,i} = I_i^{ref}/I_{control}^{ref}$ where i denotes the residue number (*red*-ile, *blue*-lys), I denotes volume of the peak and 'control' denotes a residue which does not undergo any effect of unlabeling in both selectively unlabeled and the reference sample. In this case the control residue chosen was G47. E24 and G53 were not assigned and hence are absent along with Pro

6.2.2.1 Spectral Simplification

An application of selective unlabeling for spectral simplification is in stereospecific assignments of the prochiral methyl groups of valine and isoleucine [24, 35]. Among methyl groups there is a large overlap of ^{1}H and ^{13}C chemical shifts [24, 59, 60]. This hampers the unambiguous stereospecific assignments of Val and Leu. The overlap can be alleviated if resonances from residues overlapping with those of Val and Leu are suppressed. These residues typically are Ile, Leu, Lys and Thr [24]. Selective unlabeling of these specific amino acid residues along with fractional ^{13}C labeling of other residues results in an enhanced spectral simplification. This methodology was demonstrated on

Table 6.2 Classification of amino acid types based on $^{13}C^\alpha$ and $^{13}C^\beta$ chemical shifts

Amino acid types	$^{13}C^\alpha$ and $^{13}C^\beta$ chemical shifts range (in ppm)	Group
Gly	Absence of $^{13}C^\beta$	I
Ala	$15 < \delta(^{13}C^\beta) < 24$	II
Ser	$\delta(^{13}C^\beta) < 66$	III
Thr	$\delta(^{13}C^\beta) > 66$	IV
Lys, Arg, Gln, Glu, His, Trp, Cys^{red}, Met	$24 < \delta(^{13}C^\beta) < 36$	V
Asp, Asn, Phe, Tyr, Cys^{oxd} and Leu	$36 < \delta(^{13}C^\beta) < 50$	VI
Ile	$36 < \delta(^{13}C^\beta) < 40$ $58 < \delta(^{13}C^\alpha)$	VII
Val	$30 < \delta(^{13}C^\beta) < 34$ $60 < \delta(^{13}C^\alpha)$	VIII
Pro	$30 < \delta(^{13}C^\beta) < 34$ $60 < \delta(^{13}C^\alpha)$	IX

proteins such as *Eh*CaBP (15 kDa) [24, 61, 62] and MSG (80 kDa) [35]. Selective unlabeling of Ile and Leu was not found to adversely affect the ^{13}C labeling of Val [24]. However, selective unlabeling of Leu resulted in unlabeling of Val (as discussed above in the case of 2D [^{15}N-^{1}H] HSQC).

6.2.2.2 Spectral Simplification by Isotope Filtering

Selective unlabeling is useful in NOESY based structure determination of proteins. In 3D ^{13}C- or ^{15}N-edited [^{1}H-^{1}H] NOESY a number of correlations between residues far in sequence but close in space are observed. Such 'long range' correlations are pivotal for high resolution structure determination [1]. Spectral overlaps hamper accurate analysis leading to incorrect or ambiguous assignments. One solution is to introduce ^{12}C and ^{14}N labeled amino acids in the protein via selective unlabeling and observe only those correlations which originate on ^{12}C or ^{14}N bound protons and get detected on ^{13}C or ^{15}N bound protons or vice-versa. This can be achieved by isotope filtered experiments or heteronuclear ^{14}N/^{15}N-half filters [63, 64] resulting in simplification of the NOESY spectrum thereby helping in assigning the cross peaks. Another advantage of the unlabeling method is that it removes the effects of ^{13}C relaxation on proton linewidths thereby improving the sensitivity of the NOEs observed. Typically isotope filtered experiments are performed to select and assign intermolecular NOEs [64]. However, using selective unlabeling, these experiments can be applied to select NOEs within a given protein. Figure 6.3b shows a radio frequency (r.f.) pulse scheme of a ^{12}C/^{14}N (ω_1) filtered-^{15}N edited [^{1}H-^{1}H] NOESY which implements the magnetization transfer scheme shown in Fig. 6.3a. In this experiment, ^{12}C and ^{14}N bound protons are selected in the indirect dimension (ω_1) and their NOE to ^{15}N bound protons are detected in the direct (ω_3) dimension. Using this pulse sequence a 3D spectrum was acquired for a Gln, Ile-selectively unlabeled- ^{13}C, ^{15}N labeled ubiquitin sample. A 2D $\{\omega_1$ (^{1}H), ω_3 (^{1}H)$\}$ strip plot from the 3D spectrum is shown in Fig. 6.3c wherein cross peaks from unlabeled Gln and Ile are observed in the spectrum.

An application for assignments of peaks in the NOESY spectrum was demonstrated by Vuister et al. [22] which was also one of the first applications of selective unlabeling. Subsequently, an application of the method to deuterated protein was proposed by Kelly et al. [33] in which five amino acids: Phe, Ile, Thr, Val and Tyr were selectively unlabeled against a background of ^{2}H and ^{15}N labeling in a 47 kDa homodimeric protein. Half-filter 2D and 3D ^{15}N-edited [^{1}H-^{1}H] NOESY spectra were acquired for the sample which helped to selectively detect NOEs originating from the unlabeled amino acid residues to the rest of the residues in the protein [33].

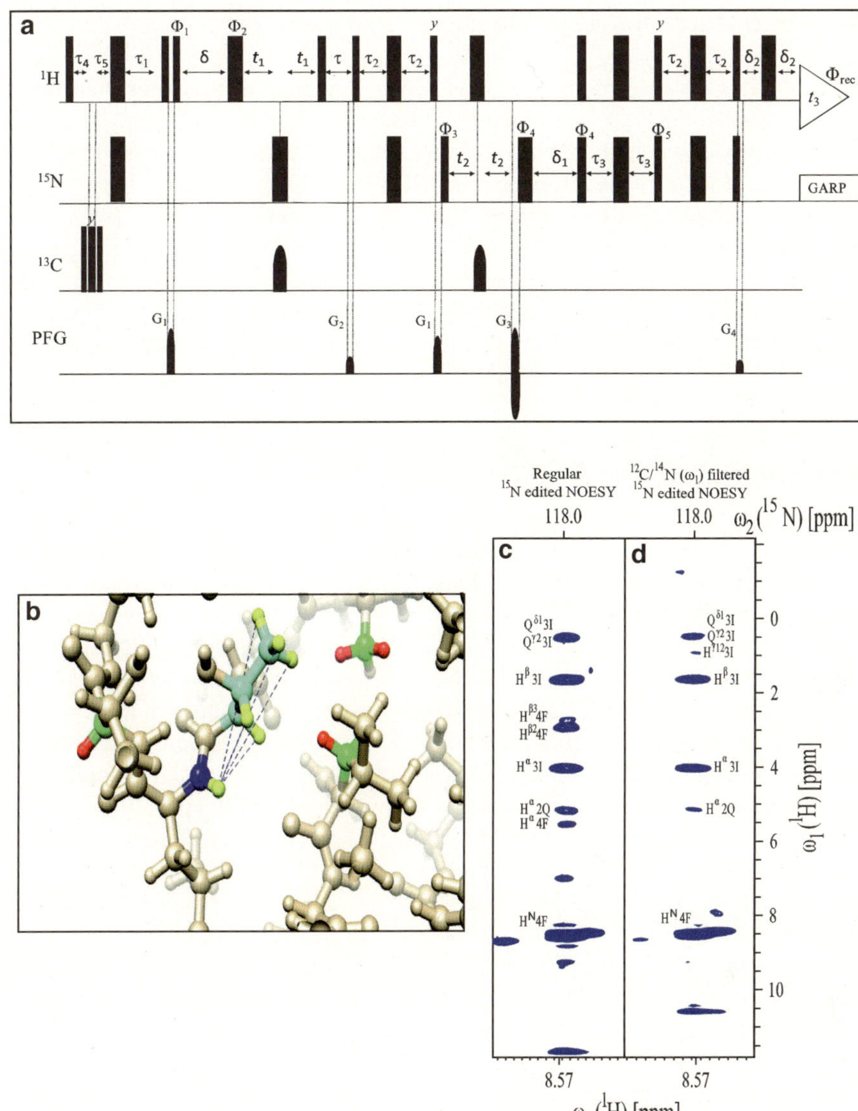

Fig. 6.3 (**a**) R.f. pulse scheme of 3D $^{12}C/^{14}N$ (ω_1) filtered-^{15}N edited [^{1}H-^{1}H] NOESY. In this experiment, NOEs to all ^{15}N bound protons originating from $^{12}C/^{14}N$ bound protons are detected as depicted in (**b**) Rectangular 90° and 180° pulses are indicated by *thin* and *thick vertical bars*, respectively, and phases are indicated above the pulses. Where no r.f. phase is marked, the pulse is applied along x. High-power 90° pulse lengths are: 9.4 μs for ^{1}H, 31 μs for ^{15}N and 11 μs for ^{13}C. The ^{1}H r.f. carrier is placed at the position of the solvent line at 4.7 ppm. The ^{13}C and ^{15}N carrier position is set to 38 ppm and 119 ppm, respectively, throughout the sequence. GARP [2] is employed to decouple ^{15}N during acquisition. The shaped 180° pulses on $^{13}C'$ are of Gaussian cascade type with duration of 256 us. All pulsed z-field gradients (PFGs) are sinc-shaped with gradient recovery delay of 200 μs. The duration and strengths of the PFGs are: G1 (1 ms, 26.8 G/cm); G2: (1 ms 16.08 G/cm); G3: (1 ms 42.88 G/ ms); G4: (1 ms 4.3 G/cm). The delays are: τ_1 ($\approx 1/2J_{NH}$)=2.7 ms, τ_2 ($\approx 1/2J_{NH}$)=2.5 ms, τ_3=2.25 ms, δ=0.5 ms and δ_1=1.5 ms, δ_2=1.2 ms, τ=120 ms. Phase cycling: ϕ_1=x, x, x, x , −x, −x, −x, −x; ϕ_2=y, y, y, y, −y, −y, −y, −y; ϕ_3=x, −x; ϕ_4=x, x, −x, −x; ϕ_5=y, y, −y, −y; ϕ_{rec}=x, −x, −x, x, −x, x, x, −x. Quadrature detection along ω_2(^{15}N) is achieved using Echo-AntiEcho mode with sensitivity enhancement (G3 is inverted with a 180° shift for ϕ_5). States-TPPI is used for quadrature of ω_1(^{1}H) dimension. (**b**) Schematic depiction of the selection of NOE's from $^{12}C/^{14}N$ bound protons to ^{15}N bound protons. (**c**) and (**d**) A representative strip plot showing comparison of the normal 3D N-edited NOESY (b1) with the 3D $^{12}C/^{14}N$ (ω_1) filtered-^{15}N edited [^{1}H-^{1}H] NOESY (c) for residue 4F of Q & I unlabeled ubiquitin protein

6.2.2.3 Spin System Identification and Resonance Assignments

The identification of spin system forms an important step in the process of protein structure determination. It also helps in automated methods for resonance assignments [23, 56, 65]. Spin system identification by selective unlabeling is accomplished in a manner similar to that carried out in selective labeling. The objective here is to identify the resonances in a 2D [^{15}N-^{1}H] HSQC belonging to a specific amino acid-type. As discussed above, this is carried out using a difference spectrum (with respect to the 2D spectrum of a uniformly labeled sample) or using the ratios of intensities described above. One of the applications of unlabeling in spin system identification is the combinatorial labeling scheme proposed by Shortle [21]. Selective unlabeling can also be used for sequence specific resonance assignments [25]. This is explained in more detail below in the section on combinatorial labeling.

6.3 Combinatorial Amino Acid Selective Labeling

One useful starting point in sequence specific resonance assignments is the assignment of the resonances observed in the different 2D and 3D NMR spectra to the corresponding amino acid type (also referred to as 'spin-system identification') [1, 2]. In general this can be achieved by making 19 samples in each of which one out of the 19 amino acid types is selectively ^{15}N-labeled (note that proline does not contain an amide proton and hence is not observed in the 2D [^{15}N, ^{1}H] HSQC spectrum). This is a tedious and time consuming approach. Hence it becomes important to develop a strategy to reduce the number of samples required by labeling more than one amino acid type in a sample. In selective labeling involving a combinatorial approach one or more protein samples are prepared with different combination of amino acids labeled selectively in a given sample [39–44, 46, 47, 50–53]. This requires proper selection of amino acids such that when taken together each amino acid has a unique labeling pattern across the different samples. This can be explained as follows.

Consider two labeling modes or states for a given amino acid (e.g., ^{15}N-labeled or unlabeled). In N samples up to 2^N-1 amino acid types can be chosen such that each amino acid can have a unique labeling pattern across the N samples. This is explained in Table 6.3 for $N=4$. For example, if the two labeling modes chosen are: ^{15}N-labeling or no labeling, then in each sample a given amino acid will either be labeled or unlabeled. This gives rise to 2^N amino acid types (indicated by numbering in the first column of Table 6.3) which can be distributed over N samples (Table 6.3a; left). It can be seen that no two amino acids have the same labeling pattern across the four samples shown in Table 6.3a. Thus based on a specific labeling pattern observed across the four samples (one spectrum is recorded per sample) the 16 amino acid types can be identified. Note that according to this calculation, one amino acid will remain unlabeled in all samples (the last amino acid shown in Table 6.3), which is of no use since its absence in all samples does not help in its assignment. Hence, (2^N-1) amino acids can be chosen with unique labeling pattern across the N samples. The number of amino acids assignable can be increased with lesser of number samples if more than two labeling modes are chosen. For instance, with three modes of labeling: ^{13}C/^{15}N, ^{15}N-only labeling and no labeling (i.e., ^{12}C/^{14}N), up to 3^N-1 amino acid types can be chosen with unique labeling pattern across the N samples. For 20 amino acid types this results in requirement for just three samples (Table 6.3).

The early implementations of combinatorial labeling involved cell-based methods [39] whereas those developed recently are exclusively based on cell-free protein synthesis. In the cell-free method, site-specifically labeled amino acids (e.g., 1-^{13}C labeled) are used during protein synthesis. The different combinatorial selective labeling approaches proposed till date can be divided into four types (Table 6.4; depending on the modes of labeling alluded to above): (1) use of ^{15}N-only labeled amino acids in different combinations, (2) use of ^{13}C, ^{15}N and ^{15}N-only labeled amino acids, (3) use of 1-^{13}C and

Table 6.3 Combinatorial labeling: (a) Using two (left) and (b) three labeling modes (right) for each amino acid type

Amino acid types[a]	Protein samples[b]			
	1	2	3	4
1	●	●	●	●
2	●	●	●	○
3	●	●	○	●
4	●	●	○	○
5	●	○	●	●
6	●	○	●	○
7	●	○	○	●
8	●	○	○	○
9	○	●	●	●
10	○	●	●	○
11	○	●	○	●
12	○	●	○	○
13	○	○	●	●
14	○	○	●	○
15	○	○	○	●
16	○	○	○	○

Amino acid types[a]	Protein samples[c]		
	1	2	3
1	●	●	●
2	●	●	◍
3	●	●	⊛
4	●	◍	●
5	●	◍	◍
6	●	◍	⊛
7	●	⊛	●
8	●	⊛	◍
9	●	⊛	⊛
10	◍	●	●
11	◍	●	◍
12	◍	●	⊛
13	◍	◍	●
14	◍	◍	◍
15	◍	◍	⊛
16	◍	⊛	●
17	◍	⊛	◍
18	◍	⊛	⊛
19	⊛	●	●
20	⊛	●	◍

[a] The numbers indicate the amino acid types that can be assigned based on a unique labeling pattern across the samples
[b] Filled and empty circles indicate the two different labeling modes (e.g., ^{15}N labeled and ^{14}N labeled)
[c] The three circles (right) indicate three different labeling modes used for each amino acid (see text)

^{15}N-labeled amino acids and (4) use of 1-^{13}C; ^{13}C, ^{15}N and ^{15}N-only labeled amino acids. The first approach yields only amino-acid type identification whereas the remaining three approaches help in sequence specific resonance assignments. The different methods are summarized in Table 6.4. Each of the method is described below in detail.

Table 6.4 Classification of different combinatorial labeling methods for spin system identification or sequence specific resonance assignments of backbone nuclei in proteins

Combinatorial schemes based on selective labeling

Labeling mode	Type of isotope labeling	Type of resonance assignments possible	References
Dual (^{15}N/^{14}N)	^{15}N labeled specific amino acid types against a background of ^{12}C/^{14}N labeling	Spin system identification	[45, 46, 48]
Dual (1-^{13}C; ^{13}C/^{15}N)	1-^{13}C and ^{13}C,^{15}N labeled specific amino acid types against a background of ^{12}C/^{14}N labeling	Sequence specific resonance assignment	[39, 40, 44, 50, 53]
Dual (^{15}N; ^{13}C/^{15}N)	^{15}N and ^{13}C,^{15}N labeled specific amino acid types against a background of ^{12}C/^{14}N labeling	Sequence specific resonance assignment	[42, 43, 47, 52]
Triple (1-^{13}C; ^{15}N; ^{13}C/^{15}N)	1-^{13}C, ^{15}N and ^{13}C,^{15}N labeled specific amino acid types against a background of ^{12}C/^{14}N labeling	Sequence specific resonance assignment	[51]
Combinatorial schemes based on selective unlabeling			
Dual (^{14}N/^{15}N)	Selective unlabeling (^{12}C/^{14}N) of specific amino acid types against a background of ^{15}N labeling	Spin system identification	[21]
Dual (^{13}C/^{15}N; ^{12}C/^{14}N)	Selective unlabeling (^{12}C/^{14}N) of specific amino acid types against a background of ^{13}C/^{15}N labeling	Sequence specific resonance assignment	[25]

6.3.1 Combinatorial Selective ^{15}N-Labeling for Spin System Identification

A combinatorial scheme involving only selective ^{15}N-labeling was proposed by Wu et al. [46]. The overall aim of the method is to achieve spin system identification by reducing the number of sample required (instead of 19) [45, 46, 48]. Five samples are made using cell-free protein synthesis because this enables the encoding of 20 amino acids in five different combinations such that each sample contains a unique combination of labeled amino acids (as shown in Table 6.3). The choice of amino acids in a given sample was based on their abundance in the protein and to minimize the overlap of peaks between different amino acid types. Those amino acid types which were present in higher numbers were kept single and not combined with other types in a sample. Each sample contained about 7–8 amino acids labeled selectively corresponding to ~30% of the total number of residues in the protein. The 2D [^{15}N-^{1}H] HSQC spectrum was thus simplified by absence of ~70% of resonances compared to complete labeling of the protein. A computational strategy was subsequently proposed to optimally select the combination of amino acid types in a given sample [54]. An analogous method involving selective unlabeling was proposed by Shortle [21]. As mentioned above, this approach helps only in assigning the resonances observed in the 2D [^{15}N-^{1}H] HSQC spectrum to the corresponding amino acid type. In order to obtain sequence specific resonance assignments, different labeling patterns such as the inclusion of ^{13}C labels is required as described below.

6.3.2   Combinatorial Methods Using Two Selective Labeling Modes (Dual Selective Labeling)

In this approach the aim is to achieve both spin-system identification and sequence specific resonance assignments. The methods are based on the idea that if two contiguous amino acid residues: i, $i+1$ (say X_i and Y_{i+1}) in the protein are labeled with ^{13}C and ^{15}N, respectively, then specific NMR experiments can be used which will selectively detect the di-peptide $X_i - Y_{i+1}$. There are two variants of this method (Table 6.4): (1) use of 1-^{13}C and ^{15}N-labeled amino acids [39, 40, 44, 53] and (2) use of $^{13}C,^{15}N$ and ^{15}N- labeled amino acids [42, 43, 47, 52]. As mentioned above, the early implementations of combinatorial selective labeling involved cell-based methods [39] whereas those developed recently are exclusively based on cell-free protein synthesis.

6.3.2.1   Combinatorial Selective Labeling Using 1-^{13}C and ^{15}N Amino Acids

In this scheme two types of labeled amino acids are utilized: (i) ^{15}N-labeled amino acids and (ii) those labeled in the carbonyl position (1-^{13}C). If a residue i (X_i) gets labeled at the 1-^{13}C position followed by a residue $i+1$ (Y_i) which is ^{15}N labeled, then 3D HNCO (or its 2D ^{15}N-$^1H^N$ projection) can be used to identify the di-peptide: $X_i - Y_{i+1}$. No other di-peptide will get detected in such a spectrum. If the dipeptide is unique in the protein primary sequence, it gets assigned sequence specifically. The combinations of amino acids with 1-^{13}C or ^{15}N-labeling can be chosen so as to maximize the uniqueness of the dipeptides. This methodology was demonstrated by Trbovic et al. for backbone assignments of membrane proteins [44]. The amino acid-types were selected using an algorithm which considered the protein primary sequence and the stretches of amino acids which could not be assigned using the conventional procedure involving triple resonance experiments [44]. Dipeptide segments: $X_i - Y_{i+1}$ with X_i labeled at 1-^{13}C and Y_i labeled with ^{15}N were identified using 2D [^{15}N-^1H] HSQC and 2D ^{15}N-$^1H^N$ projection of HNCO. The methodology was also used for resonance assignment of membrane domain structures and the combinations of amino acids were chosen using a Monte Carlo approach [53]. While the method may not result in 100% sequence specific resonance assignments, it helps in providing useful starting points which otherwise would render the resonance assignments ambiguous. Hence, it serves to supplement the conventional resonance assignment strategies.

6.3.2.2   Combinatorial Selective Labeling Using ^{15}N and $^{13}C,^{15}N$ Amino Acids

In this scheme two types of labeled amino acids are utilized: (i) ^{15}N-only labeled and (ii) ^{13}C, ^{15}N labeled. The strategy for assignment is same as mentioned above. That is, specific NMR experiments can be used to identify the di-peptide: $X_i - Y_{i+1}$ where the residue i is labeled with ^{13}C and $i+1$ is with ^{15}N. While the NMR experiments employed are primarily 2D [^{15}N-^1H] HSQC and 2D [^{15}N-^1H] projection of 3D HNCO, different variants of this approach have been proposed depending on the choice of the amino acids for selection in a given sample.

In the method proposed by Shi et al. [43] four samples were prepared. In each sample two amino acid-types were chosen with ^{13}C, ^{15}N labeling (double labeled) and one with only ^{15}N labeling. The double labeled amino acid types were chosen each from two different groups (Group I and II) while the ^{15}N labeled amino acid were selected from a third group (Group III). Groups I and II contained amino acids such that those belonging to one group could be distinguished from the other based on the $^{13}C^\alpha$ and $^{13}C^\beta$ chemical shifts. Group I contained amino acids Arg, Asp, Glu, Gln, Leu and Lys and Group II consisted of Ala, Ile, Ser, Thr and Val. In Group III less abundant amino acid types were chosen such as Cys, Gly, His, Met, Phe, Tyr, Trp. For each specifically labeled sample, different NMR

Table 6.5 Combinatorial selective labeling scheme proposed by Parker et al.

Amino acid	Control sample	Samples			
		1	2	3	4
Ala	$^{13}C/^{15}N$	$^{13}C/^{15}N$	$^{13}C/^{15}N$	$^{13}C/^{15}N$	$^{14}N/^{15}N$
Asp	$^{13}C/^{15}N$	$^{13}C/^{15}N$	$^{13}C/^{15}N$	$^{14}N/^{15}N$	$^{14}N/^{15}N$
Asn	$^{13}C/^{15}N$	$^{14}N/^{15}N$	$^{13}C/^{15}N$	$^{13}C/^{15}N$	$^{14}N/^{15}N$
Arg	$^{13}C/^{15}N$	$^{14}N/^{15}N$	$^{13}C/^{15}N$	$^{14}N/^{15}N$	$^{14}N/^{15}N$
Cys	$^{13}C/^{15}N$	$^{13}C/^{15}N$	$^{13}C/^{15}N$	$^{14}N/^{15}N$	$^{13}C/^{15}N$
Gly	$^{13}C/^{15}N$	$^{14}N/^{15}N$	$^{14}N/^{15}N$	$^{14}N/^{15}N$	$^{14}N/^{15}N$
Gln	$^{13}C/^{15}N$	$^{14}N/^{15}N$	$^{13}C/^{15}N$	$^{14}N/^{15}N$	$^{13}C/^{15}N$
Phe	$^{13}C/^{15}N$	$^{13}C/^{15}N$	$^{14}N/^{15}N$	$^{13}C/^{15}N$	$^{13}C/^{15}N$
Ile	$^{13}C/^{15}N$	$^{13}C/^{15}N$	$^{14}N/^{15}N$	$^{13}C/^{15}N$	$^{14}N/^{15}N$
Lys	$^{13}C/^{15}N$	$^{13}C/^{15}N$	$^{14}N/^{15}N$	$^{14}N/^{15}N$	$^{13}C/^{15}N$
Leu	$^{13}C/^{15}N$	$^{13}C/^{15}N$	$^{14}N/^{15}N$	$^{14}N/^{15}N$	$^{14}N/^{15}N$
Met	$^{13}C/^{15}N$	$^{14}N/^{15}N$	$^{13}C/^{15}N$	$^{13}C/^{15}N$	$^{13}C/^{15}N$
Pro	^{13}C	^{13}C	^{13}C	^{13}C	^{13}C
Ser	$^{13}C/^{15}N$	$^{14}N/^{15}N$	$^{14}N/^{15}N$	$^{13}C/^{15}N$	$^{13}C/^{15}N$
Thr	$^{13}C/^{15}N$	$^{14}N/^{15}N$	$^{14}N/^{15}N$	$^{13}C/^{15}N$	$^{14}N/^{15}N$
Val	$^{13}C/^{15}N$	$^{14}N/^{15}N$	$^{14}N/^{15}N$	$^{14}N/^{15}N$	$^{13}C/^{15}N$

spectra were acquired [43]. In addition to specific labeling of different amino acids in a sample, differential amount of labeling was also tried. In this approach both unlabeled and labeled forms of different amino acid types are added in different pre-determined ratios. The intensities of the peaks corresponding to the amino acids follow these ratios and thus help in their identification and assignments. A similar method was proposed by Staunton et al. [52] using the two labeling modes similar to that of Shi et al. [43].

A combinatorial method based on differential amount of labeling was proposed by Parker et al. [42, 47]. It is a dual labeling scheme with two modes of labeling for each amino acid type: ^{13}C, ^{15}N labeling (double labeled) or 50% ^{15}N/50% ^{14}N labeling. Five samples are prepared. One uniform ^{13}C, ^{15}N labeled sample serves as a control and in four other samples different amino acids are labeled in one of the two labeling modes (Table 6.5). If I_i is the intensity of a peak in the 2D [^{15}N-^{1}H] HSQC spectrum of one of the four selectively labeled samples, its corresponding intensity in the HSQC spectra of the other samples will either be I_i or $I_i/2$ depending on whether it is ^{13}C, ^{15}N or 50% ^{15}N/50% ^{14}N labeled. Based on this, the comparison of relative peak intensity pattern in the 2D [^{15}N-^{1}H] HSQC spectra acquired on the four selectively labeled samples helps in identifying a given amino acid type. Since four samples are prepared, this approach allows the assignment of 16 amino acid types (see Table 6.3). For sequence specific resonance assignments, a 2D HNCO spectrum is acquired on the four selectively labeled samples and instead of a relative peak intensity pattern as observed in the HSQC spectrum, the absence or presence of a peak in the spectrum across the four samples becomes unique for a given di-peptide segment. Taken together, the two spectra 2D HSQC and 2D HNCO are sufficient to identify all 16*16 amino acid pairs.

6.3.3 Combinatorial Methods Using Three Selective Labeling Modes

In a recent work by Lohr et al. [51] a combinatorial approach using three types of labeled amino acids: (i) 1-^{13}C, (ii) ^{13}C,^{15}N and (iii) ^{15}N was proposed. In this method, in addition to combinatorial selective labeling using 1-^{13}C and ^{15}N amino acids, a ^{13}C,^{15}N labeled amino acid was included. This increased the number of amino acids which can be identified while reducing the number of samples. A dipeptide segment in the protein: $X_i - Y_{i+1}$ with a specific labeling pattern on X or Y was selected using a set of six 2D NMR experiments. The amino acids used for the different samples were based on the amino acid composition of the protein being studied.

6.4   Combinatorial Methods Based on Amino Acid Selective Unlabeling

In combinatorial methods based on amino acid selective unlabeling, a set of samples are prepared in each of which specific amino acids are selectively unlabeled (i.e., $^{12}C/^{14}N$). As a result the peaks due to these amino acids are absent in the spectrum acquired. Similar to methods based on selective labeling, amino acid types in selective unlabeling are chosen based on the abundance of the amino acids in the protein and the ease with which they can be distinguished from each other. As of date, all amino acid selective unlabeling schemes proposed have been cell-based. Two of the combinatorial methods involving selective unlabeling are described below.

6.4.1   Combinatorial Selective Unlabeling for Spin System Identification

In combinatorial selective unlabeling approach proposed by Shortle [21], four proteins samples containing a combination of different ^{14}N labeled amino acids (or unlabeled amino acids) against a background of ^{13}C and ^{15}N labeling was proposed. The protein was expressed using an *E. Coli* based expression system. The amino acids were identified based on the intensity pattern (i.e., absence of the peaks) in different samples. Since the number of samples were four, up to 16 amino acids (2^4) could be assigned. Amino acids which show extensive isotope scrambling such as Glutamic acid and Aspartic acid (Table 6.1) were left out. The method was one of the earliest combinatorial methods proposed and is similar to that proposed later by Wu et al. [46].

6.4.2   Combinatorial Selective Unlabeling for Sequence Specific Resonance Assignments

A method for sequence specific resonance assignments based on selective unlabeling was recently proposed by Krishnarjuna et al. [25]. The method involves unlabeling selectively a set of amino acid types in different samples. Identification and assignment of resonances in a 2D [^{15}N, ^1H] HSQC spectrum corresponding to the selectively unlabeled amino acids is done by comparing a HSQC spectrum of the selectively unlabeled sample with that of a reference sample of the same protein which is uniformly $^{13}C/^{15}N$ labeled. The particular combination of amino acids chosen depends on their relative abundance in the protein and the relative ease with the different amino acid types can be distinguished from one another in the spectrum. Such discrimination can be made on the basis of $^{13}C^\alpha$ and $^{13}C^\beta$ chemical shifts which are well-known indicators of amino acid-types [56, 57]. The 20 amino acids can be divided into nine groups (Table 6.2; labeled I–IX) with each group having a distinct range of $^{13}C^\alpha$ and $^{13}C^\beta$ chemical shifts. Based on these considerations four samples each containing two selectively unlabeled amino acid types were prepared [25]: (i) Gln, Ile (Q,I), (ii) Arg, Asn (R,N) (iii) Phe, Val (F, V) and (iv) Lys, Leu (K, L). A uniformly ^{13}C or/and ^{15}N labelled sample was additionally prepared as a reference. These amino acids were chosen on the basis of their high abundance in proteins (80% together with Ala, Gly, Ser and Thr). With an appropriate choice of the amino acid type, isotope

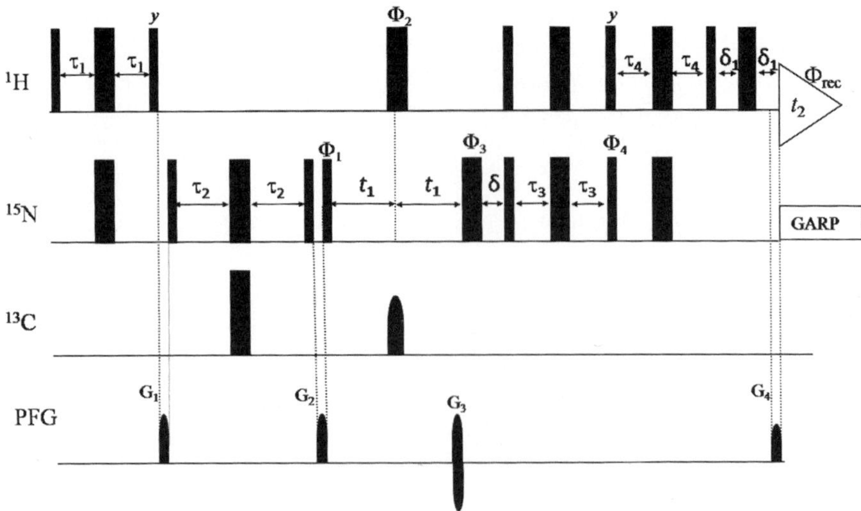

Fig. 6.4 R.f. pulse scheme of 2D $\{^{12}CO_i\text{-}^{15}N_{i+1}\}$-filtered HSQC [25]. Rectangular 90° and 180° pulses are indicated by *thin* and *thick vertical bars*, respectively, and phases are indicated above the pulses. Where no r.f. phase is marked, the pulse is applied along x. High-power 90° pulse lengths are: 8.7 μs for ^{1}H, 37 μs for ^{15}N and 15.8 μs for ^{13}C. The ^{1}H r.f. carrier is placed at the position of the solvent line at 4.7 ppm. The ^{13}C and ^{15}N carrier position is set to 176 ppm and 118.5 ppm, respectively, throughout the sequence. GARP [2] is employed to decouple ^{15}N during acquisition. The shaped 180° pulses on ^{13}C′ are of Gaussian cascade type [2] with duration of 256 μs. All pulsed z-field gradients (PFGs) are sinc-shaped with gradient recovery delay of 200 μs. The duration and strengths of the PFGs are: G1–G2 (1 ms, 26.8 G/cm); G3: (1 ms 42.9 G/cm) G4: (1 ms 4.3 G/cm). The delays are: τ_1 ($\approx 1/2J_{NH}$) = 2.3 ms, τ_2 ($\approx 1/2J_{NC'}$) = 15.0 ms, τ_3 ($\approx 1/2J_{NH}$) = 2.3 ms, δ = 1.2 ms and δ_1 = 1.2 ms. Phase cycling: $\phi_1 = x, -x$; $\phi_2 = \phi_3 = x, x, -x, -x$; $\phi_{rec} = x, -x$. Quadrature detection along ω_1(^{15}N) is achieved with sensitivity enhancement (G3 is inverted with a 180° shift for hϕ_4)

scrambling (Table 6.1) does not hamper the analysis. For instance, selective unlabeling of Ile results in significant unlabeling of Leu and Val (Table 6.1; Fig. 6.2). However, the ^{13}C$^\alpha$ and ^{13}C$^\beta$ chemical shifts of these amino acids occur in different spectral regions (Table 6.2) enabling their distinction. Thus, by avoiding selective unlabeling of amino-acid types belonging to the same group, the effect of isotope scrambling can be ignored.

In addition to spin-system identification, sequence specific resonance assignments can be carried out with the help of a 2D $\{^{12}CO_i\text{-}^{15}N_{i+1}\}$-filtered HSQC spectrum, which aids in linking the ^{1}H^N/^{15}N resonances of a selectively unlabeled residue i and $i+1$. The radio-frequency (r.f.) pulse sequence and the di-peptide selection scheme for 2D $\{^{12}CO_i\text{-}^{15}N_{i+1}\}$-filtered HSQC is shown in Fig. 6.4. The experiment works by selecting those residues ($i+1$) which are ^{15}N labeled and have an unlabeled residue as their N-terminal neighbor (i.e., residue i). This is achieved by tuning the delay periods in the pulse sequence to generate anti-phase ^{15}N$_{i+1}$-^{13}C′$_i$ magnetization which is not detectable in the 2D HSQC spectrum. If the residue i is an unlabeled amino acid, no antiphase ^{15}N$_{i+1}$-^{13}C′$_i$ magnetization is generated and ^{15}N$_{i+1}$ is observed as in a regular 2D HSQC spectrum. Thus, the resulting 2D spectrum consists of peaks corresponding only to those residues which have selectively unlabeled N-terminal neighbors.

The 2D $\{^{12}CO_i\text{-}^{15}N_{i+1}\}$-filtered HSQC is combined with 3D HNCACB and 3D CBCA(CO)NH spectrum acquired for the control sample for sequence specific resonance assignments as follows. First resonances in the 2D HSQC spectrum belonging to the selectively unlabeled amino acid residues are identified. This is done by generating a difference spectrum and/or using the method of ratios described above (Fig. 6.1). Such peaks serve as starting points. The next step is to map or establish a link between each of the ^{1}H^N/^{15}N resonances observed in the difference spectrum with the corresponding

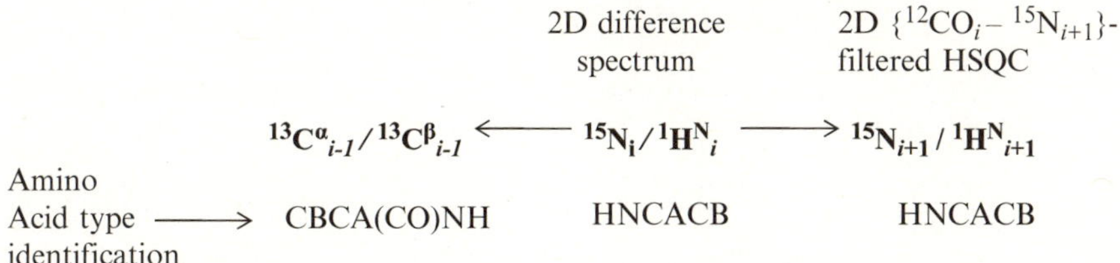

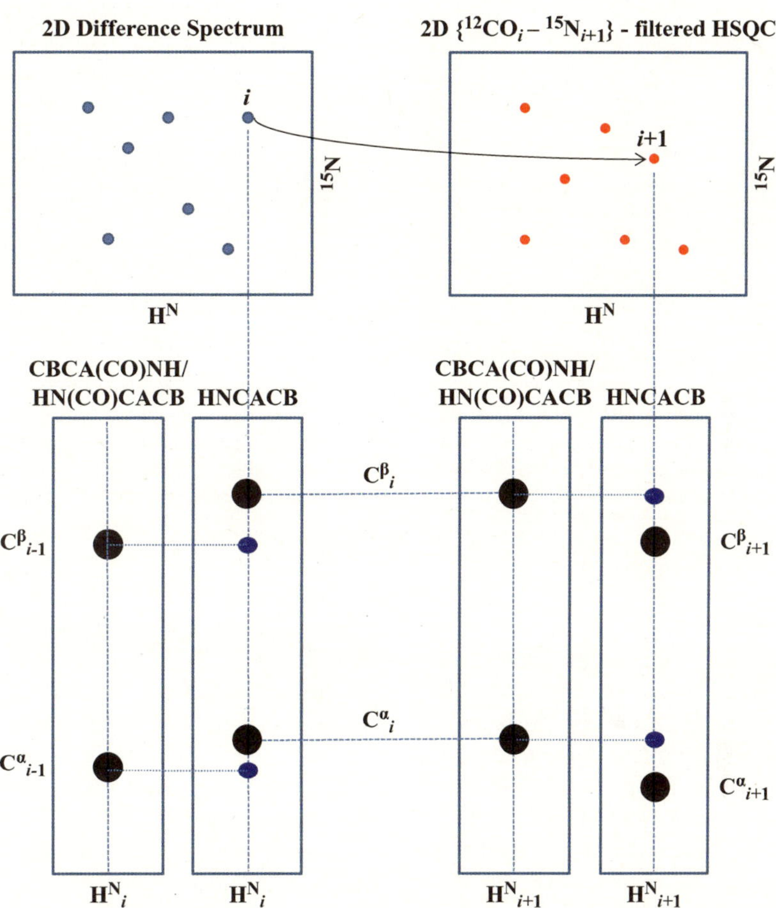

Fig. 6.5 Schematic illustration of the sequential assignment strategy used for selective unlabeling

resonances of their C-terminal neighbors ($i+1$) observed in the 2D $\{^{12}CO_i-^{15}N_{i+1}\}$-filtered HSQC. This is accomplished by an inspection of the 3D HNCACB and 3D CBCA(CO)NH spectra acquired for the reference sample as illustrated in Fig. 6.5. For a given spin-system i in the 2D difference spectrum, the spin system corresponding to $i+1$ in 2D $\{^{12}CO_i-^{15}N_{i+1}\}$-filtered HSQC is the one for which the $^{13}C^\alpha$ and $^{13}C^\beta$ chemical shifts observed in 3D CBCA(CO)NH spectrum matches that of the residue i in HNCACB (Fig. 6.5). Further, using 3D CBCA(CO)NH for a given spin-system i in the 2D difference spectrum the spin system corresponding to $i-1$ can be identified based on the $^{13}C^\alpha$ and $^{13}C^\beta$ chemical shifts (Table 6.2). This results in the identification of a tri-peptide segment from the knowledge of the amino acid types of residues: $i-1$, i and $i+1$. An alternative method to obtain the amino acid types of $i-1$ and $i+1$ is to acquire the 3D HN(CA)NH experiment [66] for the control sample.

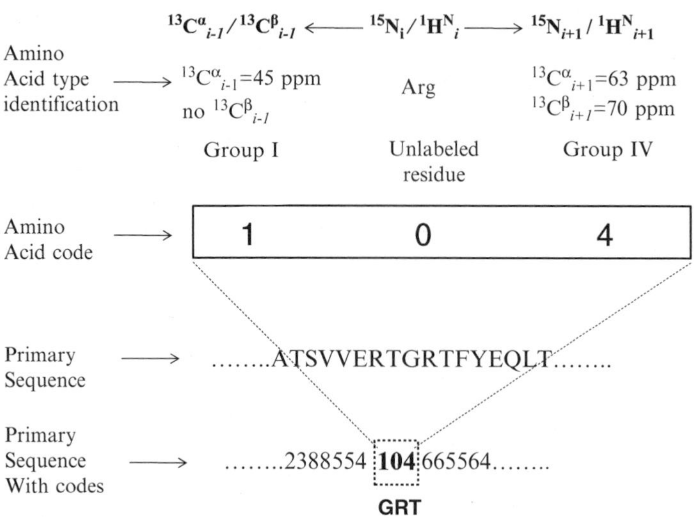

Fig. 6.6 Schematic illustration of mapping the tri-peptide segment identified using the methodology depicted in Fig. 6.5 onto the protein primary sequence. First, based on the 2D difference spectrum and 2D $\{^{12}CO_i$-$^{15}N_{i+1}\}$-filtered HSQC spectrum in concert with CBCA(CO)NH and HNCACB, the amino acid type corresponding to the tri-peptide segment ($i-1, i, i+1$) is identified (see Fig. 6.5). Each residue of the tripeptide segment is assigned a code based on the group to which it belongs (Table 6.2). The selectively unlabeled residue is assigned an unique code '0' because its amino acid type is known exactly. Next, the protein primary sequence is converted into an array of codes for each amino acid based again on the group to which each amino acid belongs in Table 6.2. The tripeptide segment identified above is then mapped on to the primary sequence wherever the tri-peptide code matches that in the primary sequence

This experiment directly correlates the chemical shifts of $^{15}N_i$ and ^1H_i with those of residues $i-1$ and $i+1$ [66]. Once $^{15}N_{i-1/i+1}$ and $^1H_{i-1/i+1}$ are identified using 3D $\underline{HN}$(CA)NH, the amino acid-types corresponding to $i-1/i+1$ can be identified based on their respective $^{13}C^\alpha$ and $^{13}C^\beta$ chemical shifts observed in 3D HNCACB. The tri-peptide thus identified using one of the above methods can then be mapped on to the protein primary sequence for sequence specific resonance assignments. The mapping is carried out as depicted in Fig. 6.6. A unique number code is assigned to amino acids in each of the nine categories shown in Table 6.2 except for the residue being selectively unlabeled, which is assigned a separate code '0' (given that its type is known exactly). Thus, all amino acid types in a given category get the same code. The different codes assigned are: Ala-1; Gly-2; Ser-3; Thr-4; Lys, Arg, Gln, Glu, His, Trp, Cys^{red}, Met-5; Asp, Asn, Phe, Tyr, Cys^{oxd} and Leu-6; Ile-7; Val-8; Pro-9. Next the protein primary sequence is converted into a new sequence of codes using the same codes as assigned to the amino acids in the different groups (Fig. 6.6). The tri-peptide segment (containing three codes) is then mapped on to the primary sequence of codes to identify its location for sequence specific resonance assignments. This is illustrated in Fig. 6.7 for a particular tripeptide in ubiquitin. Since the method relies on the presence of ^{15}N, ^1H on the residue $i+1$ following a residue i which is unlabeled, the following types of tri-peptide segments cannot be assigned: (i) the segment in which the immediate C-terminal neighbor of the central (unlabeled) residue is a Pro (i.e., $i+1\equiv$Pro), (ii) the segment which contains both residues i and $i + 1$ as selectively unlabeled and (iii) the residue is the last amino acid in the polypeptide chain (i.e., located at the C-terminal end). Compared to the other methods which assign di-peptides, the combinatorial selective unlabeling method described above helps in assigning tri-peptide segments which are more uniquely mapped on the sequence. Once the tri-peptides are assigned, the remaining residues in the protein can be assigned in di-peptide pairs, which earlier could not be assigned due to degeneracy.

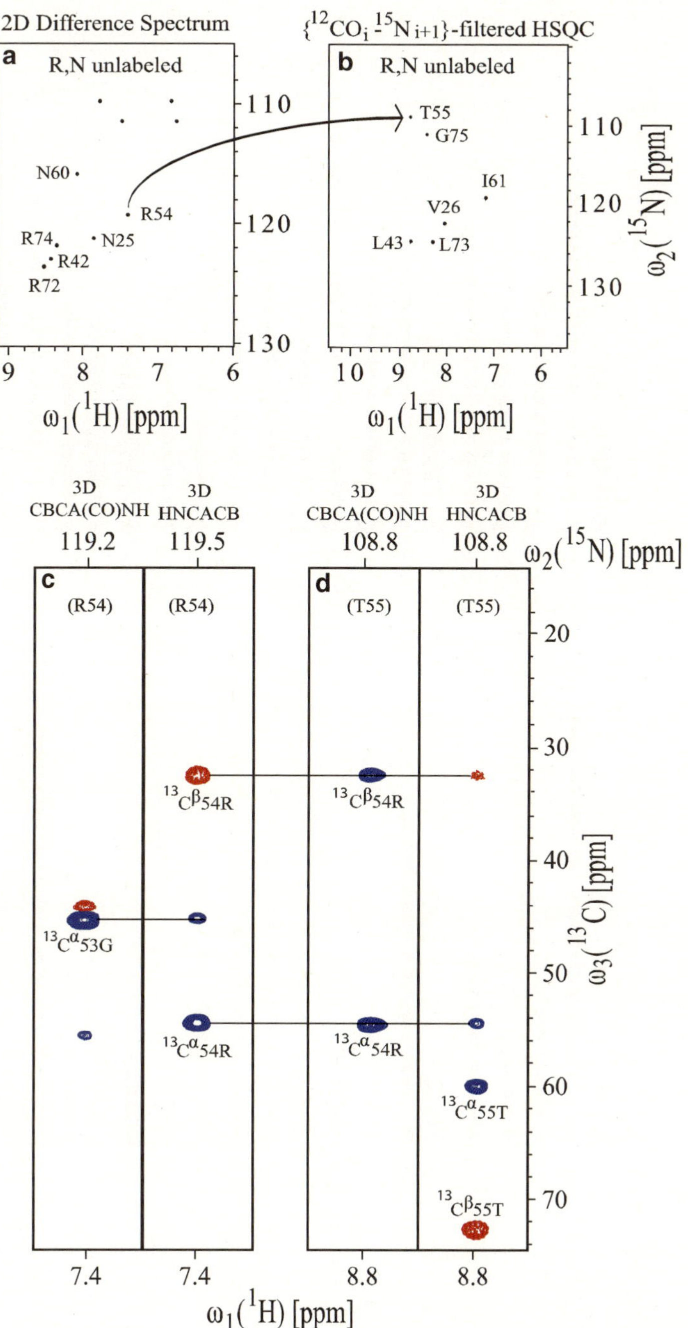

Fig. 6.7 Illustration of sequence specific resonance assignment of a tri-peptide segment in Ubiquitin. Given a peak in the 2D difference spectrum of (R, N) selectively unlabeled sample, (e.g., R54), its C-terminal neighbour ($i+1$) is identified using the procedure shown in Fig. 6.5 and its amino acid type is assigned using $^{13}C^\alpha$ and $^{13}C^\beta$ chemical shift in 3D HNCACB. The amino acid type of residue $i-1$ is assigned like-wise from 3D CBCA(CO)NH and the tri-peptide segment (containing the amino acid type information) is then mapped on to the protein sequence shown for sequence specific assignments

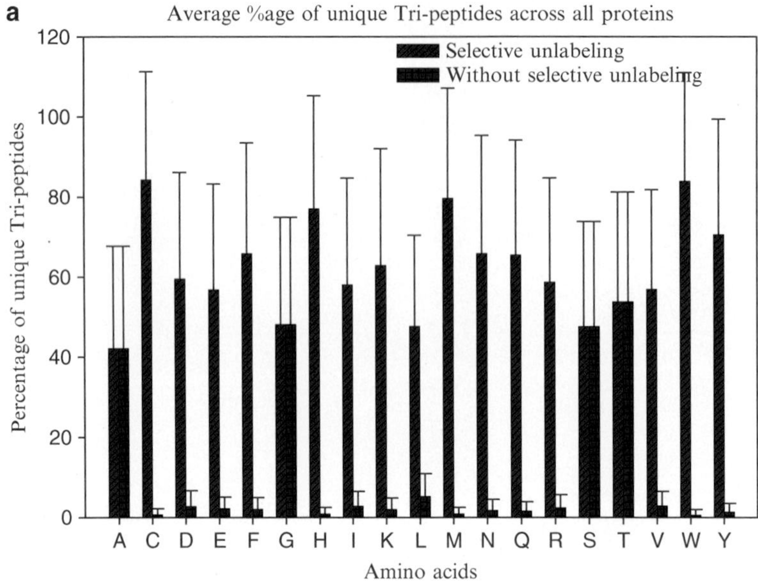

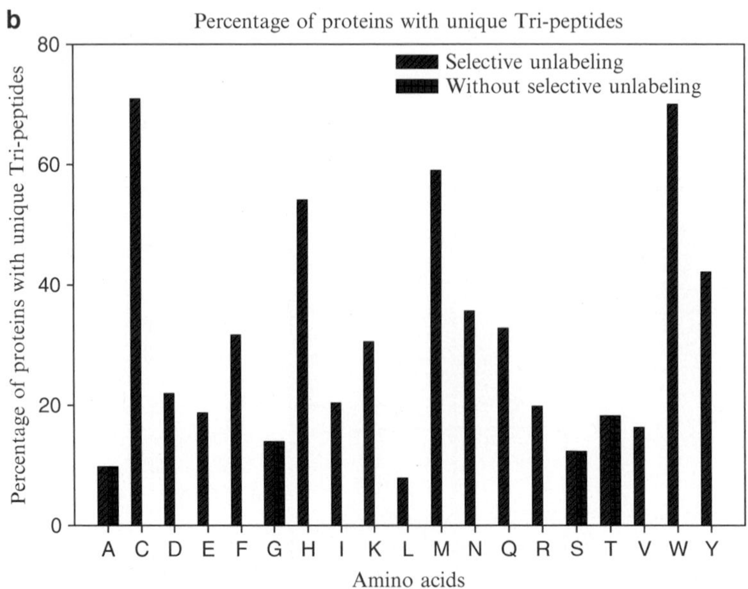

Fig. 6.8 Statistical analysis of the uniqueness of tri-peptide segments in proteins rendered with selective unlabeling. 161792 proteins from UniRef50 were considered for analysis. (**a**) The *bars* show the percentage of proteins which have unique tri-peptide sequences containing a given amino acid residue in the central position. To calculate the uniqueness of such tri-peptide segments, the residue preceding and following the central residue is given a number code according to the group to which it belongs (Table 6.2) and the central residue is given a unique code. The entire polypeptide chain (also converted to a sequence of codes) is then searched for multiple occurrences of such tri-peptide segments. (**b**) The average percentage of unique tri-peptide segments in a protein centred on a given amino acid type. The average and standard deviation was measured considering all the proteins (161,792). When considering the selective unlabeling approach (shown in *dark red*), the central residue in a tri-peptide segment is given a unique code whereas the two residues around it are given codes according to the respective groups (Table 6.2) to which they belong. While considering uniform labelling (or non-selective labelling) as shown in *blue*, the central residue in a tri-peptide is given the code according to the group to which it belongs. Thus the increase in uniqueness due to selective unlabeling arises from the identification of the amino acid being selectively unlabeled (the central residue)

We have carried out a statistical analysis of ~160,000 non-homologues primary sequences to estimate the extent to which such tri-peptide segments can be placed uniquely in a protein sequence (Fig. 6.8). In many proteins the uniqueness of the tri-peptide segment is high if the central amino acid type in the tripeptide segment is known by selective unlabeling. In a given protein the information of amino acid type significantly reduces the multiple occurrences of tri-peptide segments. For instance, on an average 65% of tri-peptide segments containing lysine as the central residue were unique given the knowledge of its amino acid type (Fig. 6.8). On the contrary, in the absence of identification of lysine only ~5% of the tripeptide sequences are unique. Note that the uniqueness is based on the code system described above based on categorizing the amino acids into different groups (Table 6.2). Taken together, the combinatorial selective unlabeling approach not only helps in reducing the search space for sequential assignments (with 2D $\{^{12}CO_i\text{-}^{15}N_{i+1}\}$-filtered HSQC), but also increases the uniqueness of the tri-peptide segment identified, resulting in unambiguous assignments. The advantage with selective unlabeling is more in the case of Group V and Group VI amino acid residues (Table 6.2), which are composed of a large number of amino acid types.

The number of assignable residues can be increased and the number of samples reduced if more than two amino acids are selectively unlabeled in the same sample. However, no two amino acids from the same group should be selectively unlabeled in one sample owing to the fact that their identity is purely based on the $^{13}C^{\alpha}$ and $^{13}C^{\beta}$ chemical shifts. Thus, for example two samples can be prepared: (1) one containing the amino acids R, N, G, S, T and A and (2) containing K. The combination of multiple samples also helps to verify the overlapping assignments.

6.5   Conclusions

Amino acid selective labeling and unlabeling are very important tools in protein structure determination. In the case of large proteins and/or proteins with severe chemical shift overlap such as intrinsically disordered/membrane proteins the process of sequence specific resonance assignments becomes a tedious and time consuming task. In such cases selective identification of amino acid residue by selective labeling or unlabeling improves the efficiency of assignments. Selective identification of amino acids helps not only in obtaining starting points for carrying out assignments, they also help directly in sequence specific resonance assignment. This is achieved either using single or a group of amino acid types labeled/unlabeled selectively in a given sample. The method of selective unlabeling is an inexpensive approach and has the advantage of providing assignments for residues preceding and following the amino acid type being unlabeled. It is possible to devise new experiments which can extend this methodology to identify residues further away (i.e., $i+2$, $i-2$) from the unlabeled site thereby providing larger segments of sequentially connected residues which can map uniquely onto the protein primary sequence.

Acknowledgments   The facilities provided by NMR Research Centre at IISc supported by Department of Science and Technology (DST), India is gratefully acknowledged. HSA acknowledges support from DST-SERC and DAE-BRNS research awards. GJ acknowledges fellowship from Council of Scientific and Industrial Research (CSIR). We thank Dr. John Cort, Pacific Northwest National Laboratory, for providing the Ubiquitin plasmid.

References

1. Wüthrich K (1986) NMR of proteins and nucleic acids. Wiley, New York
2. Cavanagh J, Fairbrother WJ, Palmer AG, Skelton NJ (1996) Protein NMR spectroscopy. Academic, San Diego
3. McDermott AE (2004) Structural and dynamic studies of proteins by solid-state NMR spectroscopy: rapid movement forward. Curr Opin Struct Biol 14:554–561

4. Marassi F, Opella SJ (1998) NMR structural studies of membrane proteins. Curr Opin Struct Biol 8:640–648

5. Sanders CR, Sonnichsen FD (2006) Solution NMR of membrane proteins: practice and challenges. Magn Reson Chem 44:S24–S40

6. Dyson HJ, Wright PE (1998) Equilibrium NMR studies of unfolded and partially folded proteins. Nat Struct Biol 5:499–503

7. Markley JL, Putter I, Jardetzky O (1968) High resolution nuclear magnetic resonance spectra of selectively deuterated staphylococcal nuclease. Science 161:1249–1251

8. Muchmore DD, McIntosh LP, Russell CB, Anderson DE, Dahlquist FW (1989) Expression and nitrogen-15 labelling of proteins for proton and nitrogen-15 nuclear magnetic resonance. Methods Enzymol 177:44–73

9. McIntosh LP, Dahlquist FW (1990) Biosynthetic incorporation of ^{15}N and ^{13}C for assignment and interpretation of nuclear magnetic resonance spectra of proteins. Q Rev Biophys 23:1–38

10. Dötsch V, Matsuo H, Wagner G (1996) Amino-acid type identification for deuterated proteins with a b-carbon edited HNCOCACB experiment. J Magn Reson B 112:95–100

11. Dötsch V, Oswald RE, Wagner G (1996) Amino acid type selective triple resonance experiments. J Magn Reson B 110:107–111

12. Dötsch V, Wagner G (1996) Editing of amino acid type in CBCA(CO)NH experiments based on the $^{13}C^b$ -$^{13}C^g$ coupling. J Magn Reson B 111:310–313

13. Schubert M, Schmalla M, Schmieder P, Oschkinat H (1999) MUSIC in triple-resonance experiments: amino acid type-selective ^{15}N/^{1}H correlations. J Magn Reson 141:34–43

14. Schubert M, Oschkinat H, Schmieder P (2001) MUSIC and aromatic residues: amino acid type-selective backbone ^{15}N/^{1}H correlations, part III. J Magn Reson 153:186–192

15. Schubert M, Oschkinat H, Schmieder P (2001) Amino acid type-selective backbone ^{15}N/^{1}H correlations for Arg and Lys. J Biomol NMR 20:379–384

16. Schubert M, Oschkinat H, Schmieder P (2001) MUSIC, selective pulses and tuned amino acid type-selective ^{15}N/^{1}H correlations, part II. J Magn Reson 148:61–72

17. Schubert M, Labudde D, Leitner D, Oschkinat H, Schmieder P (2005) A modified strategy for sequence specific assignment of protein NMR spectra based on amino acid type selective experiments. J Biomol NMR 31:115–128

18. Barnwal RP, Rout AK, Atreya HS, Chary KVR (2008) Identification of C-terminal neighbours of amino acid residues without an aliphatic C-13(gamma) as an aid to NMR assignments in proteins. J Biomol NMR 41:191–197

19. Goto NK, Kay LE (2000) New developments in isotope labeling strategies for protein solution NMR spectroscopy. Curr Opin Struct Biol 10:585–592

20. Ohki S, Kainosho M (2008) Stable isotope labeling methods for protein NMR. Prog NMR Spectrosc 53:208–226

21. Shortle D (1994) Assignment of amino acid type in ^{1}H-^{15}N correlation spectra by labeling with ^{14}N-amino acids. J Magn Reson B105:88–90

22. Vuister GW, Kim SJ, Wu C, Bax A (1994) 2D and 3D NMR-study of phenylalanine residues in proteins by reverse isotopic labeling. J Am Chem Soc 116:9206–9210

23. Atreya HS, Chary KVR (2000) Amino acid selective 'unlabelling' for residue-specific NMR assignments in proteins. Curr Sci 79:504–507

24. Atreya HS, Chary KVR (2001) Selective 'unlabeling' of amino acids in fractionally C-13 labeled proteins: an approach for stereospecific NMR assignments of CH3 groups in Val and Leu residues. J Biomol NMR 19:267–272

25. Krishnarjuna B, Jaipuria G, Thakur A, D'Silva P, Atreya HS (2011) Amino acid selective unlabeling for sequence specific resonance assignments in proteins. J Biomol NMR 49:39–51

26. Vinarov DA, Lytle BL, Peterson FC, Tyler EM, Volkman BF, Markley JL (2004) Cell-free protein production and labeling protocol for NMR-based structural proteomics. Nat Methods 1:149–153

27. Apponyi MA, Ozawa K, Dixon NE, Otting G (2008) Cell-free protein synthesis for analysis by NMR spectroscopy. Methods Mol Biol 426:257–268

28. Tugarinov V, Hwang PM, Kay LE (2004) Nuclear magnetic resonance spectroscopy of high-molecular-weight proteins. Annu Rev Biochem 73:107–146

29. Tugarinov V, Kay LE (2004) An isotope labeling strategy for methyl TROSY spectroscopy. J Biomol NMR 28:165–172

30. Katzen F, Chang G, Kudlicki W (2005) The past, present and future of cell-free protein synthesis. Trends Biotechnol 23:150–156

31. Kainosho M, Torizawa T, Iwashita Y, Terauchi T, Ono AM, Güntert P (2006) Optimal isotope labelling for NMR protein structure determinations. Nature 440:52–57

32. Hong M, Jakes K (1999) Selective and extensive ^{13}C labeling of a membrane protein for solid-state NMR investigation. J Biomol NMR 14:71–74

33. Kelly MJS, Krieger C, Ball LJ, Yu Y, Richter G, Schmieder P, Bacher A, Oschkinat H (1999) Application of amino acid type-specific 1H and 14N labeling in a 2H-, 15N-labeled background to a 47 kDa homodimer: potential for NMR structure determination of large proteins. J Biomol NMR 14:79–83

34. Mohan PMK, Barve M, Chatteljee A, Ghosh-Roy A, Hosur RV (2008) NMR comparison of the native energy landscapes of DLC8 dimer and monomer. Biophys Chem 134:10–19
35. Tugarinov V, Kay LE (2004) Stereospecific NMR assignments of prochiral methyls, rotameric states and dynamics of valine residues in malate synthase G. J Am Chem Soc 126:9827–9836
36. Mukherjee S, Mustafi SM, Atreya HS, Chary KVR (2005) Measurement of (1)J(N-i, C-i(alpha)), (1)J(N-i, C '(i-1)), (2)J(N-i, C-alpha (i-1)), (2)J (H-N (i), C ' (i-1)) and (2)J(H-N (i), C-alpha (i)) values in C-13/N-15-labelled proteins. Magn Reson Chem 43:326–329
37. Hilty C, Fernandez C, Wider G, Wüthrich K (2002) Side chain NMR assignments in the membrane protein OmpX reconstituted in DHPC micelles. J Biomol NMR 23:289–301
38. Turano P, Lalli D, Felli IC, Theil EC, Bertini I (2010) NMR reveals pathway for ferric mineral precursors to the central cavity of ferritin. Proc Natl Acad Sci USA 107:545–550
39. Kainosho M, Tsuji T (1982) Assignment of the three methionyl carbonyl carbon resonances in *streptomyces* subtilisin inhibitor by a carbon-13 and nitrogen-15 double-labeling technique. A new strategy for structural studies of proteins in solution. Biochemistry 21:6273–6279
40. Yabuki T, Kigawa T, Dohmae N, Takio K, Terada T, Ito Y, Laue ED, Cooper JA, Kainosho M, Yokoyama S (1998) Dual amino acid-selective and site-directed stable isotope-labeling of the human c-Ha-Ras protein by cell free synthesis. J Biomol NMR 11:295–306
41. Weigelt J, van Dongen M, Uppenberg J, Schultz J, Wikstrom M (2002) Site-selective screening by NMR spectroscopy with labeled amino acid pairs. J Am Chem Soc 124:2446–2447
42. Parker MJ, Aulton-Jones M, Hounslow AM, Craven CJ (2004) A combinatorial selective labeling method for assignment of backbone amide resonances. J Am Chem Soc 126:5020–5021
43. Shi J, Pelton JG, Cho HS, Wemmer DE (2004) Protein signal assignments using specific labeling and cell-free synthesis. J Biomol NMR 28:235–247
44. Trbovic N, Klammt C, Koglin A, Lohr F, Bernhard F, Dotsch V (2005) Efficient strategy for the rapid backbone assignment of membrane proteins. J Am Chem Soc 127:13504–13505
45. Ozawa K, Wu PSC, Dixon NE, Otting G (2006) 15N-labelled proteins by cell-free protein synthesis: strategies for high-throughput NMR studies of proteins and protein-ligand complexes. FEBS J 273:4154–4159
46. Wu PSC, Ozawa K, Jergic S, Su XC, Dixon NE, Otting G (2006) Amino-acid type identification in ¹⁵N-HSQC spectra by combinatorial selective ¹⁵N-labeling. J Biomol NMR 34:13–21
47. Craven CJ, Al-Owais M, Parker MJ (2007) A systematic analysis of backbone amide assignments achieved via combinatorial selective labelling of amino acids. J Biomol NMR 38:151–159
48. Xun Y, Tremouilhac P, Carraher C, Gelhaus C, Ozawa K, Otting G, Dixon NE, Leippe M, Grotzinger J, Dingley AJ, Kralicek AV (2009) Cell-free synthesis and combinatorial selective ¹⁵N-labeling of the cytotoxic protein amoebapore A from *Entamoeba histolytica*. Protein Expr Purif 68:22–27
49. Sobhanifar S, Reckel S, Junge F, Schwarz D, Kai L, Karbyshev M, Lohr F, Bernhard F, Dotsch V (2010) Cell-free expression and stable isotope labelling strategies for membrane proteins. J Biomol NMR 46:33–43
50. Hefke F, Bagaria A, Reckel S, Ullrich SJ, Dotsch V, Glaubitz C, Guntert P (2011) Optimization of amino acid type-specific ¹³C and ¹⁵N labeling for the backbone assignment of membrane proteins by solution- and solid state NMR with the UPLABEL algorithm. J Biomol NMR 49:75–84
51. Lohr F, Reckel S, Karbyshev M, Connolly PJ, Abdul-Manan N, Bernhard F, Moore JM, Dotsch V (2012) Combinatorial triple-selective labeling as a tool to assist membrane protein backbone resonance assignment. J Biomol NMR 52:197–210
52. Staunton D, Schlinkert R, Zanetti G, Colebrook SA, Campbell ID (2006) Cell-free expression and selective isotope labelling in protein NMR. Magn Reson Chem 44:S2–S9
53. Maslennikov I, Klammt C, Hwang E, Kefala G, Okamura M, Esquivies L, Mors K, Glaubitz C, Kwiatkowski W, Jeon YH, Choe S (2010) Membrane protein structure of three classes of histidine kinase receptors by cell free expression and rapid NMR analysis. Proc Natl Acad Sci USA 107:10902–10907
54. Sweredoski MJ, Donovan KJ, Nguyen BD, Shaka AJ, Baldi P (2007) Minimizing the overlap problem in protein NMR: a computational framework for precision amino acid labeling. Bioinformatics 23:2829–2835
55. Cai ML, Huang Y, Sakaguchi K, Clore GM, Gronenborn AM, Craigie R (1998) An efficient and cost-effective isotope labeling protocol for proteins expressed in Escherichia coli. J Biomol NMR 11:97–102
56. Atreya HS, Sahu SC, Chary KVR, Govil G (2000) A tracked approach for automated NMR assignments in proteins (TATAPRO). J Biomol NMR 17:125–136
57. Atreya HS, Chary KVR, Govil G (2002) Automated NMR assignments of proteins for high throughput structure determination: TATAPRO II. Curr Sci 83:1372–1376
58. Rasia RM, Brutscher B, Plevin MJ (2012) Selective isotopic unlabeling of proteins using metabolic precursors: application to NMR assignment of intrinsically disordered proteins. Chembiochem 13:732–739
59. Barnwal RP, Atreya HS, Chary KVR (2008) Chemical shift based editing of CH3 groups in fractionally C-13-labelled proteins using GFT (3,2)D CT-HCCH-COSY: stereospecific assignments of CH3 groups of Val and Leu residues. J Biomol NMR 42:149–154

60. Jaipuria G, Thakur A, D'Silva P, Atreya HS (2010) High-resolution methyl edited GFT NMR experiments for protein resonance assignments and structure determination. J Biomol NMR 48:137–145
61. Sahu SC, Atreya HS, Chauhan S, Bhattacharya A, Chary KVR, Govil G (1999) Letter to the editor: sequence-specific ^{1}H, ^{13}C and ^{15}N assignments of a calcium binding protein from Entamoeba histolytica. J Biomol NMR 14:93–94
62. Atreya HS, Sahu SC, Bhattacharya A, Chary KVR, Govil G (2001) NMR derived solution structure of an EF-hand calcium-binding protein from Entamoeba histolytica. Biochemistry 40:14392–14403
63. Otting G, Wüthrich K (1990) Heteronuclear filters in two-dimensional [1H,1H]-NMR spectroscopy: combined use with isotope labelling for studies of macromolecular conformation and intermolecular interactions. Q Rev Biophys 23:39–96
64. Breeze AL (2000) Isotope-filtered NMR methds for the study of biomolecular structure and interactions. Prog NMR Spectrosc 36:323–372
65. Baran MC, Huang YJ, Moseley HNB, Montelione GT (2004) Automated analysis of protein NMR assignments and structures. Chem Rev 104:3541–3555
66. Chandra K, Jaipuria G, Shet D, Atreya HS (2011) Efficient sequential assignments in proteins with reduced dimensionality 3D HN(CA)NH. J Biomol NMR 52:115–126

Part II
Nucleic Acids

Chapter 7
Isotope Labeling and Segmental Labeling of Larger RNAs for NMR Structural Studies

Olivier Duss, Peter J. Lukavsky, and Frédéric H.-T. Allain

Abstract NMR spectroscopy has become substantial in the elucidation of RNA structures and their complexes with other nucleic acids, proteins or small molecules. Almost half of the RNA structures deposited in the Protein Data Bank were determined by NMR spectroscopy, whereas NMR accounts for only 11% for proteins. Recent improvements in isotope labeling of RNA have strongly contributed to the high impact of NMR in RNA structure determination. In this book chapter, we review the advances in isotope labeling of RNA focusing on larger RNAs. We start by discussing several ways for the production and purification of large quantities of pure isotope labeled RNA. We continue by reviewing different strategies for selective deuteration of nucleotides. Finally, we present a comparison of several approaches for segmental isotope labeling of RNA. Selective deuteration of nucleotides in combination with segmental isotope labeling is paving the path for studying RNAs of ever increasing size.

7.1 Introduction

RNA has become widely recognized not only as protein coding information carrier but also as an important regulator of gene expression such as riboswitches, miRNAs or large non-coding RNAs [1]. It has been shown that with increasing complexity of the organism the extent of non protein coding DNA is increasing drastically [2]. In humans, 98.8% of the DNA is not coding for proteins but is mainly transcribed into RNA. The ratio between the number of different proteins and RNAs in the cell is in strong contrast to the ratio of structures deposited in the Protein Data Bank. As of April 2012 only 1.1% of the structures are RNA molecules, whereas proteins account for 92.6%. This discrepancy is due to difficulties in crystallizing and phasing RNA for X-ray crystallography. The strong negative charges on the RNA backbone as well as the dynamic nature of the RNA molecules often impede crystallization. NMR spectroscopy, on the other hand, has proven to have high potential to determine RNA structures. The percentage of RNA structures elucidated by NMR (45%) is almost equal to the ones elucidated by X-ray crystallography (52%). Small RNAs (<40 nucleotides) can be studied using uniformly ^{1}H, ^{13}C, ^{15}N-labeled nucleotides with different nucleotide specific labeling schemes [3–5]. However, larger biologically relevant RNA structures are difficult to tackle due to the severe spectral overlap observed in NMR spectra of RNA [6]. Furthermore, strong relaxation of

O. Duss • P.J. Lukavsky (✉) • F.H. Allain (✉)
Institute for Molecular Biology and Biophysics, Swiss Federal Institute of Technology Zürich,
ETH Hönggerberg, CH-8093 Zürich, Switzerland
lpeter@mol.biol.ethz.ch, allain@mol.biol.ethz.ch

H.S. Atreya (ed.), *Isotope Labeling in Biomolecular NMR*, Advances in Experimental Medicine and Biology 992,
DOI 10.1007/978-94-007-4954-2_7, © Springer Science+Business Media Dordrecht 2012

important RNA resonances due to dipolar interactions (mostly ^{1}H-^{13}C in ^{13}C-labeled nucleotides or ^{1}H-^{1}H in the sugar or the pyrimidine H5-H6 proton pair) limited the studies of larger RNAs. Improvements in isotope labeling of RNA, especially site-specific deuteration and segmental labeling of RNA, have opened the avenue for studying RNA molecules of ever increasing size by NMR spectroscopy. NMR spectroscopy is therefore expected to become more and more important in the elucidation of RNA structures and their complexes with other nucleic acids, small molecules or proteins.

In this book chapter, we discuss several aspects of isotope labeling of RNA focusing on larger RNAs. We start by comparing the different methods for the production of RNA that are chemical synthesis, in vitro enzymatic transcription and in vivo RNA production. We also point out different possibilities that allow obtaining RNAs with homogenous termini. In a second part, we present different approaches to purify RNA in both a native and denaturing manner. The third part focuses on several strategies for selective deuteration/protonation of nucleotides leading to a strong reduction of the spectral overlap and also to a reduction of transverse relaxation. The review ends with an extensive description of different strategies for segmental isotopic labeling of RNA that are becoming essential for studying RNAs of moderate to large size by NMR spectroscopy, especially in combination with measurements of residual dipolar couplings, paramagnetic relaxation enhancement [5, 7], electron paramagnetic resonance (our unpublished results) and small angle X-ray scattering [8].

7.2 RNA Synthesis

RNA can be synthesized in three different ways depending on the length and the requirements for isotope labeling: chemical synthesis, *in vitro* enzymatic transcription and *in vivo* production of RNA (see Fig. 7.1a).

7.2.1 *Chemical Synthesis*

Chemical synthesis is the method of choice for preparing small RNAs, as *in vitro* enzymatic synthesis of RNAs smaller than ten nucleotides has been reported not to be successful [9], except in one case [10]. Chemical synthesis of RNA is reported for RNAs up to 80 nucleotides [11–13]. However, low yields and high costs for larger RNAs make the chemical synthesis suitable only for short RNAs (<20 nts). A unique advantage of chemical synthesis is the possibility of introducing modified nucleotides at desired positions. For example, introduction of thiouridines at specific positions in an RNA allows the attachment of nitroxide spin-labels for measuring paramagnetic relaxation enhancement [14]. Moreover, a protocol for synthesizing short RNAs, that are selectively ^{13}C-labeled on sugar carbons has been developed [15] and used to solve the structure of several protein-RNA complexes [16–20], but unfortunately isotope labeled phosphoamidites are not yet commercially available. Introduction of modified nucleotides into larger RNAs can be achieved by enzymatic ligation of small synthetic RNAs (see last section in this chapter) [21, 22].

7.2.2 *In Vitro Enzymatic Transcription*

In vitro enzymatic transcription using SP6, T3 or T7 RNA polymerases is the most widely used method for the production of RNAs larger than 12 nucleotides [23–26]. The possibility of incorporating commercially available ^{13}C- and/or ^{15}N- labeled, perdeuterated or even partially deuterated nucleoside triphosphates (NTPs) allows production of RNAs suitable for heteronuclear multidimensional NMR

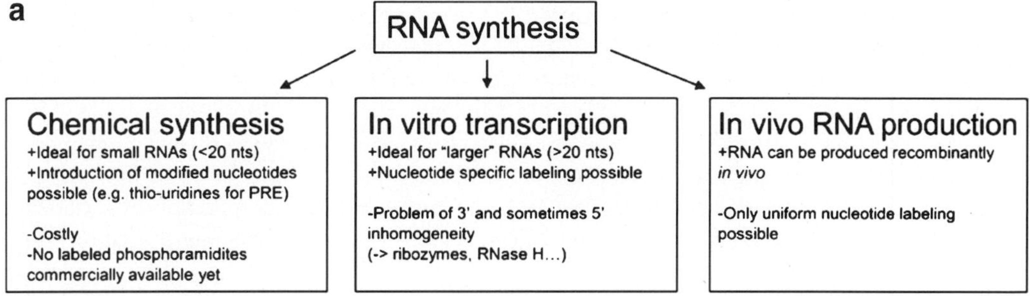

b

How to get homogenous 5' and 3' termini

5' hammerhead ribozyme in cis	No sequence requirement at 5'-end, generates 5'-hydroxyl, can also be used cis or trans in 3'-end but requires UX (X≠G) preceding cleavage site	Ferre-D'Amare et al. (1996) Shields et al. (1999)
3' VS or HDV ribozyme in trans or cis	3' Varkud satellite ribozyme (VS) cuts after any base N≠C and hepatitis delta virus ribozyme (HDV) has no sequence requirement. Ribozyme cleavage generates a 2',3'-cyclic phosphate	Ferre-D'Amare et al. (1996) Walker et al. (2003)
Sequence specific RNase H cleavage	Sequence specific RNase H cleavage can be used both 5' (generating a 5'-phosphate) and 3' (generating a 3'-hydroxyl)	Lapham et al. (1996)
DNAzymes	The RNA substrate is cut between a purine and a pyrimidine. DNAzymes can be used both 5' (generating a 5'-hydroxyl) and 3' (generating a 2',3'-cyclic phosphate)	Pyle et al. (2000)
C2'-methoxyl modified DNA template strand	DNA template modified with C2'-methoxyls at last two nucleotides of 5'-terminus dramatically reduces non-templated nucleotide addition by T7 polymerase	Kao et al. (1999)

c

Methods for RNA purification

PAGE (polyacrylamide gel electrophoresis)	Single nucleotide resolution up to 30 nts for preparative scale, time-consuming, low yields, requires denaturing of RNA	Price et al. (1998)
Non-denaturing size-exclusion or anion-exchange FPLC	Separation of monomeric from multimeric RNA species, purification of RNA from proteins, abortive transcription products and DNA	Easton et al. (2010) McKenna et al. (2007)
Denaturing anion-exchange HPLC	Fast and efficient method for purification of RNA from proteins, abortive transcription products and DNA, almost single nucleotide resolution up to 40 nts	Duss et al. (2010) Shields et al. (1999)
Affinity chromatography	Affinity tag is attached to RNA, which can be cleaved by either an activated ribozyme or by RNase H cleavage or a DNAzyme	Batey et al. (2007) Walker et al. (2008)

Fig. 7.1 RNA production. (**a**) RNA synthesis can be performed either by chemical synthesis, *in vitro* transcription or *in vivo* RNA production. (**b**) Methods for generating homogenous 5'- and 3'-termini. (**c**) Methods for preparative RNA purification (Adapted with permission from Dominguez et al. [5])

[27–31]. T7 RNA polymerase can be obtained commercially or produced in-house by overexpressing a His-tagged T7 RNA polymerase in *E. coli* [32]. Transcription reactions should be first optimized on small scale reactions by changing concentrations of $MgCl_2$, DNA, NTPs and T7 RNA polymerase and testing the influence of the addition of pyrophosphatase and/or guanine monophosphate (GMP). The best condition can be scaled-up to a large scale reaction of for example 10 ml, which typically yields around 500 nmol of RNA. Unlabeled and uniformly labeled NTPs are commercially available but can also be produced in-house [28]. Transcription using T7 RNA polymerase can be performed from chemically synthesized double-stranded DNA templates or from linearized plasmids. Since only the 18 nts T7 promoter on the top-strand is sufficient for transcription, the same top-strand can be used for any transcription. However, it has been observed specially for structured RNAs that higher yields are obtained when fully double-stranded DNA is used [33]. The first nucleotide, which is incorporated, must be a guanine. Transcription efficiency is highly dependent on the starting six nucleotides. Excellent starting sequences are native starting sequences of the T7 RNA polymerase, such as

GGGAGA, GGGAUC, GGCAAC or GGCGCU [23]. Besides the 5′ sequence requirements, another drawback of T7 *in vitro* transcription is the 3′ and 5′ inhomogeneity. More than 30% of untemplated 5′ nucleotides have been observed for sequences starting with 4–5 consecutive guanines, whereas a sequence starting with GCG showed no detectable 5′ inhomogeneity [34]. More severe is the 3′ inhomogeneity, where up to six additional nucleotides can be added. An overview of several methods to overcome 5′ and 3′ inhomogeneity is presented in Fig. 7.1b.

The problem of 5′ and 3′ inhomogeneity can be circumvented by incorporation of a ribozyme sequence in *cis*, which cleaves co-transcriptionally leading to an homogenous 5′-hydroxyl or a 2′,3′-cyclic phosphate end, respectively (see Fig. 7.6d step 1) [31, 35]. Concerning 5′-inhomogeneity, hammerhead ribozymes are interesting because they have no sequence requirements [26]. When placed 5′ to the RNA of interest, they allow cleavage of the RNA with $MgCl_2$ as cofactor almost to completion [26, 36, 37]. Concerning 3′-inhomogeneity, the hepatitis delta virus (HDV) RNA ribozyme, that has no sequence requirements [38, 39], or the *Neurospora* Varkud satellite (VS) ribozyme that has minimal sequence requirements (VS will cut efficiently after any nucleotide other than cytosine) can be efficiently used [36, 40]. It has been shown that hammerhead ribozymes [41] and VS ribozymes [36] can be added in *trans* thereby saving isotope labeled NTPs that otherwise would be used to produce the ribozyme incorporated in *cis* (see Fig. 7.6d step 1).

In addition to ribozymes, DNAzymes have been developed by *in vitro* evolution as engineering tools [42–44]. The 10–23 family of DNAzymes cleaves between a purine and a pyrimidine, which is the only sequence requirement. Cleavage results in a 5′-hydroxyl group and a 2′,3′-cyclic phosphate similarly to small ribozymes. Moreover, it has been shown that RNAs can be cleaved sequence-specifically by RNase H, when the RNA of interest is hybridized with a 2′-*O*-methyl-RNA/DNA chimera [45, 46]. In contrast to ribozyme and DNAzyme mediated RNA cleavage, RNase H produces 5′-monophosphates and 3′-hydroxyl groups (see Fig. 7.6d step 2). Another approach is the use of a DNA template strand for transcription, in which the two 5′ nucleotides are modified with C2′-methoxyls. This dramatically reduced 3′-end inhomogeneities [47].

7.2.3 In Vivo Production of RNA

Dardel and co-workers developed an *in vivo* method using a tRNA scaffold to protect the RNA from cellular RNases for the production of milligram quantities of RNA for NMR studies [48, 49]. The tRNA scaffold can be removed either by DNAzymes or by sequence-specific RNase H cleavage [42, 44–46]. Using this method, a reasonable amount of RNA for NMR studies (0.8 μmol) was obtained from 2 l of *E. coli* culture grown on $^{15}N/^{13}C$-labeled medium.

7.3 RNA Purification

RNA obtained by *in vitro* enzymatic or *in vivo* transcription must be purified from proteins (T7 RNA polymerase, pyrophosphatase) and abortive transcription products (a large number of smaller oligoribonucleotides of two to six nucleotides in length are generated during transcription due to abortive initiation events) as well as from unused NTPs. In addition, RNA with one or two additional nucleotides arising from untemplated nucleotide addition at the 3′ end must be removed, when a homogenous RNA is required. An overview of different purification methods is presented in Fig. 7.1c.

Denaturing polyacrylamide gel electrophoresis (PAGE) is the most commonly used purification method for large quantities of RNA needed for NMR spectroscopy. Single nucleotide resolution for preparative scales is typically achieved for RNAs up to 30 nucleotides. However, this procedure is laborious and suffers from low recovery yields, especially with larger RNAs. Additionally, PAGE requires the RNA to be denatured and refolded after purification, which might lead to aggregation and dimerization of the RNA [50]. Furthermore, the RNA is not free of low-molecular-weight acrylamide contaminants, which might interact with RNA and also compromise NMR spectral analysis [51]. Therefore, different chromatographic methods have been developed to purify RNA. Frederick and coworkers proposed purifying RNA by non-denaturing anion-exchange chromatography [52]. Depending on the salt type of the elution buffer (NaCl, CsCl or $MgCl_2$) they could separate RNAs with different conformations. Recently, Lukavsky and co-workers showed that weak anion-exchange fast protein liquid chromatography (FPLC) under non-denaturing conditions can be used to separate the desired RNA product from the T7 RNA polymerase, unincorporated rNTPs, small abortive transcripts and the plasmid DNA template [53]. Rapid purification of homogeneous RNAs can be achieved by using *trans*-acting hammerhead ribozymes in combination with anion-exchange high performance liquid chromatography (HPLC) chromatography at high temperature (90°C) [41]. In our laboratory, we are using an anion exchange HPLC under denaturing conditions (6 M Urea) at high temperature (85°C) allowing separation of RNAs to almost single nucleotide resolution up to 40 nucleotides. These harsh denaturing conditions also allow the separation of an RNA from a long DNA splint used in splinted ligation [37]. The eluted RNA is subsequently liberated from urea using butanol extraction [54]. Certain biologically relevant RNAs might fold into different conformations or might form multimers, which can be separated by purifying them under non-denaturing conditions using size-exclusion FPLC [50, 51, 55]. In addition to reverse-phase HPLC [56], the use of affinity chromatography has been described [57–60]. Batey and Kieft developed a sophisticated approach, where an affinity tag is attached to the 3′-end of the RNA by a glucosamine-6-phosphate activated (*glmS*) ribozyme [59]. The affinity tag is based on two RNA stem-loops having high affinity for the MS2 coat protein fused to a 6×His-tagged MBP, which binds to a Ni^{2+}-affinity column. Elution of the RNA can be achieved by activating the ribozyme with addition of GlcN6P. Affinity purification based on aptamer tags binding Sephadex or Streptavidin have also been proposed [48, 60].

Depending on the purification method, the RNA can either be eluted directly into NMR buffer or needs to be exchanged into a suitable buffer. Buffer exchange can be performed by dialysis or by washing and concentrating the RNA with ultracentrifugation with an appropriate molecular weight cut-off (MWCO). Dialysis bags and ultracentrifugation filter devices with 1,000 Da MWCO are commercially available and are appropriate for RNAs produced by *in vitro* transcription. The RNAs can be lyophilized and subsequently resuspended into NMR buffer. Typical NMR buffers for RNA are 10–50 mM sodium phosphate at pH=5.5–6.

7.4   Deuteration Strategies for Larger RNAs

Small RNAs can be studied using uniformly ^1H, ^{13}C, ^{15}N-labeled nucleotides with different nucleotide specific labeling schemes [3–5]. However, once the size of RNA oligonucleotides exceeds 40 nucleotides, NMR spectroscopic studies suffer from two major problems, which make unambiguous resonance assignments increasingly difficult [61]. First, this is caused by extensive overlap of the six ribose protons, of which five resonate within a narrow window of 1 ppm. Often, it is also impossible to gain additional resolution from the attached ribose carbon atoms, which also resonate in defined, but again narrow spectral regions. In turn, this leads to unresolvable resonance overlap in through-bond and through-space 3D-type NMR experiments and thus makes resonance assignments impossible. Second, larger RNA molecules are often elongated and tumble slowly in solution and thus ^1H-^{13}C dipolar relaxation is more pronounced. This results in increased proton and carbon linewidths, which additionally aggravate the existing resonance overlap and often broaden resonances beyond detection.

For larger proteins, partial or full deuteration helps to overcome short transverse relaxation times caused by dipolar ^{1}H-^{1}H and ^{1}H-^{13}C interactions and thus allows for efficient magnetization transfer along the protein backbone and into the side chains. This approach makes resonance assignments even for very large proteins feasible [62]. In larger RNA molecules, on the other hand, short transverse relaxation times are mainly caused by dipolar ^{1}H-^{13}C interactions and not so much by dipolar ^{1}H-^{1}H interactions due to the much lower proton density in RNA as compared to proteins [63]. Moreover, resonance assignments of RNA heavily rely on 3D ^{1}H-^{13}C correlation NMR spectra and therefore full deuteration is not an option for resonance assignments of large RNAs. For these reasons, different approaches have been developed to overcome the problem of resonance overlap in large RNAs using site-specific ribose or base labeling.

7.4.1 Site-Specific Ribose Deuteration

RNA consists of four types of residues, namely adenosine, guanosine, cytidine, and uridine, which differ by the nature of their base, but all contain the same ribose moiety. Hence, most extensive overlap in large RNAs is observed for the ribose protons H2'-H5'/H5'', which typically resonate between 4 and 5 ppm as well as their attached carbon atoms C2'-C5', which resonate in specific, but narrow spectral regions from 60 to 85 ppm [64]. To resolve this overlap, selective labeling methods are required rather than uniform or residue-type specific labeling schemes as used for proteins.

Several site-specific labeling schemes have been developed for the ribose moiety of RNA: Site-specific deuteration in the ribose ring is used for spectral simplification of the crowded ribose region. Moreover, a combination of uniform or selective ^{13}C-labeling with site-specific deuteration allows for spectral editing and helps to reduce the dipole-dipole interaction induced relaxation for improved efficiency of ^{1}H-^{13}C correlation experiments. While chemical synthesis would allow to precisely control the placement of specific isotopic labels in the ribose ring, this method requires expensive labeled precursors, multistep synthesis and often results in poor overall yields. Enzymatic synthesis with enzymes from the pentose phosphate pathway, on the other hand, can use glucose with a variety of labeling schemes and thus is more cost effective and results in higher overall yields. This method has been pioneered by the Williamson lab and some of these different possible labeling schemes have been employed to determine structures of large RNAs [29, 65].

Starting from fully deuterated glucose with or without ^{13}C-labeling, enzymes from the pentose pathway are used to prepare 5-phospho-D-ribosyl α-1-pyrophosphate, which can be linked to the corresponding bases using enzymes from the nucleotide biosynthesis and salvage pathways. In addition, specific enzymes can be used in the first reaction steps to back-exchange the H1' and/or H2' position in the resulting ribose moiety, while H3', H4', H5' and H5'' positions remain deuterated (Fig. 7.2a). This labeling scheme leads to great spectral simplification in the crowded ribose–ribose and ribose–base regions in 2D NOESY spectra, but retains important structural information. The sugar pucker can still be determined from H1'-H2' cross peaks in 2D TOCSY and DQF COSY spectra and the cross peaks intensities of sequential H1'-base and H2'-base connectivities are indicative of canonical A-form or non-canonical RNA conformation. If combined with ^{13}C-labeling, additional resolution of the H1' and H2' protons is gained from the attached carbon atoms in 3D ^{1}H-^{13}C correlation NMR experiments. In addition, this labeling scheme allows editing of C2'-H2' correlations, which can overlap with C3'-H3' correlations and thus makes unambiguous assignments possible, when through-bond HCCH-type experiments fail due to short transverse relaxation times [65].

The Butcher group has successfully applied this labeling scheme to determine the structure of a 30 kDa GAAA tetraloop receptor complex [66]. Using uniform as well as nucleotide-specific ^{13}C, ^{15}N-labeling together with 3D NOESY and 2D filter/edited NOESY experiments, it was not possible to assign many NOEs in the ribose region which crucially defined the interaction interface. Selective deuteration of the ribose H3'-H5'/H5'' protons together with pyrimidine base H5 deuteration, on

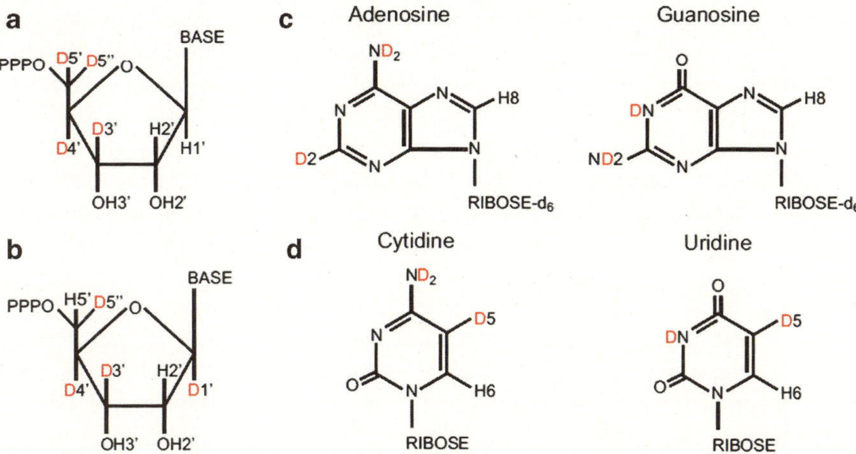

Fig. 7.2 Site-specific ribose and base labeling schemes. (**a**) Ribose moiety protonated in the H1′ and H2′ positions and deuterated in the H3′, H4′, H5′/H5″ positions from the Williamson group [65]. Deuterated positions are highlighted in *red*. This labeling scheme is also commercially available for all four nucleotides. (**b**) Ribose moiety protonated in the H2′ and H5′ positions and deuterated in the H1′, H3′, H4′, H5″ positions from the Wijmenga group [30]. Deuterated positions are highlighted in *red*. (**c**): Adenosine and guanosine nucleotides with the purine base moiety selectively protonated in the H8 position and a perdeuterated ribose moiety prepared from perdeuterated rNTPs following procedures described in [31]. Deuterated positions are highlighted in *red*. (**d**) Cytidine and uridine nucleotides with the pyrimidine base moiety selectively deuterated in the C5 position and a fully protonated ribose moiety prepared from fully protonated rNTPs following procedures described in [68]. Deuterated positions are highlighted in *red*

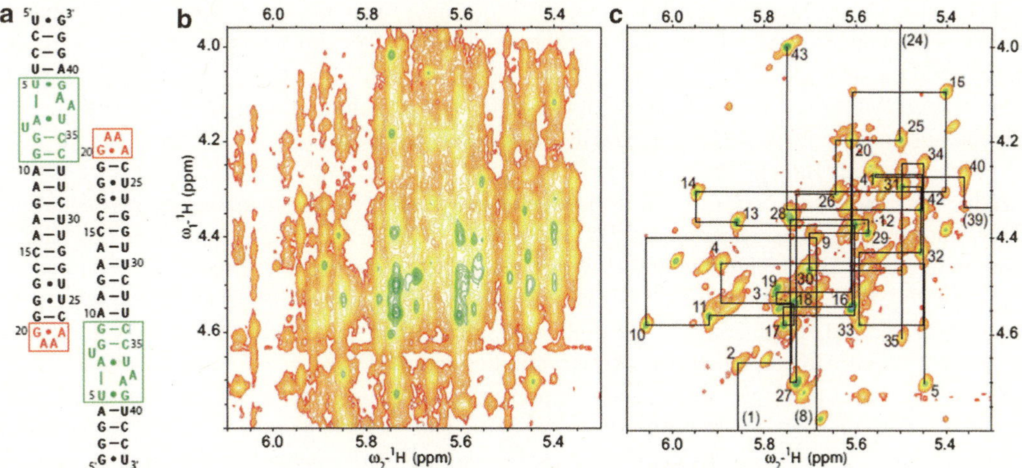

Fig. 7.3   Relief of spectral crowding by site-specific ribose deuteration. (**a**) Secondary structure of a 30 kDa GAAA tetraloop-receptor RNA dimer. The helical regions are shown in *black*, the tetraloop is shown in *red*, and the receptor in *green*. Numbering according to [66]. (**b, c**) Region of the 2D NOESY spectrum obtained for the fully protonated in (**b**) or selectively ribose deuterated in (**c**) tetraloop-receptor RNA dimer. Deuteration of the ribose H3′, H4′, H5′ and H5 positions (see Fig. 7.2a) reliefs the severe overlap in the ribose – base region of the 2D NOESY spectrum and allows sequential assignments through H2′ – base connectivities (Reprinted with permission from Davis et al. [66])

the other hand, allowed to assign many NOEs in otherwise crowded spectral regions (Fig. 7.3) and also helped to differentiate H2′ and H3′ chemical shifts, which were overlapped even in ^{13}C-edited NOESY spectra. The authors also point out a minor disadvantage of this labeling scheme which results from the lack of NOEs from H4′ and H5′/H5″ protons, which often yield important structural information in non-canonical regions of the RNA molecule, such as bulges and loops [66]. But in many cases, non-A-form torsion angles or unusual stacking interactions in these regions result in unique ^{13}C and ^{1}H chemical shifts for these resonances and thus allow assignments even without selective labeling.

A partial deuteration scheme for the ribose moiety has also been described by the Wijmenga group. Again, starting from fully deuterated and ^{13}C -labeled glucose, stereochemical H-exchange is used to selectively back-exchange the H2′ and H5′ position in the resulting ribose moiety, while the H1′, H3′, H4′ and H5″ positions remain deuterated (Fig. 7.2b) [30]. This labeling scheme preserves important structural information from H2′-base connectivities, the H2′-H5′ intra-sugar NOE contacts can provide sugar pucker information and sequential H2′-H5′ NOE contacts constrain the RNA backbone. The main advantage is that in addition, important torsion angle information is accessible through $J_{H5'-C3'}$ coupling (γ torsion angle) and $J_{C4'P}$ and $J_{H5'P}$ couplings (β torsion angle). This scheme has been successfully demonstrated on a 31 nucleotide RNA, but still awaits an application to larger RNA systems [30].

7.4.2 Site-Specific Base Deuteration

Spectral simplification of the ribose–base region in NOESY spectra can also be achieved by site-specific deuteration of the purine and pyrimidine base moieties. The purine C8 and the pyrimidine C5 position of 5′ rNMPs can be deuterated using bisulfite modification under basic conditions before conversion to 5′ rNTPs [67, 68]. Likewise, protons can be selectively introduced at these positions in deuterated NMPs (Fig. 7.2c, d). The purine C8 position can also be efficiently exchanged in rNTPs rather than rNMPs under basic conditions without significant hydrolysis and degradation. Incubation of rATP or rGTP with triethylamine (TEA) at 60°C for 5 days or 24 h, respectively, leads to essentially complete exchange in the C8 position and the resulting rNTPs can be used directly for *in vitro* transcription after removal of volatile TEA by lyophilization [31]. Exchange at the pyrimidine C6 position of rCMP or rUMP is less straightforward, but can be achieved with up to 70% efficiency in alkaline solution with DMSO before the rNMPs start to decompose [68].

Purine C8 deuteration has been used prior to ^{13}C,^{15}N-labeling of RNA to distinguish adenine from guanine H8 protons and both of them from adenine H2 protons [67]. Deuteration of the C5 position of pyrimidines eliminates the strong H5-H6 crosspeak in NOESY and TOCSY spectra and by labeling RNA with either C5-deuterated uracil or cytosine their H6 protons can be discriminated through the absence of H5-H6 correlations for the C5-deuterated pyrimidine (Fig. 7.2d). In addition, this is advantageous for spectral simplification in large RNAs, since strong crosspeaks arising from the H5-H6 protons can obscure important sequential connectivities in the H1′ – base region of 2D and 3D NOESY spectra. The Lukavsky group has recently employed this labeling scheme to unambiguously assign the 48 nts *K10* transport signal RNA (Fig. 7.4a) [69]. This RNA displayed a high sequence redundancy containing 14 Watson-Crick A:U base pairs and a total of 18 adenosine and 19 uridine residues. Nevertheless, almost all adenine C2-H2 and C8-H8 crosspeaks were resolved in ^{1}H-^{13}C correlation NMR experiments, while uracil C6-H6 correlations showed a high degree of overlap (Fig. 7.4b), which could not be resolved in 3D NOESY-HSQC spectra. Likewise, unambiguous resonance assignments using homonuclear 2D NOESY spectra were also compromised by the prominent H5-H6 crosspeaks from the 19 uridine and 6 cytidine residues (Fig. 7.4c). Site-specific deuteration of the pyrimidine H5 positions helped to edit for the H1′–base connectivities, which allowed unambiguous, sequential assignments of a run of five consecutive uridine residues (Fig. 7.4d) [69]. In addition, this labeling scheme eliminates the nearby ^{1}H,^{1}H dipolar coupling partner of the pyrimidine H6 proton and subsequently leads to line narrowing of NOESY crosspeaks involving the pyrimidine H6 protons.

The Summers group makes extensive use of different combinations of perdeuterated, fully protonated and partially deuterated rNTPs to achieve assignments of very large RNA systems. These labeling

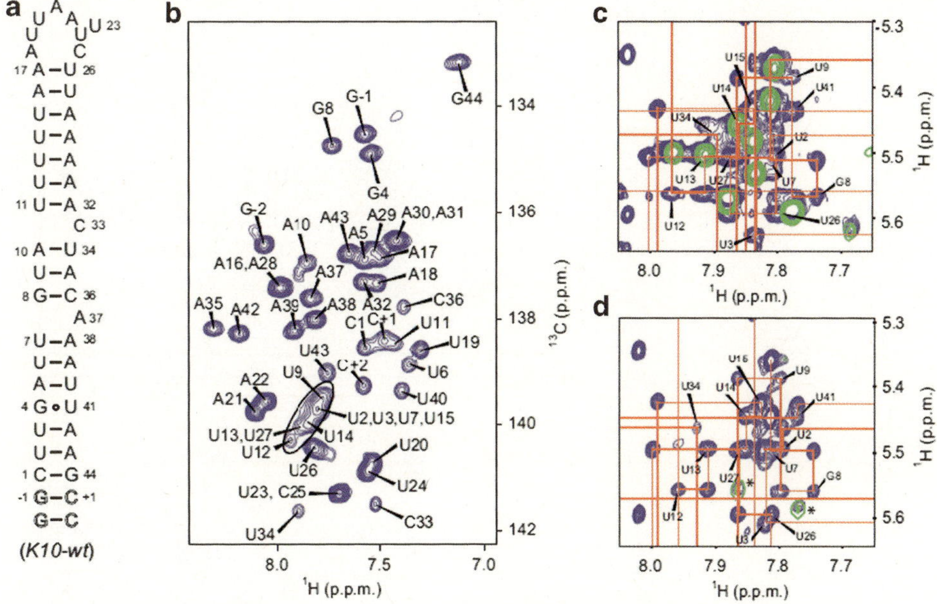

Fig. 7.4 RNA labelling with C5 deuterated pyrimidines simplifies 2D NOESY spectra. (**a**) Secondary structure of wild-type *K10* TLS RNA. Numbering according to [69]. Outlined nucleotides were added to improve transcription efficiency. The three helical segments are separated by single nucleotide bulges (C33 and A37). (**b**) ^{1}H,^{13}C-ct-HSQC (constant-time heteronuclear single quantum coherence) spectrum of aromatic pyrimidine C6-H6 and purine C8-H8 correlations in ^{13}C, ^{15}N-labeled *K10-wt* RNA. Most aromatic resonances (labeled according to **a**) display excellent chemical shift dispersion, except for several uracil C6-H6 resonances, which show severe chemical shift overlap in both the carbon and proton dimensions (indicated by an *oval*). (**c**) Overlay of a homonuclear TOCSY (*green*) and NOESY (*purple*) spectrum of unlabeled *K10-wt* RNA presenting a crowded region of the aromatic-anomeric walk. Strong NOESY cross-peaks from pyrimidine H5-H6 protons (also indicated as *green TOCSY peaks*) obscure part of the aromatic-anomeric walk indicated with *orange lines* (e.g. U13 or U14). Intraresidue correlations are labeled with residue numbers according to (**a**). (**d**) Overlay of a homonuclear TOCSY (*green*) and NOESY (*purple*) spectrum of site-specifically pyrimidine C5 deuterated *K10-wt* RNA showing the same part of the aromatic-anomeric walk as in (**c**). Strong NOESY cross peaks (*purple*) from pyrimidine H5-H6 protons are now absent due to the site-specific pyrimidine C5 deuteration and only two residual TOCSY peaks (*green*) are visible as indicated by an *asterix*. The "pyrimidine H5-H6 crosspeak-free" aromatic-anomeric walk is indicated with *orange lines* (Reprinted with permission from Bullock et al. [69])

schemes are often combined with segmental labeling using techniques outlined in the next section [31]. Unambiguous assignments for the 101 nts core encapsidation signal of the moloneymurine leukemia virus (MoMuLV) RNA were accomplished by structure determination of individual stem-loops complemented with conventional 3D and 4D ^{13}C-edited NOESY experiments using nucleotide-specific ^{13}C,^{15}N-labeling on the entire 101 nts RNA [70, 71]. The assignments were completed with 2D NOESY spectra collected on four different nucleotide-specifically protonated samples with the remaining nucleotides being perdeuterated. This labeling scheme allows editing of intra-nucleotide NOEs for the protonated residue type and identifies inter-nucleotide NOEs between protonated residues if they are adjacent in sequence or close in space as a result of tertiary interactions [70].

For the assignments of the 132 nts double kissing hairpin of the MoMuLV 5′ untranslated region (Fig. 7.5a), this approach was extended using RNA samples with specifically C6/C8-protonated and perdeuterated or C8-deuterated and fully protonated nucleotides in combination with perdeuterated or fully protonated nucleotides (Fig. 7.2c) [72]. For instance, a 2D NOESY spectrum collected on an RNA sample containing perdeuterated uridine and cytosine, fully protonated guanosine and C8-protonated, perdeuterated adenosine edits the following structural information: intra ribose – base

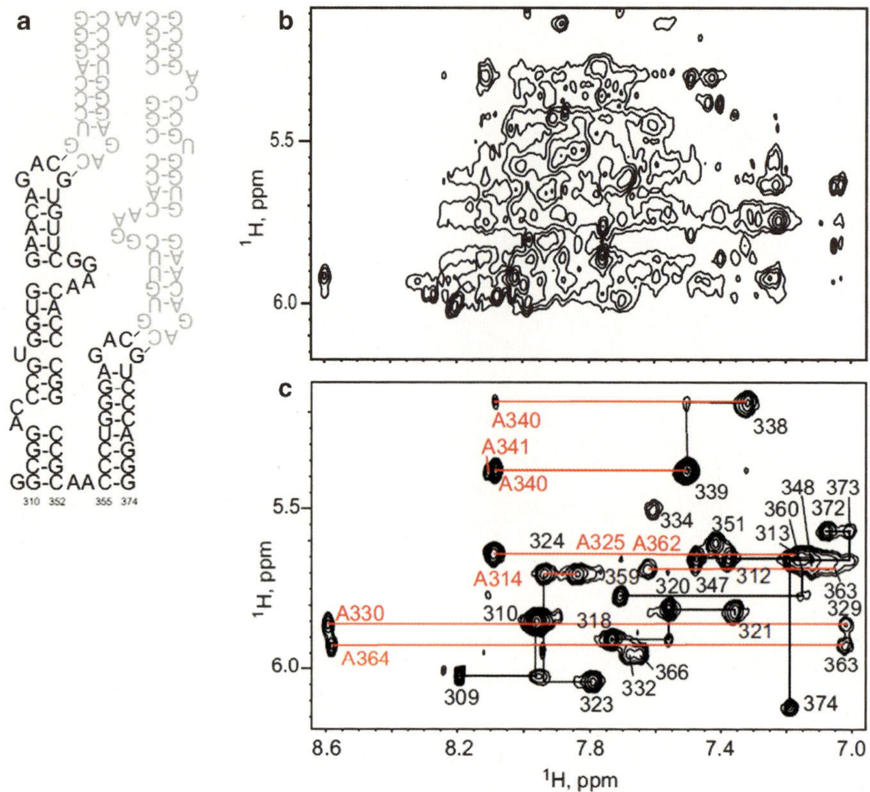

Fig. 7.5 H8 enrichment of perdeuterated purines simplifies 2D NOESY spectra. (**a**) Secondary structure of the stem loops C and D of the MoMuLV 5′UTR. Numbering according to [72]. (**b**) 2D NOESY spectrum of the fully protonated stem loop CD dimer displays severe resonance overlap in the H1′ – base region. (**c**) 2D NOESY spectrum of the stem loop CD dimer transcribed using fully protonated GTP, H8-protonated, perdeuterated ATP, and perdeuterated CTP and UTP. *Black lines* denote inter-guanosine H1′ – base connectivities and *red lines* denote adenosine H8 to guanosine H1′ connectivities (For detailed explanation see text. Reprinted with permission from Miyazaki et al. [72])

crosspeaks for all guanosines, inter ribose – base crosspeaks for sequential guanosines ($G_{i\text{-ribose}}$ to $G_{i+1\text{-H8}}$), inter ribose – base crosspeaks for sequential guanosine-adenosine pairs ($G_{i\text{-ribose}}$ to $A_{i+1\text{-H8}}$) and sequential H8 – H8 NOEs for all guanosines and adenosines (for example see Fig. 7.5b, c). Depending on the chemical shift dispersion of the aromatic protons and the number of different sequential nucleoside pairs, several combinations are required to obtain full assignments of a large RNA. In the case of the 132 nts double kissing hairpin, six different labeling combinations were required to obtain 1,248 unique NOE-derived ^{1}H-^{1}H distance restraints. Together with additional restraints which aid to maintain A-form conformation in helices, this yielded a well-defined structure of this large RNA with an r.m.s.d of 0.4 Å for the final ensemble of 20 structures [72]. This represents the largest RNA determined by solution NMR spectroscopy to date.

Many of the presented labeling schemes can be implemented using commercially available rNTPs. Fully protonated, fully deuterated, ^{15}N-, ^{13}C- or ^{13}C,^{15}N-labeled rNTPs and rNMPs can be purchased and converted into rNTPs selectively protonated or deuterated at various base positions. The attractive labeling scheme from the Williamson group with deuterated H3′, H4′ and H5′/H5″ positions in the ribose moiety and the H5 position in pyrimidine bases is also commercially available and other labeling schemes might be available upon request [29, 65].

7.5 Segmental Isotope Labeling for Large RNAs

Although signal overlap can be substantially reduced using selectively labeled nucleotides or RNAs containing nucleotide-specific labeling schemes, larger RNA structures are difficult or impossible to tackle due to the tremendous spectral overlap [6]. Segmental isotopic labeling of RNA is therefore essential to study RNAs of moderate to large size by NMR spectroscopy. Firstly, it allows verifying whether a small structural RNA element such as a hairpin that can be studied by standard NMR methods retains the same structure as in the larger, biological active RNA. One can, for example, ligate a small isotopically labeled fragment produced by chemical synthesis or in vitro transcription to a larger unlabeled fragment. Using this approach, Puglisi and co-workers could show that a 25 kDa RNA domain adopts the same structure in isolation as found in the context of the entire 100 kDa natural RNA [73, 74]. Very recently, the group of Summers segmentally isotope labeled 29 nts at the 3′ end of the intact 230 kDa HIV-1 5′-leader [75] and they could detect structures that regulate HIV-1 genome packaging. Secondly, segmental isotope labeling is reducing the number of resonances in ^{13}C- or ^{15}N-edited correlation spectra and therefore the spectral overlap. Properly choosing the segmental isotope labeling strategy allows the measurements of residual dipolar couplings and paramagnetic relaxation enhancement of enough RNA resonances to extract structural information (our unpublished results and references [7, 73]).

RNA segmental labeling can be also very useful for other biophysical techniques. Site-specific incorporation of heavy atoms into internal positions within longer RNAs showed to have high potential for solving phases in X-ray crystallography [21, 76]. Similarly, single molecule experiments with RNA often require the incorporation of modified nucleotides at specific positions within a long RNA to study its structure or folding [22]. Finally, we could show using segmental labeling that introduction of two nitroxyl radical tags into specific positions of a longer RNA allows the measurement of long-range distances by pulse electron paramagnetic resonance (our unpublished results). Therefore, enzymatic ligation between a short synthetic RNA containing modified nucleotides and longer fragments produced by *in vitro* transcription is expected to become a method of choice for studying biologically important RNAs (unpublished results and reference [77]). However, methods for incorporating modified nucleotides into internal positions of longer RNAs (>100 nts) for structural studies or segmental isotope labeling for NMR structural studies remained very time-consuming, costly as the yield they provide is low and not always applicable because of the sequence-dependence of most protocols [31, 35]. In the following sections, we will first describe the principle and shortly present several methods for segmental isotope labeling by emphasizing their advantages and disadvantages. Then we will describe our recently developed alternative approach for segmental labeling of RNA by which we can obtain very rapidly (5–7 days) high amounts (up to ten-fold higher than previously reported) of segmentally labeled RNAs without sequence requirements [37]. We will finish this chapter by providing a detailed protocol of our recently published method.

7.5.1 General Principle

Basically, two or several RNA fragments that are unlabeled, uniformly isotope labeled or are containing modified or specifically labeled nucleotides at specific positions in the RNA sequence are combined and ligated to obtain the full length RNA of interest. The RNA fragments can be produced by chemical synthesis, *in vitro* enzymatic transcription or *in vivo* depending on their required property (see first section). The RNA fragments have to fulfill certain requirements such that they can be ligated without self-ligation or ligation in the wrong sequential order. RNA ligation can be performed by T4 RNA or T4 DNA ligase [78–80] or by using a deoxyribozyme that catalyzes RNA ligation [81]. Both T4 RNA and T4 DNA ligases require a 5′-monophosphate on the donor fragment and 3′-hydroxyl on the acceptor fragment at the site of ligation (see Fig. 7.6d step3), whereas the deoxyribozyme catalyzes a ligation

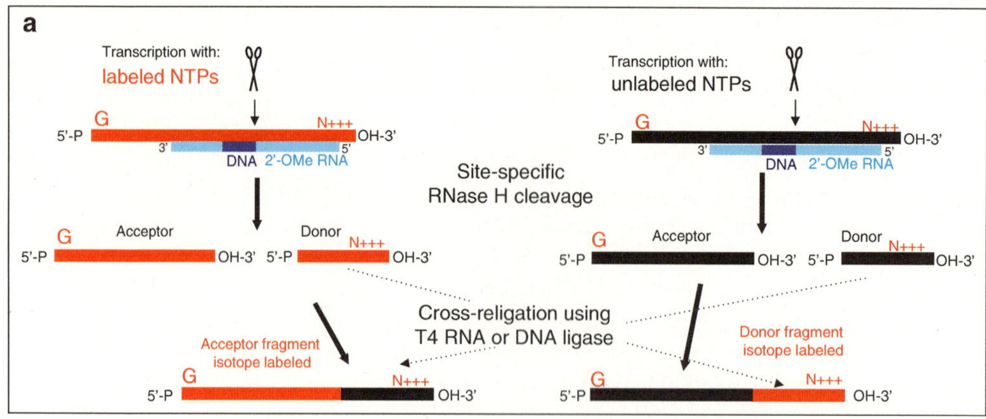

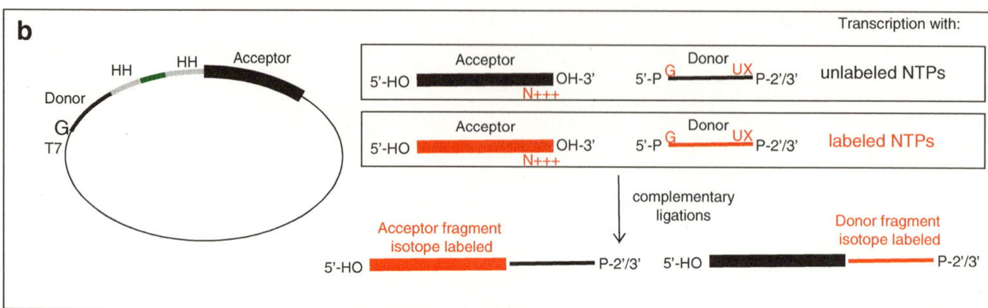

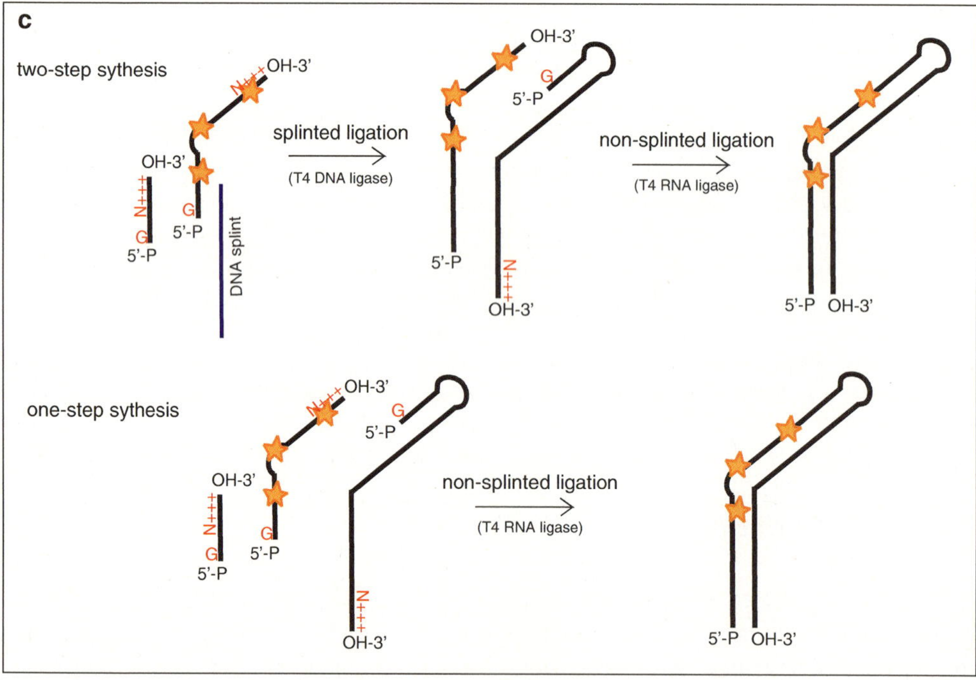

Fig. 7.6 Different approaches for segmental isotope labeling of RNA. The approaches by Xu et al. (**a**), Tzakos et al. (**b**), Nelissen et al. (**c**) and Duss et al. (**d**) are explained in the main text. An overview of the four methods is presented in (**e**). (**a**–**d**) The isotope labeled material is highlighted in *red*, the unlabeled material in *black*. The G or the UX at the 5′- or the 3′-end of certain fragments indicate sequence requirements. The G at the 5′-end of certain RNA fragments represents a good transcription start site. N+++ stands for an inhomogenous 3′-end. P-2′/3′ indicates a 2′/3′-cyclic phosphate. In the 2′-O-methyl RNA/DNA chimeras (**a, d**) the DNA is in *dark blue* and the 2′-O-methyl RNA in *light blue*. *Scissors* indicate RNase H cleavage sites. The *orange stars* in (**c**) represent selectively deuterated uridines. (**d**) The correctly protected termini of both acceptor and donor fragments are encircled in *green*. E: To compare the different approaches the yield was calculated to be the amount obtained for two segmental isotopically labeled samples, in which only one fragment is isotopically labeled starting from one 20 ml labeled transcription reaction (4.5 mM in each NTP):

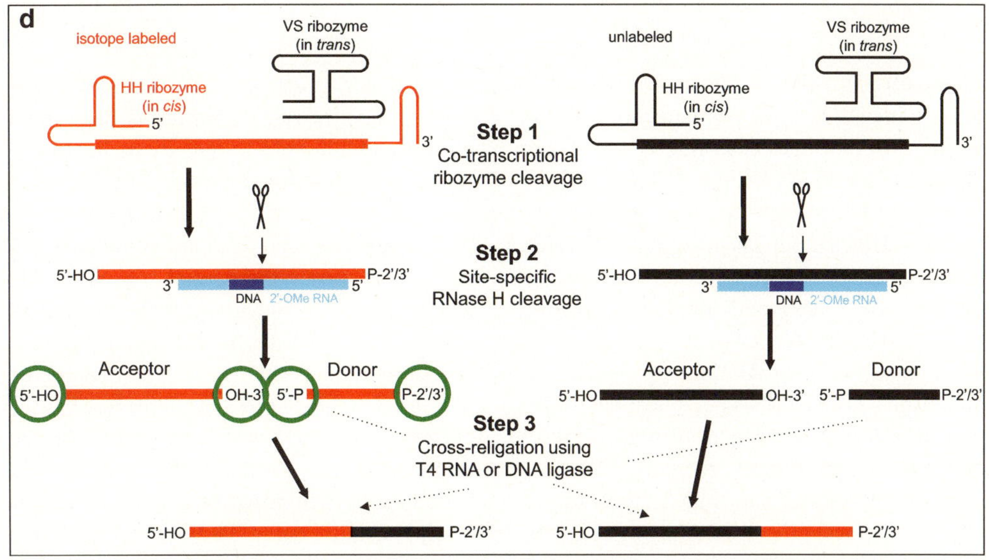

Method	Sequence requirements	Correct fragment termini	Fragment homogeneity	# of labeled transcription reactions required	Multiple segmental isotope labeling possible	Yield and dependence of yield on sequence of fragments to be ligated	Time needed [days]
Xu et al. 1996	**Yes** (G at 5' of RNA)	**No**	**No**	**One**	**Yes**	**4-50 nmol** (dependent on the 5'-sequence of the full-length RNA)	14-17
Tzakos et al. 2007	**Yes** (3' segment must start with G and requires U at penultimate position)	**Yes**	**No** (Non-native nucleotides can be introduced at ligation site)	**One**	**No**	**20/ 22 nmol** (dependent on sequence 3' to ligation site)	12-14
Nelissen et al. 2008	**Yes** (all segments must start with G and must be able to form target-like structure for proper ligation)	**No**	**No** (Non-native nucleotides can be introduced at ligation sites)	**Several** (One per segment to be ligated)	**Yes** (not feasible for >3 segments)	**< 30 nmol** (2 fragments) (dependent on 5'-sequence of all fragments and on ability of fragments to form target-like structure)	9-11
Duss et al. 2010	**No**	**Yes**	**Yes**	**One**	**Yes**	**100-300 nmol** (2 fragments) **50-270 nmol** (4 fragments) (barely sequence dependent)	5-7 / 7-9

Fig. 7.6 (continued) For the approach of Duss et al., the yields for the two-piece ligation (**d**, one RNase H cleavage and one ligation reaction) or for the four-piece ligation (Fig. 7.7; two RNase H cleavages and one ligation reaction) were determined by starting with each 400 nmol of labeled and 400 or 1,200 nmol of unlabeled RNA (obtained from one 20 ml labeled and one 20 or 60 ml unlabeled transcription reaction for the two-fragments or the four-fragments ligation, respectively) and determining the final yield assuming typical yields (after purification) for RNase H cleavage of 50–90% and T4 DNA/RNA ligation of 50–85%. In the approach of Nellisen et al., they start from 665 nmol of isotopically labeled fragment and obtain 173 nmol of *one* segmentally labeled RNA (using the more efficient two-pot ligation), in which only the middle fragment is isotopically labeled. It is not mentioned from which volume of transcription reaction this was obtained. Considering the fact that PAGE purification was used to purify the fragments and that the yield of the fragments is RNA sequence dependent, it can be assumed that the 665 nmols of labeled RNA fragment were obtained from >60 ml transcription reaction (at a concentration of 4.5 mM for each NTP). To obtain *two* segmentally labeled samples, in which a different segment is labeled, 120 ml transcription reaction would be required to obtain 173 nmol of each sample. Therefore, they would obtain <30 nmol of two segmentally isotope labeled RNAs from one 20 ml labeled transcription reaction. The time requirement is without cloning and template plasmid production, however it includes the time needed for small scale optimization reactions required for a new system under study. If not mentioned by the authors, 1 day for each small scale optimization (transcription, RNase H cleavage or ligation) and 3–4 days for PAGE purification is assumed (**d** and **e** are adapted with permission from Duss et al. [37])

reaction with a 5′-triphosphate on the donor fragment and with a 3′-hydroxyl on the acceptor fragment. Ligation with DNA ligase, which recognizes a nicked double-stranded substrate, is performed by annealing a DNA oligonucleotide or a 2′-O-methyl-RNA/DNA chimera to the site of ligation [78]. Unlike DNA ligase, RNA ligase requires a single-stranded site of ligation. Preferentially, the acceptor and the donor are brought together by base-pairing such that the site of ligation is in a hairpin loop [37, 80, 82]. However, it has been shown that RNA ligase can also be used in combination with a DNA oligonucleotide annealing with the site of ligation designed to mimic the natural substrate of RNA ligase [83]. To prevent self-ligation or ligation of the fragments in the improper sequential order (especially using T4 RNA ligase), the acceptor fragment should contain a hydroxyl group both at its 5′ and 3′ ends, whereas the donor fragment should have a monophosphate at the ligation site and a monophosphate or a 2′,3′-cyclic phosphate at the 3′-end (see Fig. 7.6d step3). As discussed in the first section, RNAs obtained by in vitro transcription contain a 5′-terminus with a tri-phosphate and an inhomogenous 3′-hydroxyl terminus. Ribozymes engineered at the 3′-end producing homogenous 2′,3′-cyclic phosphates can thus be used to generate 3′-ends of both acceptor and donor fragment, whereupon the acceptor 3′-end has to be further dephosphorylated using T4 polynucleotide kinase (PNK), which has 3′-phosphatase activity [31, 75, 84, 85]. Hammerhead ribozymes located 5′ to the acceptor fragment generate the correct 5′-hydroxyl end, whereas the 5′ donor end generated by a hammerhead ribozyme has to be phosphorylated by T4 PNK. To generate two samples, in which one segment is labeled and the second one is unlabeled, two unlabeled and two labeled transcription reactions must be performed. In addition, the use of T4 PNK is an additional costly step, which also requires an additional purification.

In the last 15 years, several groups came up with elegant approaches to improve the efficiency and practicability of segmental isotope labeling of RNA. In the following sections, we will present their principles and discuss their advantages and disadvantages.

7.5.2 Previous Methods

Crothers and co-workers showed that sequence-specific RNase H cleavage of an unlabeled and a labeled RNA can be followed by direct cross re-ligation of a labeled with an unlabeled fragment using T4 DNA ligase (see Fig. 7.6a) [86]. This method allows for multiple segmental isotope labeling and requires only one labeled and one unlabeled transcription reaction. However, this method has 5′ sequence requirements in that the 5′-end has to be a good transcription start site. Therefore, the final yield depends on the 5′-end. Furthermore, the method suffers from 3′-inhomogeneities. Because the four termini of the two fragments to be ligated are not correctly protected ligation can only be performed using splinted T4 DNA ligation as non-splinted ligation would lead to undesired side-products (self-ligation, ligation in the wrong sequential order, multimerization, circularization).

Lukavsky and co-workers used a plasmid encoding the 3′ donor fragment followed by a hammerhead ribozyme, which is connected by a flexible linker to a second hammerhead ribozyme preceding the 5′ acceptor fragment yielding a terminal 3′-hydroxyl after transcription (see Fig. 7.6b) [73, 87, 88]. If the transcription reaction is primed with GMP, both fragments are correctly protected for ligation with T4 RNA ligase (or T4 DNA ligase), the 5′ acceptor fragment being protected in both 5′- and 3′-ends with hydroxyl groups and the 3′ donor fragment being protected by a 5′-phosphate and a 2′,3′-cyclic phosphate. An additional advantage is that this approach requires only one labeled and one unlabeled transcription reaction. The drawbacks of this technique are that a G is present at the 3′ of the ligation site and a U at the penultimate position of the RNA. Furthermore, transcription can potentially generate an inhomogenous 3′-end of the acceptor fragment leading to possible incorporation of additional nucleotides at the site of ligation, especially when using RNA ligase. Finally, this technique does not allow for multiple segmental isotope labeling. It is therefore not possible to label a fragment within an RNA.

Wijmenga and co-workers introduced selectively deuterated uridine residues into the central position of a 20 kDa RNA (see Fig. 7.6c) [89]. The three fragments (two unlabeled and one transcribed with selectively deuterated, uniformly $^{13}C/^{15}N$-labeled UTP and otherwise unlabeled NTPs) were transcribed separately. Subsequently, the three fragments were either ligated in a two-step protocol first with T4 DNA ligase followed by T4 RNA ligation or by a one-step protocol using only T4 RNA ligase. The ligation sites have to be designed in such a way that the three fragments fold into the target or target-like structure for preventing undesired side-product formation (self-ligation, ligation in the wrong sequential order, multimerization, circularization) as the termini are not correctly protected. Furthermore, the different fragments have to be purified by PAGE to prevent or minimize the introduction of non-native nucleotides at the sites of ligation. As the three fragments are obtained from three separate transcription reactions every fragment has to start with a guanosine and the final yield strongly depends on the specific starting sequence of the several fragments. As demonstrated for this specific example, multiple segmental isotope labeling is possible. However, its practicability is limited to cases for which the fragments can fold into target-like structures and start with a guanosine. A similar approach, using both T4 DNA and RNA ligase for ligating multiple chemically synthesized RNA fragments, has allowed introducing site-specific 2′-methylseleno labels into long RNAs [21, 82].

Very recently, the site-specific introduction of a single isotopically labeled guanosine residue into a long RNA has been presented. This method consists of two simple enzymatic reactions. Firstly, a group I self-splicing intron transfers an isotopically labeled GMP into an RNA of interest generating a 5′-residue labeled fragment. This 3′-fragment is then ligated to an unlabeled 5′ fragment using T4 DNA splinted ligation [90].

A comparison of the different methods for segmental isotope labeling of RNA summarizing their strong and weak points is shown in Fig. 7.6e.

The techniques presented in this section resulted in a so far very limited use of segmental isotope labeling in NMR studies of RNA [74, 75]. Reasons are:

- the reported low yields (less than 30 nmol of segmentally labeled RNA were obtained starting from 20 ml labeled transcription reactions (see Fig. 7.6e))
- the sequence requirements at the sites of ligation (such as the need of a guanosine 3′ to the ligation site [88]) or that all [89] or some [86] fragments must start with a guanosine
- the requirement that the different fragments to be ligated need to fold into a target-like structure for efficient ligation [89]
- the tedious purification steps by PAGE
- the need to use additional enzymes such as T4 polynucleotide kinase for engineering the 3′- and 5′-termini [31, 75]
- the inability or difficultly to introduce labeled fragments within and not only at the end of an RNA

In the next section, we will discuss an approach for segmental isotope labeling of RNA that we recently published [37]. This approach allows for multiple segmental isotope labeling and does not suffer from sequence requirements. High yields are obtained within minimal preparation times paving the path for studying RNAs of biologically relevant size.

7.5.3 Fast, Efficient and Sequence-Independent Method for Flexible Multiple Isotope Labeling of RNA

7.5.3.1 Principle

This approach is based on the transcription of two full-length RNAs with identical sequence, one isotopically labeled and one unlabeled (Fig. 7.6d) [37]. The transcribed RNAs are flanked at the 5′-end by a hammerhead (HH) ribozyme in *cis* and at the 3′-end by the minimal sequence required by the *Neurospora* Varkud satellite ribozyme for cleavage in *trans* [36] (step 1 in Fig. 7.6d) or an HDV ribozyme having no sequence requirements when placed at the 3′-end [38, 39]. Both ribozymes (in *cis* or in *trans*) cleave co-transcriptionally leading to two homogenous termini, a 5′-hydroxyl and 2′/3′-cyclic phosphate for the full-length RNA. After purification, the two transcribed RNAs are cleaved site-specifically by RNase H using a guide 2′-*O*-methyl-RNA/ DNA splint yielding an acceptor fragment (5′-fragment) with two hydroxyl termini and a donor fragment (3′-fragment) with a phosphate at its 5′-end and a cyclic 2′/3′-phosphate at its 3′-end [45] (step 2 in Fig. 7.6d). After separation of the two fragments of each cleavage reaction, the subsequent two cross-religations between the labeled fragment and the unlabeled fragment using T4 RNA or T4 DNA ligase results in two segmentally labeled RNAs, in which either the 5′- or the 3′-fragment is isotopically labeled [86] (step 3 in Fig. 7.6d). Each reaction step is followed by a fast and efficient denaturing anion-exchange HPLC purification followed by an n-butanol extraction or a dialysis to remove urea and salts. For a two-piece ligation, 5–7 days are required in total, whereas 2–3 days are needed for step 1 (1 day transcription optimization, 1–2 days large scale transcription and purification), 1.5–2 days for step 2 (0.5–1 day RNase H cleavage optimization, 1 day large scale cleavage and purification) and 1.5–2 days for step 3 (0.5–1 day ligation optimization, 1 day large scale ligation and purification).

One main advantage of this method compared to others is that one obtains from only one labeled transcription reaction two homogenous fragments, which are correctly engineered at all four termini (Fig. 7.6d after step 2). This is essential to obtain the highest possible yields during the ligation step (Fig. 7.6d step 3) since no self-ligation or ligation in the wrong sequential order is possible and only the correct product can be obtained. Furthermore, there are no sequence restrictions on the identity of the fragments. This method is flexible in that it allows not only ligating two differentially labeled fragments but also allows to introduce any labeled fragment *within* an unlabeled RNA via three- or more-piece ligations. The method was applied on a 72 nts non-coding RNA containing four stem-loops, for which four different NMR samples could be obtained with one of the four stem-loops isotopically labeled at the time (see Fig. 7.7) [37, 91]. For generating these four samples, only one labeled transcription reaction was required.

7.5.3.2 Protocol

Vector Construction and Plasmid Amplification

1. DNA construct design: The construct used for co-transcriptional cleavage of the target RNA (step 1 Fig. 7.6d) is composed of a T7 promoter, an optimal transcription starting sequence (for example GGGAUC, see Milligan et al. 1987 [23]), a HH ribozyme [36], the coding sequence of the target RNA, a minimal sequence for recognizing the VS RNA in *trans* [40] and finally a BamHI restriction site for linearization. If the target RNA has a similar length like the HH ribozyme additional nucleotides (such as an MS2 stem-loop structure) can be included preceding the 5′ HH ribozyme to extend the length of the fragment containing the HH ribozyme. This allows for an optimal separation of the target RNA from the 5′HH during anion-exchange HPLC. Instead of a VS ribozyme (which has minimal sequence requirements by cutting after any nucleotide other than a cytosine) a hepatitis delta virus (HDV) ribozyme can be used at the 3′ end. The HDV ribozyme has no sequence requirements when placed at the 3′ end. To check that both HH and VS ribozyme are predicted to fold properly M-fold [92] should be used to check the secondary structure of the entire transcribed RNA.
2. Synthesis of insert by PCR: The full insert (for example 300 nts) is obtained by three consecutive PCR reactions. A first PCR reaction is performed using only one overlapping primer pair and no template. After purification, this first PCR product is used as template for a second PCR reaction using primers extending on both ends. After a second purification, this extended primer PCR is repeated once by again using the product of the previous reaction as template for the next PCR to finally obtain the insert sequence.

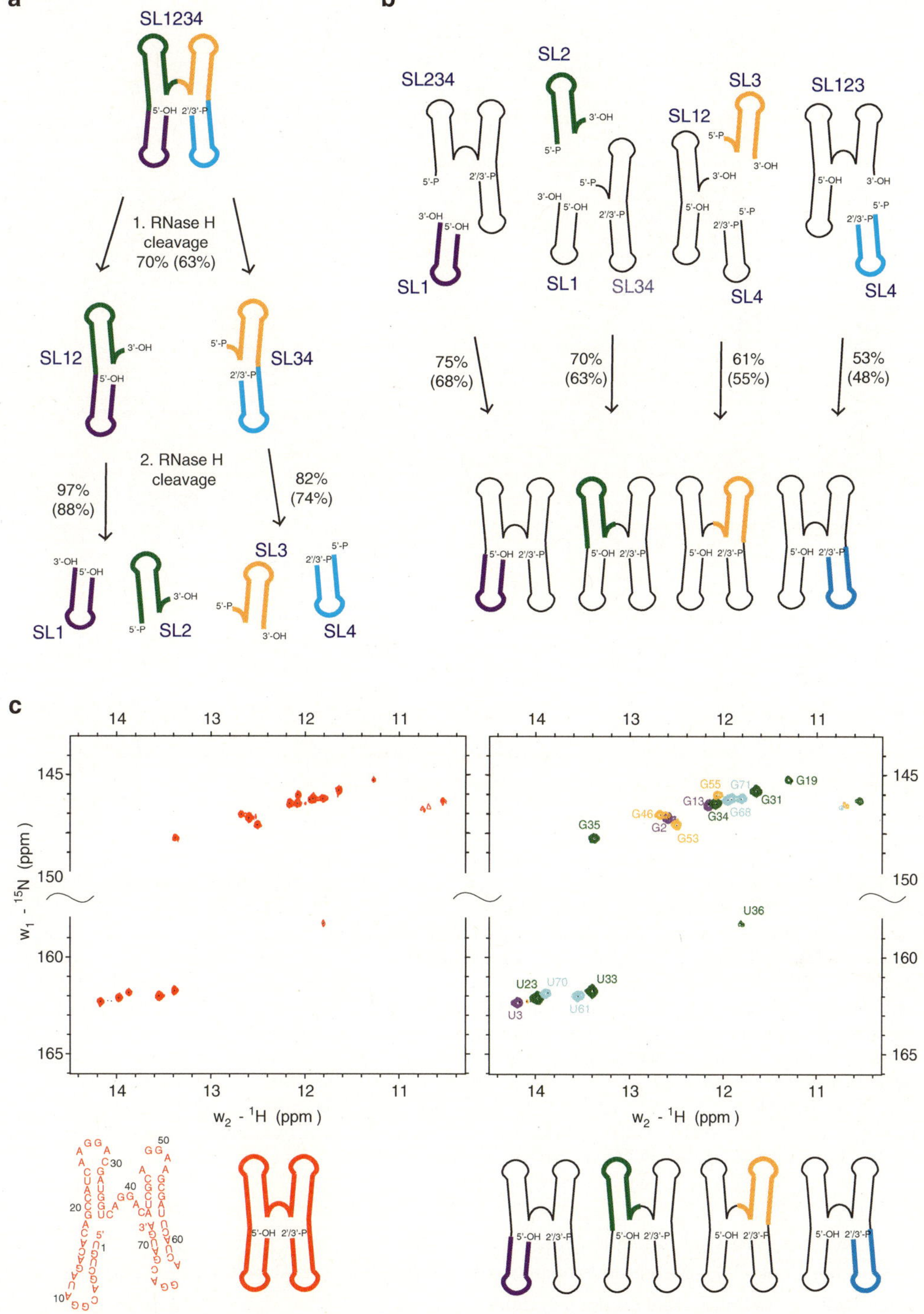

Fig. 7.7 Principle, reaction efficiencies and NMR evidence for isotope labeling of each stem-loop of the RsmZ RNA separately. (**a**) Sequence-specific RNase H cleavages to obtain all four isotopically labeled stem-loop fragments. The yields of the cleavage reactions before HPLC purification are indicated, the values in brackets are expressing the yield after purification. The different stem-loops are colored (SL1: *magenta*, SL2: *green*, SL3: *orange*, SL4: *cyan*). (**b**) Splinted T4 DNA ligase mediated ligations of isotope labeled (*in color*) and unlabeled (*in black*) fragments. The unlabeled fragments were obtained in a similar way as the labeled fragments. (**c**) NMR evidence for the successful segmental isotope labeling of each stem-loop separately. ¹H-¹⁵N-HSQC NMR spectrum of the uniformly ¹⁵N-labeled RsmZ RNA (*left*) and overlay of the ¹H-¹⁵N-HSQC NMR spectra of the four segmentally labeled RsmZ RNAs with each stem-loop labeled separately (*right*). The spectra were recorded on a Bruker 600 MHz spectrometer at 10°C (Reprinted with permission from Duss et al. [37])

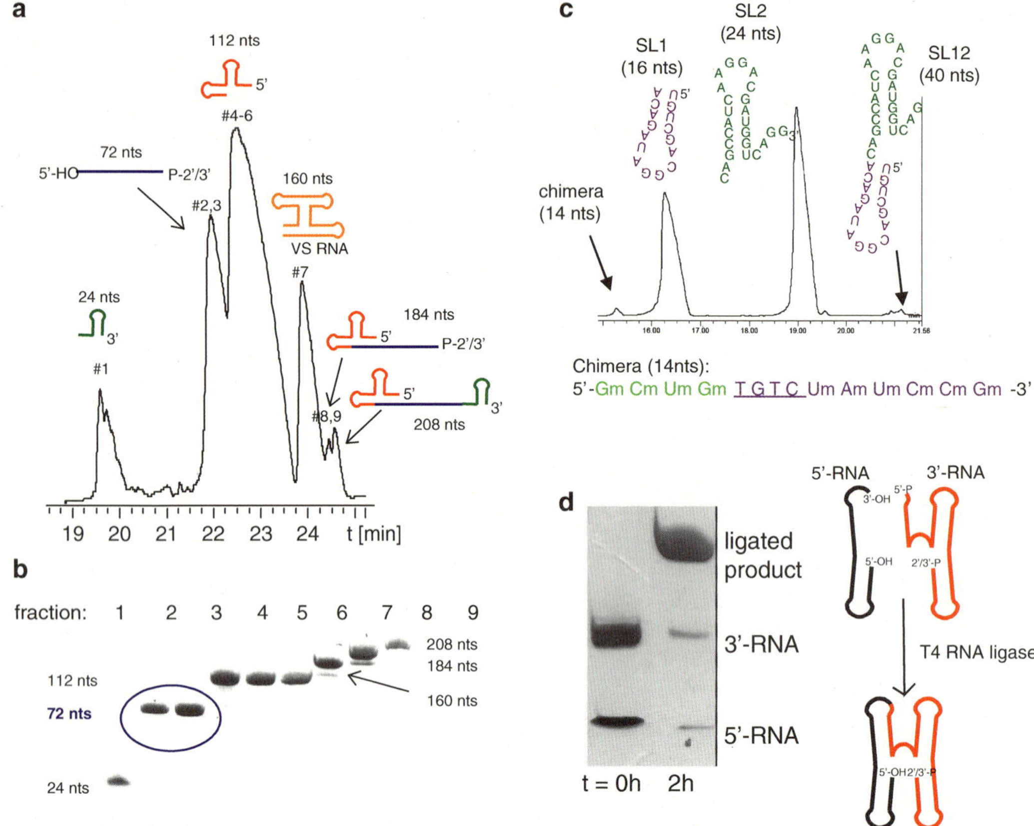

Fig. 7.8 **Protocol for multiple segmental isotope labeling of RNA.** (**a**, **b**) Co-transcriptional ribozyme cleavage and its purification. (**a**) Denaturing anion-exchange HPLC profile of a 10 ml transcription mix (which corresponds to 200 nmol of a 72 nts RNA product after purification) and (**b**) analytical 16% denaturing PAGE gel of the corresponding elution fractions. The different fragments obtained by co-transcriptional ribozyme cleavage are shown on the *top* of their corresponding peak (*blue*: target 72nts RNA, *red*: hammerhead ribozyme, *green*: 24 nts VS stem-loop sequence required for VS ribozyme cleavage in *trans*, *orange*: VS ribozyme). (**c**) Preparative scale (120 nmol) denaturing anion-exchange HPLC profile of fragments obtained by site-specific RNase H cleavage of a 40 nts RNA (SL12) to obtain SL1 (16 nts) and SL2 (24 nts) using only 5% chimera (14 nts). The different fragments obtained by RNase H cleavage are shown on the *top* of their corresponding peak. The chimera used for this cleavage is shown on the *bottom*. Its nucleotides are *colored* according to the stem-loops they are hybridizing to. The nucleotides in the chimeras are either DNA (*underlined*) or 2′-*O*-methyl-RNA (Am, Cm, Gm, Um). (**d**) Analytical 16% denaturing PAGE gel of a non-splinted ligation reaction of a 29 nts RNA (5′-RNA) and a 43 nts RNA (3′-RNA) using T4 RNA ligase. In the ligation scheme the unlabeled RNA is in *black* and the labeled RNA in *red* (Adapted with permission from Duss et al. [37])

3. The insert is double digested and purified.
4. The purified double digested insert is ligated with a doubly digested pUC19 plasmid and subsequently transformed into appropriate competent cells.
5. After sequencing, the plasmid is amplified by performing a large plasmid purification to obtain typically 5–8 mg of DNA (QIAGEN plasmid Giga kit).
6. The plasmid is linearized using 0.2 U BamHI/μg of DNA doing an overnight digestion. No further purification is required for transcription.
7. The DNA sequence coding for the VS ribozyme RNA [40] is obtained with the same extended primer PCR as described above, is also cloned into a pUC19 vector, amplified using a large plasmid purification and finally linearized.

RNA Purification

After every reaction step (transcription, RNase H cleavage and T4 RNA or DNA ligation; see below) an RNA purification is required (see Fig. 7.8).

1. The purification is performed by anion-exchange chromatography on a preparative Dionex DNAPac PA-100 column (22 × 250 mm) at 85°C. Flow rate: 20 ml/min; eluent A: 12.5 mM Tris–HCl (pH = 8.0), 6 M urea; eluent B: 12.5 mM Tris–HCl (pH = 8.0), 0.5 M NaClO$_4$, 6 M urea; detection at 260 nm; 30–75% B gradient within 18 min.
2. Fractions containing the purified RNA are determined by 16% urea acrylamide gels
3. Fractions are liberated from urea and desalted by dialysis against water or by n-butanol extraction of the aqueous phase until RNA precipitation [54]. The RNA precipitate is redissolved into 1 ml of water and precipitated with a 30–50 ml of n-butanol followed by centrifugation. The last step is repeated twice more. The final precipitate is resuspended in a few hundred microliters of water and freeze-dried overnight. The lyophilized RNA is dissolved into an appropriate buffer.

RNA Transcription and Co-transcriptional Ribozyme Cleavage (Step 1 Fig. 7.6d)

1. Transcription yields and ribozyme cleavage efficiencies are optimized on 40 µl small scale reactions with changing concentrations of MgCl$_2$, plasmid DNA, NTPs and T7 polymerase and testing the influence of the addition of pyrophosphatase and/or GMP.
2. The best condition is scaled-up to a large scale reaction of 5–20 ml. A typical reaction contains 42.5 mM MgCl$_2$, 4.5 mM of each NTP, 33 ng/µl linearized plasmid, 10 µM separately transcribed VS RNA and 1.7 µM *in-house* produced T7 Polymerase [32] in a transcription buffer containing 40 mM Tris–HCl pH = 8.0, 1 mM spermidine, 0.01% Triton X-100 and 5 mM DTT. After 4–6 h of transcription it is usually sufficient to heat the reaction mix to 65°C for 15 min to complete the ribozyme cleavage. However, it might be necessary to perform several cycles of thermal cycling if the ribozyme cleavage efficiencies are unsatisfactory. Finally, the reaction is stopped with the addition of 100 mM EDTA pH = 8.0.
3. After filtration of the RNA using a 0.22 µm filter the RNA is purified by anion-exchange HPLC followed by n-butanol extraction (see Fig. 7.8a, b).

Note: The labeled NTPs are commercially available but can also be prepared *in-house* according to the protocol from Batey et al. [28].

Sequence-Specific RNase H Cleavage (Step 2 Fig. 7.6d)

1. Sequence-specific RNase H cleavage is performed by annealing a 2′-*O*-methyl-RNA/DNA chimera to the site of ligation [93]. An example of a chimera is shown in Fig. 7.8c.
2. The best conditions for cleavage are determined by small scale reactions (typically 500 pmol RNA in 15 ul reaction volume) mainly optimizing the RNase H enzyme concentration (NEB, or in-house produced [49]) and the ratio between the RNA and the 2′-*O*-methyl-RNA/DNA chimera. The reaction temperature and the reaction time do usually not have a big influence on the cleavage specificities. Most reactions are conducted for 1 h at 37°C. However, some reactions require cleavage at 4°C to minimize unspecific cleavage. Most RNase H reactions are performed using only 5% stoichiometric amount of 2′-*O*-methyl-RNA/ DNA chimera. However, some reactions are less sensitive to unspecific cleavage when using stoichiometric amounts of chimera.
3. The best conditions are scaled up to a large scale reaction (20–200 nmol). The only parameter that cannot be scaled up is the RNase H enzyme concentration that has to be down-scaled ten times compared to the small scale reaction to prevent potential unspecific cleavage. A typical reaction to cleave

200 nmol of RNA is performed in 6 ml volume containing 33 μM RNA, 1.65 μM chimera, 80 nM *in-house* produced RNase H in 50 mM Tris–HCl pH = 7.5, 100 mM NaCl and 10 mM MgCl$_2$.

4. The reactions are directly loaded onto the anion-exchange HPLC followed by n-butanol extraction and lyophilization (see Fig. 7.8c).

RNA Ligation Using T4 RNA and DNA Ligase (Step 3 Fig. 7.6d)

1. Non-splinted T4 RNA based ligations are first performed on small scale reactions (typically 400 pmol RNA fragments in 10 ul reaction volume) mainly optimizing the T4 RNA enzyme concentration (NEB) and testing the addition of BSA (see Fig. 7.8d).
2. Splinted T4 DNA based small scale ligation reactions (typically 200 pmol RNA fragments in 20 ul reaction volume) are performed by optimizing the T4 DNA enzyme concentration (NEB, fermentas or *in-house*), the reaction time, the reaction temperature and testing the influence of PEG-4000. The DNA splints, which are added in a 1–1.2-fold excess in respect to the RNA fragments, are usually annealed to the RNA fragments prior to ligation. However, we found that depending on the secondary structure of the RNA and the DNA splints annealing is not required.
3. The best reaction conditions are scaled up for the large scale reactions. A typical large scale ligation reaction (e.g. 100 nmol of RNA fragments) using T4 RNA ligase is 40 μM in both RNA fragments, 1× in NEB ligation buffer (50 mM Tris–HCl pH = 7.8, 1 mM ATP, 10 mM MgCl$_2$, 10 mM DTT), 1× in BSA using 5U T4 RNA ligase per nmol of RNA to be ligated. The reaction is performed for 2 h at 37°C.
4. A typical large scale ligation reaction (e.g. 100 nmol of RNA fragments) using T4 DNA ligase is 10 μM in RNA fragments, 15 μM in DNA splint oligo, 10% in PEG-4000, 40 mM Tris–HCl pH = 7.8, 0.5 mM ATP, 10 mM MgCl$_2$, 10 mM DTT, 50U T4 DNA ligase (fermentas) per nmol of RNA to be ligated or 2 μM final concentration of *in-house* produced T4 DNA ligase. The reaction is performed for 2–6 h at 37°C.
5. The reactions are subjected to HPLC purification followed by n-butanol extraction and lyophilization.

NMR Spectroscopy

The lyophilized RNAs are typically dissolved into 250 ul Buffer containing 10 mM sodium phosphate at pH = 6.0 containing 10% ^{2}H$_2$O with RNA concentrations of 40–500 μM.

7.6 Conclusions and Outlook

Although almost half of the RNA structures deposited in the PDB database were determined by NMR spectroscopy only ten of them are >20 kDa and only 4 >30 kDa (April 2012). The limitations in isotope labeling of RNA are a main factor that prevented solving more RNA structures of larger molecular weight. The severe spectral overlap of RNA resonances and the strong relaxation of the sugar or the H5-H6 protons make RNA structure determination of larger size difficult or impossible. Recent improvements in strategies for selective deuteration/protonation of nucleotides that can be incorporated into RNAs have helped to reduce the spectral overlap but also to decrease broadening of the NMR resonances due to proton-proton relaxation. On the other hand, segmental isotope labeling of RNA is decreasing the spectral overlap providing the basis for studying RNAs of larger size. Specially, isotope labeling of RNA will gain in importance in verifying if a small RNA domain has the same structure in isolation as in the context of the larger biologically relevant RNA. Using a modular approach, it will allow in future to target structures of large RNAs or protein-RNA complexes. Initially, the structures of small RNA domains will be solved by conventional NMR methods or X-ray crystallography. Then, the chemical shifts of the small domains will be compared to the chemical shifts of the domains in context of the large RNA or protein-RNA complexes using segmental isotope labeling of the large RNA. The small RNA domains with identical chemical shifts will then be taken as building blocks for the large RNA. Finally, the relative positions of the different domains with respect to each other will be determined using residual couplings (RDCs), paramagnetic relaxation enhancement (PRE) and other non NMR methods. As the measurement of RDCs and PREs relies on the observation of non-overlapped, well-separated resonances, both segmental isotope labeling of RNA and the incorporation of selectively deuterated NTPs will be necessary to study larger RNAs. Using segmental isotope labeling of RNA in combination with measurements of RDCs and measurements of long-range distances by electron paramagnetic resonance, allowed us to solve the solution structure of a 70 kDa tetra-molecular complex containing a 72 nucleotides RNA bound to three protein dimers (Duss and Allain, unpublished results). This would not have been possible without a highly efficient method for multiple segmental isotope labeling of RNA [37]. We expect that more structures in solution of high molecular weight RNA and protein-RNA complexes will be determined in the future following the same approach.

Acknowledgements   We thank Christophe Maris for setting up the HPLC and Christine von Schroetter for the production of isotopically labeled NTPs. This work was supported by the SNF-NCCR Iso-lab, the SNF grant Nr. 3100A0-118118 (to F.A.) and the grant from HFSP [RGP0024/2008] (to P.J.L.).

References

1. Sharp PA (2009) The centrality of RNA. Cell 136:577–580
2. Mattick JS (2011) The central role of RNA in human development and cognition. FEBS Lett 585:1600–1616
3. Varani G, Aboulela F, Allain FHT (1996) NMR investigation of RNA structure. Prog Nucl Magn Reson Spectrosc 29:51–127
4. Wijmenga SS, van Buuren BNM (1998) The use of NMR methods for conformational studies of nucleic acids. Prog Nucl Magn Reson Spectrosc 32:287–387
5. Dominguez C, Schubert M, Duss O, Ravindranathan S, Allain FHT (2011) Structure determination and dynamics of protein-RNA complexes by NMR spectroscopy. Prog Nucl Magn Reson Spectrosc 58:1–61
6. Lukavsky PJ, Puglisi JD (2005) Structure determination of large biological RNAs. Methods Enzymol 394:399–416
7. Tzakos AG, Grace CRR, Lukavsky PJ, Riek R (2006) NMR techniques for very large proteins and RNAs in solution. Annu Rev Biophys Biomol Struct 35:319–342
8. Zuo XB, Wang JB, Foster TR, Schwieters CD, Tiede DM, Butcher SE, Wang YX (2008) Global molecular structure and interfaces: refining an RNA: RNA complex structure using solution X-ray scattering data. J Am Chem Soc 130:3292–3293
9. Hudson BP, Martinez-Yamout MA, Dyson HJ, Wright PE (2004) Recognition of the mRNA AU-rich element by the zinc finger domain of TIS11d. Nat Struct Mol Biol 11:257–264
10. Perez-Canadillas JM (2006) Grabbing the message: structural basis of mRNA 3′UTR recognition by Hrp1. EMBO J 25:3167–3178
11. Ohtsuki T, Vinayak R, Watanabe Y, Kita K, Kawai G, Watanabe K (1996) Automated chemical synthesis of biologically active tRNA having a sequence corresponding to Ascaris suum mitochondrial tRNA(Met) toward NMR measurements. J Biochem 120:1070–1073
12. Scaringe SA, Wincott FE, Caruthers MH (1998) Novel RNA synthesis method using 5′-O-silyl-2′-O-orthoester protecting groups. J Am Chem Soc 120:11820–11821
13. Pitsch S, Weiss PA (2002) Chemical synthesis of RNA sequences with 2′-O-[(triisopropylsilyl)oxy]methyl-protected ribonucleoside phosphoramidites. Curr Protoc Nucleic Acid Chem Chapter 3:Unit 3.8
14. Ramos A, Varani G (1998) A new method to detect long-range protein-RNA contacts: NMR detection of electron-proton relaxation induced by nitroxide spin-labeled RNA. J Am Chem Soc 120:10992–10993
15. Wenter P, Reymond L, Auweter SD, Allain FH, Pitsch S (2006) Short, synthetic and selectively 13C-labeled RNA sequences for the NMR structure determination of protein-RNA complexes. Nucleic Acids Res 34:e79

16. Oberstrass FC, Auweter SD, Erat M, Hargous Y, Henning A, Wenter P, Reymond L, Amir-Ahmady B, Pitsch S, Black DL, Allain FH (2005) Structure of PTB bound to RNA: specific binding and implications for splicing regulation. Science 309:2054–2057

17. Auweter SD, Fasan R, Reymond L, Underwood JG, Black DL, Pitsch S, Allain FH (2006) Molecular basis of RNA recognition by the human alternative splicing factor Fox-1. EMBO J 25:163–173

18. Oberstrass FC, Lee A, Stefl R, Janis M, Chanfreau G, Allain FH (2006) Shape-specific recognition in the structure of the Vts1p SAM domain with RNA. Nat Struct Mol Biol 13:160–167

19. Skrisovska L, Bourgeois CF, Stefl R, Grellscheid SN, Kister L, Wenter P, Elliott DJ, Stevenin J, Allain FH (2007) The testis-specific human protein RBMY recognizes RNA through a novel mode of interaction. EMBO Rep 8:372–379

20. Dominguez C, Fisette JF, Chabot B, Allain FH-T (2010) Structural basis of G-tract recognition and encaging by hnRNP F quasi RRMs. Nat Struct Mol Biol 17:853–861

21. Hobartner C, Rieder R, Kreutz C, Puffer B, Lang K, Polonskaia A, Serganov A, Micura R (2005) Syntheses of RNAs with up to 100 nucleotides containing site-specific 2′-methylseleno labels for use in X-ray crystallography. J Am Chem Soc 127:12035–12045

22. Rieder R, Lang K, Graber D, Micura R (2007) Ligand-induced folding of the adenosine deaminase A-riboswitch and implications on riboswitch translational control. Chembiochem 8:896–902

23. Milligan JF, Groebe DR, Witherell GW, Uhlenbeck OC (1987) Oligoribonucleotide synthesis using T7 RNA-polymerase and synthetic DNA templates. Nucleic Acids Res 15:8783–8798

24. Gurevich VV, Pokrovskaya ID, Obukhova TA, Zozulya SA (1991) Preparative in vitro messenger-RNA synthesis using Sp6 and T7 RNA-polymerases. Anal Biochem 195:207–213

25. Pokrovskaya ID, Gurevich VV (1994) In-vitro transcription – preparative RNA yields in analytical scale reactions. Anal Biochem 220:420–423

26. Price SR, Ito N, Oubridge C, Avis JM, Nagai K (1995) Crystallization of RNA-protein complexes.1. Methods for the large-scale preparation of RNA suitable for crystallographic studies. J Mol Biol 249:398–408

27. Nikonowicz EP, Sirr A, Legault P, Jucker FM, Baer LM, Pardi A (1992) Preparation of C-13 and N-15 labeled RNAs for heteronuclear multidimensional NMR-studies. Nucleic Acids Res 20:4507–4513

28. Batey RT, Battiste JL, Williamson JR (1995) Preparation of isotopically enriched RNAs for heteronuclear NMR. Methods Enzymol 261:300–322

29. Scott LG, Tolbert TJ, Williamson JR (2000) Preparation of specifically H-2- and C-13-labeled ribonucleotides. Methods Enzymol 317:18–38

30. Cromsigt J, Schleucher J, Gustafsson T, Kihlberg J, Wijmenga S (2002) Preparation of partially H-2/C-13-labelled RNA for NMR studies. Stereo-specific deuteration of the H5″ in nucleotides. Nucleic Acids Res 30:1639–1645

31. Lu K, Miyazaki Y, Summers MF (2010) Isotope labeling strategies for NMR studies of RNA. J Biomol NMR 46:113–125

32. Price RP, Oubridge C, Varani G, Nagai K (1998) Preparation of RNA: protein complexes for X-ray cristallography and NMR. In: Smith CWJ (ed) RNA-protein interactions: a practical approach. Oxford University Press, Oxford, pp 37–74

33. Gallo S, Furler M, Sigel RKO (2005) In vitro transcription and purification of RNAs of different size. Chimia 59:812–816

34. Pleiss JA, Derrick ML, Uhlenbeck OC (1998) T7 RNA polymerase produces 5′ end heterogeneity during in vitro transcription from certain templates. RNA 4:1313–1317

35. Dayie KT (2008) Key labeling technologies to tackle sizeable problems in RNA structural biology. Int J Mol Sci 9:1214–1240

36. Ferre-D'Amare AR, Doudna JA (1996) Use of cis- and trans-ribozymes to remove 5′ and 3′ heterogeneities from milligrams of in vitro transcribed RNA. Nucleic Acids Res 24:977–978

37. Duss O, Maris C, von Schroetter C, Allain FHT (2010) A fast, efficient and sequence-independent method for flexible multiple segmental isotope labeling of RNA using ribozyme and RNase H cleavage. Nucleic Acids Res 38:e188

38. Schurer H, Lang K, Schuster J, Morl M (2002) A universal method to produce in vitro transcripts with homogeneous 3′ ends. Nucleic Acids Res 30:e56

39. Walker SC, Avis JM, Conn GL (2003) General plasmids for producing RNA in vitro transcripts with homogeneous ends. Nucleic Acids Res 31:e82

40. Guo HC, Collins RA (1995) Efficient trans-cleavage of a stem-loop RNA substrate by a ribozyme derived from neurospora VS RNA. EMBO J 14:368–376

41. Shields TP, Mollova E, Marie LS, Hansen MR, Pardi A (1999) High-performance liquid chromatography purification of homogenous-length RNA produced by trans cleavage with a hammerhead ribozyme. RNA 5:1259–1267

42. Santoro SW, Joyce GF (1997) A general purpose RNA-cleaving DNA enzyme. Proc Natl Acad Sci USA 94:4262–4266

43. Santoro SW, Joyce GF (1998) Mechanism and utility of an RNA-cleaving DNA enzyme. Biochemistry 37:13330–13342

44. Pyle AM, Chu VT, Jankowsky E, Boudvillain M (2000) Using DNAzymes to cut, process, and map RNA molecules for structural studies or modification. Methods Enzymol 317:140–146

45. Inoue H, Hayase Y, Iwai S, Ohtsuka E (1987) Sequence-dependent hydrolysis of RNA using modified oligonucleotide splints and RNase H. FEBS Lett 215:327–330

46. Lapham J, Crothers DM (1996) RNase H cleavage for processing of in vitro transcribed RNA for NMR studies and RNA ligation. RNA 2:289–296

47. Kao C, Zheng M, Rudisser S (1999) A simple and efficient method to reduce nontemplated nucleotide addition at the 3′ terminus of RNAs transcribed by T7 RNA polymerase. RNA 5:1268–1272

48. Ponchon L, Dardel F (2007) Recombinant RNA technology: the tRNA scaffold. Nat Methods 4:571–576

49. Ponchon L, Beauvais G, Nonin-Lecomte S, Dardel F (2009) A generic protocol for the expression and purification of recombinant RNA in *Escherichia coli* using a tRNA scaffold. Nat Protoc 4:947–959

50. Kim I, Mckenna SA, Puglisi EV, Puglisi JD (2007) Rapid purification of RNAs using fast performance liquid chromatography (FPLC). RNA 13:289–294

51. Lukavsky PJ, Puglisi JD (2004) Large-scale preparation and purification of polyacrylamide-free RNA oligonucleotides. RNA 10:889–893

52. Anderson AC, Scaringe SA, Earp BE, Frederick CA (1996) HPLC purification of RNA for crystallography and NMR. RNA 2:110–117

53. Easton LE, Shibata Y, Lukavsky PJ (2010) Rapid, nondenaturing RNA purification using weak anion-exchange fast performance liquid chromatography. RNA 16:647–653

54. Cathala G, Brunel C (1990) Use of n-butanol for efficient recovery of minute amounts of small RNA fragments and branched nucleotides from dilute solutions. Nucleic Acids Res 18:201

55. McKenna SA, Kim I, Puglisi EV, Lindhout DA, Aitken CE, Marshall RA, Puglisi JD (2007) Purification and characterization of transcribed RNAs using gel filtration chromatography. Nat Protoc 2:3270–3277

56. Murray JB, Collier AK, Arnold JRP (1994) A general purification procedure for chemically synthesized oligoribonucleotides. Anal Biochem 218:177–184

57. Cheong HK, Hwang E, Lee C, Choi BS, Cheong C (2004) Rapid preparation of RNA samples for NMR spectroscopy and X-ray crystallography. Nucleic Acids Res 32:e84

58. Kieft JS, Batey RT (2004) A general method for rapid and nondenaturing purification of RNAs. RNA 10:988–995

59. Batey RT, Kieft JS (2007) Improved native affinity purification of RNA. RNA 13:1384–1389

60. Walker SC, Scott FH, Srisawat C, Engelke DR (2008) RNA affinity tags for the rapid purification and investigation of RNAs and RNA-protein complexes. Methods Mol Biol 488:23–40

61. Allain FH, Varani G (1997) How accurately and precisely can RNA structure be determined by NMR? J Mol Biol 267:338–351

62. Tugarinov V, Muhandiram R, Ayed A, Kay LE (2002) Four-dimensional NMR spectroscopy of a 723-residue protein: chemical shift assignments and secondary structure of malate synthase g. J Am Chem Soc 124:10025–10035

63. Marino JP, Diener JL, Moore PB, Griesinger C (1997) Multiple-quantum coherence dramatically enhances the sensitivity of CH and CH2 correlations in uniformly C-13-labeled RNA. J Am Chem Soc 119:7361–7366

64. Lukavsky PJ, Puglisi JD (2001) RNAPack: an integrated NMR approach to RNA structure determination. Methods 25:316–332

65. Tolbert TJ, Williamson JR (1997) Preparation of specifically deuterated and C-13-labeled RNA for NMR studies using enzymatic synthesis. J Am Chem Soc 119:12100–12108

66. Davis JH, Tonelli M, Scott LG, Jaeger L, Williamson JR, Butcher SE (2006) RNA helical packing in solution: NMR structure of a 30 kDa GAAA tetraloop-receptor complex (vol 351, pg 371, 2005). J Mol Biol 360:742

67. Varani G, Tinoco I (1991) RNA structure and NMR-spectroscopy. Q Rev Biophys 24:479–532

68. Nikonowicz EP (2001) Preparation and use of H-2-labeled RNA oligonucleotides in nuclear magnetic resonance studies. Nucl Magn Reson Biol Macromol Pt A 338:320–341

69. Bullock SL, Ringel I, Ish-Horowicz D, Lukavsky PJ (2010) A′-form RNA helices are required for cytoplasmic mRNA transport in Drosophila. Nat Struct Mol Biol 17:703–709

70. D'Souza V, Dey A, Habib D, Summers MF (2004) NMR structure of the 101-nucleotide core encapsidation signal of the Moloney murine leukemia virus. J Mol Biol 337:427–442

71. D'Souza V, Summers MF (2004) Structural basis for packaging the dimeric genome of Moloney murine leukaemia virus. Nature 431:586–590

72. Miyazaki Y, Irobalieva RN, Tolbert BS, Smalls-Mantey A, Iyalla K, Loeliger K, D'Souza V, Khant H, Schmid MF, Garcia EL, Telesnitsky A, Chiu W, Summers MF (2010) Structure of a conserved retroviral RNA packaging element by NMR spectroscopy and cryo-electron tomography. J Mol Biol 404:751–772

73. Kim I, Lukavsky PJ, Puglisi JD (2002) NMR study of 100 kDa HCV IRES RNA using segmental isotope labeling. J Am Chem Soc 124:9338–9339

74. Lukavsky PJ, Kim I, Otto GA, Puglisi JD (2003) Structure of HCV IRES domain II determined by NMR. Nat Struct Biol 10:1033–1038

75. Lu K, Heng X, Garyu L, Monti S, Garcia EL, Kharytonchyk S, Dorjsuren B, Kulandaivel G, Jones S, Hiremath A, Divakaruni SS, LaCotti C, Barton S, Tummillo D, Hosic A, Edme K, Albrecht S, Telesnitsky A, Summers MF (2011) NMR detection of structures in the HIV-1 5′-leader RNA that regulate genome packaging. Science 334:242–245
76. Serganov A, Keiper S, Malinina L, Tereshko V, Skripkin E, Hobartner C, Polonskaia A, Phan AT, Wombacher R, Micura R, Dauter Z, Jaschke A, Patel DJ (2005) Structural basis for Diels-Alder ribozyme-catalyzed carbon-carbon bond formation. Nat Struct Mol Biol 12:218–224
77. Akiyama BM, Stone MD (2009) Assembly of complex RNAs by splinted ligation. Methods Enzymol 469:27–46
78. Moore MJ, Query CC (2000) Joining of RNAs by splinted ligation. Methods Enzymol 317:109–123
79. Frilander MJ, Turunen JJ (2005) RNA ligation using T4 DNA ligase. In: Hartmann RK, Bindereif A, Schön A, Westhof E (eds) Handbook of RNA biochemistry. WILEY-VCH Verlag GmBH & Co, Weinheim, pp 36–52
80. Persson T, Willkomm DK, Hartmann RK (2005) T4 RNA ligase. In: Hartmann RK, Bindereif A, Schön A, Westhof E (eds) Handbook of RNA biochemistry. WILEY-VCH Verlag GmBH & Co, Weinheim, pp 53–74
81. Purtha WE, Coppins RL, Smalley MK, Silverman SK (2005) General deoxyribozyme-catalyzed synthesis of native 3′-5′ RNA linkages. J Am Chem Soc 127:13124–13125
82. Lang K, Micura R (2008) The preparation of site-specifically modified riboswitch domains as an example for enzymatic ligation of chemically synthesized RNA fragments. Nat Protoc 3:1457–1466
83. Stark MR, Pleiss JA, Deras M, Scaringe SA, Rader SD (2006) An RNA ligase-mediated method for the efficient creation of large, synthetic RNAs. RNA 12:2014–2019
84. Ohtsuki T, Kawai G, Watanabe Y, Kita K, Nishikawa K, Watanabe K (1996) Preparation of biologically active Ascaris suum mitochondrial tRNAMet with a TV-replacement loop by ligation of chemically synthesized RNA fragments. Nucleic Acids Res 24:662–667
85. Ohtsuki T, Kawai G, Watanabe K (1998) Stable isotope-edited NMR analysis of Ascaris suum mitochondrial tRNAMet having a TV-replacement loop. J Biochem 124:28–34
86. Xu J, Lapham J, Crothers DM (1996) Determining RNA solution structure by segmental isotopic labeling and NMR: application to Caenorhabditis elegans spliced leader RNA 1. Proc Natl Acad Sci USA 93:44–48
87. Tzakos AG, Easton LE, Lukavsky PJ (2006) Complementary segmental labeling of large RNAs: economic preparation and simplified NMR spectra for measurement of more RDCs. J Am Chem Soc 128:13344–13345
88. Tzakos AG, Easton LE, Lukavsky PJ (2007) Preparation of large RNA oligonucleotides with complementary isotope-labeled segments for NMR structural studies. Nat Protoc 2:2139–2147
89. Nelissen FH, van Gammeren AJ, Tessari M, Girard FC, Heus HA, Wijmenga SS (2008) Multiple segmental and selective isotope labeling of large RNA for NMR structural studies. Nucleic Acids Res 36:e89
90. Kawahara I, Haruta K, Ashihara Y, Yamanaka D, Kuriyama M, Toki N, Kondo Y, Teruya K, Ishikawa J, Furuta H, Ikawa Y, Kojima C, Tanaka Y (2012) Site-specific isotope labeling of long RNA for structural and mechanistic studies. Nucleic Acids Res 40:e7
91. Schubert M, Lapouge K, Duss O, Oberstrass FC, Jelesarov I, Haas D, Allain FH (2007) Molecular basis of messenger RNA recognition by the specific bacterial repressing clamp RsmA/CsrA. Nat Struct Mol Biol 14:807–813
92. Zuker M (2003) Mfold web server for nucleic acid folding and hybridization prediction. Nucleic Acids Res 31:3406–3415
93. Hayase Y, Inoue H, Ohtsuka E (1990) Secondary structure in formylmethionine tRNA influences the site-directed cleavage of ribonuclease H using chimeric 2′-O-methyl oligodeoxyribonucleotides. Biochemistry 29:8793–8797

Part III
Metabolomics

Chapter 8
Isotope Enhanced Approaches in Metabolomics

G.A. Nagana Gowda, Narasimhamurthy Shanaiah, and Daniel Raftery

Abstract The rapidly growing area of "metabolomics," in which a large number of metabolites from body fluids, cells or tissue are detected quantitatively, in a single step, promises immense potential for a number of disciplines including early disease diagnosis, therapy monitoring, systems biology, drug discovery and nutritional science. Because of its ability to detect a large number of metabolites in intact biological samples reproducibly and quantitatively, nuclear magnetic resonance (NMR) spectroscopy has emerged as one of the most powerful analytical techniques in metabolomics. NMR spectroscopy of biological samples with isotope labeling of metabolites using nuclei such as ^{2}H, ^{13}C, ^{15}N and ^{31}P, either *in vivo* or *ex vivo,* has dramatically improved our ability to identify low concentrated metabolites and trace important metabolic pathways. Considering the somewhat limited sensitivity and high complexity of NMR spectra of biological samples, efforts have been made to increase sensitivity and selectivity through isotope labeling methods, which pave novel avenues to unravel biological complexity and understand cellular functions in health and various disease conditions. This chapter describes current developments in isotope labeling of metabolites *in vivo* as well as *ex vivo,* and their potential metabolomics applications.

8.1 Metabolomics

Metabolomics, also sometimes referred to as metabonomics or metabolic profiling, is defined as "the quantitative measurement of the dynamic multiparametric metabolic response of living systems to pathophysiological stimuli or genetic modifications" [1]. In metabolomics, a large number of small-molecule metabolites (typically less than 1,000 Da) from body fluids, cells or tissue samples are detected quantitatively in a single step, and then analyzed typically with multivariate statistical methods to yield information that is essential for systems biology, drug discovery, toxicology, food and nutrition sciences and other studies. Metabolites in biological systems represent the end products of genes, transcripts and proteins functions, and therefore the study of concentrations and fluxes of such metabolites provides extremely important information for understanding the composition and function of biochemical networks. Hence, metabolomics may lead to solutions to many important questions related to human disease diagnosis, prognosis and therapeutic development [2–8].

G.A.N. Gowda (✉) • N. Shanaiah • D. Raftery
Department of Anesthesiology and Pain Medicine, Mitochondria
and Metabolism Center, University of Washington, Seattle, WA 98109, USA
e-mail: ngowda@uw.edu

H.S. Atreya (ed.), *Isotope Labeling in Biomolecular NMR*, Advances in Experimental Medicine and Biology 992,   147
DOI 10.1007/978-94-007-4954-2_8, © Springer Science+Business Media Dordrecht 2012

The importance of metabolomics for biomarker discovery stems from the fact that a modest change in enzyme activity can cause substantial perturbations in the metabolic profile; identification of such perturbations represents a sensitive measure of the onset of many diseases. In addition, metabolomics enables the establishment of biological variation among individuals, and understanding such variations is important for achieving the goal of 'personalized medicine', which represents a new treatment procedure tailored to the individual's needs. In principle, this personalized approach could be based on the metabolic fingerprint of the patient [9]. A major advantage of the metabolomics approach over other related *'omic'* areas such as genomics, transcriptomics and proteomics is its association with high throughput measurements, both quantitatively and reproducibly, using non-invasive or minimally invasive approaches, and its minimal requirements regarding sample preparation procedures. Several hundreds of metabolites can be characterized in parallel using modern analytical techniques such as high-resolution NMR spectroscopy, mass spectrometry (MS), or even Fourier transform infrared spectroscopy (FTIR) and electrochemical detection arrays, that provide efficient methods for monitoring altered biochemistry. However, because of the wealth of information they provide, the primary analytical methods that have been widely employed in the field of metabolomics are NMR, which we focus in this chapter, and MS. After a brief review of standard NMR methods used in metabolmics, we focus on isotope enhanced methods.

8.1.1 NMR Spectroscopy in Metabolomics

NMR spectroscopy is advantageous for metabolomics studies because it requires little or no sample preparation; is rapid, non-destructive, and non-invasive; and provides highly reproducible and easily quantifiable data. Moreover, NMR spectra can be reliably assigned to specific metabolic species, based on their chemical shifts and multiplet patterns; thus, NMR provides a wealth of information on the identity and quantity of a large number of metabolites in parallel from a single experiment [7, 10–13]. Independent of molecular identity, a particular type of nuclei (^{1}H, ^{2}H, ^{13}C, ^{15}N, ^{31}P, etc.) are detected at the same sensitivity in one NMR experiment, and the absolute quantification of different metabolites can therefore be measured with a single internal or external standard. With advanced high-throughput NMR methodology, up to 200 samples can be measured within a day with the assistance of flow-injection probes and automated liquid handlers [7].

8.1.2 Biological Samples Used in Metabolomics

To date, blood serum/plasma and urine are the most studied biofluids in the area of metabolomics. This is because they both contain hundreds to thousands of detectable metabolites and can be obtained minimally or non-invasively. Blood maintains a normal homeostasis in the human body by constant regulatory mechanisms. It perfuses all living cells and hence is presumed to carry vital information on virtually every cell in the human body. Hence metabolic profiling of blood serum or plasma enables global visualization of the metabolic status. The NMR spectrum of blood serum/plasma includes both narrow lines from small-molecule metabolites and broad lines from macromolecules such as proteins and lipids. A variety of spectral editing methods are often used to suppress the macromolecular components effectively and detect the small molecules selectively. Because of the relatively low concentrations of proteins and high concentrations of low molecular weight metabolites in urine, metabolomics study of urine is relatively simple in terms of the sample preparation and the NMR experiments. However, a major problem with urine results from its high salt concentration and varying pH across a sample set. In addition, variability in concentration due to dilution effects leads to spectra with significantly different integrated intensities (although normalization can compensate for this effect).

Metabolic profiling of urine is particularly useful for studying organ toxicity and drug metabolism in animal models using a variety of drugs or toxins [14, 15]. Other fluids such as cerebrospinal fluid, bile, seminal fluid, amniotic fluid, synovial fluid, gut aspirate, and saliva have also been studied [16–20]. Studies have also involved intact tissues and its lipid and aqueous extracts [21]. Metabolic studies of cells including yeast [22], bacterium, tumor cells, and tissue spheroids have also been reported [23, 24]. Specifically, metabolic profiling of intact tissue has gained increased interest for understanding the molecular basis of diseases [25–27]. This is because, biomarkers due to pathophysiological stress are considered to be highly concentrated in the tissue due to their close association with the pathological source, such as tumors. The rich metabolic profile of tissue is thought to be particularly useful for guiding the detection of biomarkers in relatively easily accessible biofluids. The latest technological advancements in NMR have reduced the required sample quantity to a few mg such that even the biopsy tissue is sufficient to obtain good-quality NMR spectra with resolution that is often comparable to solution-state spectra. Detailed procedures to collect, store, and prepare biofluids or tissue samples for NMR analysis have been provided as guidelines for metabolomics applications [28, 29].

8.1.3 NMR Experiments Used in Metabolomics

8.1.3.1 Common NMR Experiments

A variety of NMR experiments are used to analyze complex samples such as biofluids, tissue and cells. 1D NOESY (nuclear Overhauser enhancement spectroscopy) and CPMG (Carr-Purcell-Meiboom-Gill) are the two most commonly used experiments. The 1D NOESY is particularly useful for samples with inherently narrow line shapes such as urine, while the CPMG experiment is often used for samples such as blood serum/plasma to suppress the broad background signals from macromolecules such as proteins and lipids. Since these signals are generally not of interest, the CPMG sequence [7, 30] attenuates them and improves the signal-to-noise ratio (SNR) of spectra of small-molecule metabolites. The number of spin echoes and the delay time between the pulses are critical for effective suppression of the broad signals without appreciably losing the sensitivity for the metabolites of interest. Alternatively, based on the large difference between the diffusion coefficients of small and macro molecules, their signals can be separated using diffusion ordered spectroscopy (DOSY) experiment. DOSY provides an alternative method for profiling small molecule metabolites by eliminating the signals of macromolecules [31–33].

Water signal suppression is common to most NMR experiments in metabolomics since most biological samples comprise water, abundantly. A variety of water suppression sequences including the one pulse sequence with presaturation, presaturation utilizing relaxation gradients and echoes (PURGE), excitation sculpting and the combinations of gradient and weak *rf* pulses such as the WET sequence, have been developed and used [31, 34–36]. However, the most commonly used and simplest sequence uses presaturation that involves a continuous low-power irradiation at the water resonance position (~4.8 ppm) during the relaxation delay of usually a few seconds [37]. Incorporation of the presaturation pulse into the 1D NOESY sequence is useful to obtain flat and undistorted baseline for the NMR spectrum.

8.1.3.2 Other NMR Experiments

1D ^{1}H NMR spectra obtained using common pulse sequences such as NOESY or CPMG with water suppression are generally complex with analyte signal intensities ranging over three orders of magnitude. In a typical ^{1}H NMR spectrum, most signals are crowded into a region of ~10 ppm and the resulting spectral overlap impedes identification and quantification of certain metabolites of interest. Alternative methods have been used to overcome such limitations.

Selective TOCSY

The 1D selective TOCSY (total correlation spectroscopy) methodology was explored for the analysis of metabolites in biofluids and was shown to detect metabolites quantitatively, even at concentrations much lower than those of the major components [38–40]. In a 1D selective TOCSY experiment, protons signals of interest are selectively excited by a shaped soft pulse; the magnetization is then transferred to all sequentially coupled spin system through J-couplings. This targeted approach suppresses unwanted and intense signals from other molecules significantly, and thus alleviates any dynamic-range problem and enables detection of molecules at low concentrations.

2D NMR

2D NMR experiments improve spectral resolution and information content compared to 1D NMR, which can be very helpful for identifying unknown metabolites. Extensive use of such experiments has been limited, however, due to their longer acquisition times, larger data sizes and increased complexity in data analysis. Nevertheless, due to increased interest in unraveling the complexity of biological samples, the use of two-dimensional NMR methods is on the rise [11, 31, 41–48]. Commonly used 2D NMR experiments include 2D J-resolved spectroscopy, correlation spectroscopy (COSY), total correlation spectroscopy (TOCSY), heteronuclear single quantum coherence spectroscopy (HSQC), and heteronuclear multiple bond correlation spectroscopy (HMQC). 2D J-resolved spectroscopy separates the chemical shifts and spin-spin couplings (J-coupling) along two frequency axes [46, 49]. The projection of this experiment is a simplified 1D spectrum; the simplification is achieved by removing the multiplicity associated with J-couplings and retaining only the chemical shift information.

Chromatography Resolved NMR Spectroscopy

There is a growing interest to couple NMR with high pressure liquid chromatography (HPLC). This combination offers access to low-concentration metabolites and hence promises numerous possibilities for metabolomics applications. One area that has highly benefited from this approach is pharmaceutical research and development [50]. Reverse phase columns commonly used in HPLC suffer from poor chromatographic resolution and long elution times (>30 min each run). Recent developments in hydrophilic interaction liquid chromatography (HILIC) columns offer significant improvements to the separation of polar metabolites [51]. Moreover, developments in ultrahigh pressure liquid chromatography (UPLC) have enabled significant improvements in chromatographic resolution, reductions in separation time and enhancement in detection limits (by three- to five-folds) [52].

8.1.4 Advanced NMR Methods

The inherently low sensitivity of NMR poses a major challenge for identifying low concentration metabolites in complex biological samples. Numerous efforts have been focused on improving the resolution and sensitivity to alleviate the limitations of NMR. Strong magnets and cryogenic probes have significantly lowered the NMR detection limit [53]. The use of cryogenic probes can significantly reduce the level of thermal noise resulting in an increase in the signal-to-noise ratio, typically by four-fold. In addition, advancements in micro-coil probe technology has shown that small-diameter detection coils have better signal-to-noise ratios and are particularly useful for mass limiting samples [54–56]. While saddle coils are widely used in conventional NMR, metabolomics applications are shown to benefit from solenoidal micro-coils because solenoidal coil enhances sensitivity by factor of two- to three-fold [55, 56]. Using pure compounds, such coils have been demonstrated to significantly lower the sample volume as well as the mass detection limit. More recently, a variety of micro-coil probes with different sample volumes and multiple coils have been designed and constructed, which are potentially useful for profiling metabolites in biofluids with limited sample quantities [57–60].

8.2 Isotope Labeling

While the standard and advanced methods discussed above have benefited the field of metabolomics, stable isotope labeling methods provide numerous capabilities and potential to take NMR-based metabolomics to a new level. Both NMR and MS based methods use isotope labeling for applications that include metabolites flux analysis, identification and quantitation of low concentration metabolites in complex biological samples. The unique advantage of using isotope enhanced NMR for metabolomics is that it enables tracing of biochemical pathways and the derivation of metabolic fluxes accurately in animal and human subjects. It allows unraveling interconnectivity of pathways and provides vital clues to disease mechanisms. Moreover, the isotope labeling approach, *ex vivo*, enables detection of a large number of (nearly 200) metabolites with improved resolution and sensitivity, which is unprecedented from NMR point of view. The use of mixtures of metabolites containing light and heavy isotopes is a popular method currently for relative and absolute quantitation of metabolites using MS methods [61–64], while the addition of isotope labeled internal standards has been used for MS quantitation for decades. Broadly, isotope labeling methods in NMR-based metabolomics are used in two major applications. One area focuses on metabolic pathway and flux analysis, and the other provides enhancement of resolution and sensitivity for complex mixture analysis.

8.2.1 *In Vivo Isotope Labeling for Metabolic Pathway and Flux Analysis*

Metabolic pathway and flux analysis have significantly evolved as a consequence of the combination of stable isotopes and sensitive detection using advanced analytical methods. Flux analysis allows a detailed understanding of metabolic pathways, enables quantification of all intracellular metabolite levels and determination of the rates at which metabolites are produced or consumed. Although both NMR and MS methods are used in metabolic flux analysis, NMR is especially powerful as it allows both identification and quantification of metabolites in intact biological samples, and determination of positional isotopomer distributions derived from precursors enriched with stable isotopes [65–69] (Fig. 8.1).

Metabolite concentrations within a cell are controlled by regulation through homeostasis and a large number of metabolites are associated with many metabolic pathways. Hence global analysis of such metabolites does not enable a clear understanding of the biosynthesis of metabolites. For example, pyruvate can be obtained from several precursors and it is itself a precursor of many other metabolites [70–74]. Changes in pyruvate's concentration can therefore be due to any or all perturbations in such pathways. Similarly, lactate is secreted at a very high rate in cancer cells and accounts for significant fraction of the glucose consumed by these cells [73]. However, the total lactate concentration represents both the amount of glycolysis as well as amino acid oxidation. Therefore, it is not possible to quantify the contributions of different pathways based on the total concentrations of pyruvate or lactate.

The cells supplied with ^{13}C-labeled glucose enable measurements of labeled glucose consumption as well as the concentrations of the fractions of the metabolites produced from labeled glucose. Thus, for lactate, two forms – one containing ^{12}C at the methyl position and the other containing ^{13}C derived solely from the glycolysis of ^{13}C labeled glucose- can be clearly seen in the NMR spectra [75, 76]. In the same way, a number of additional labeled metabolites can be used to quantitatively measure their rate of consumption as well as the fractions of labeled metabolites produced. A number of stable isotopes such

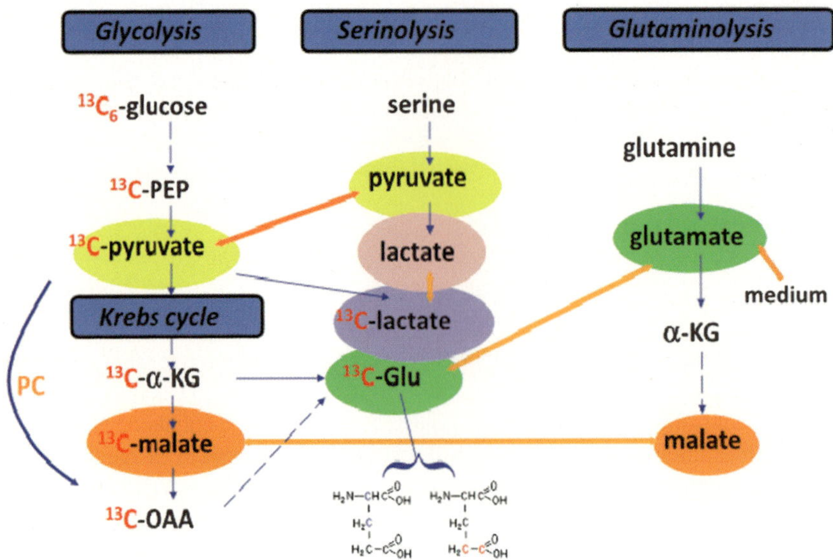

Fig. 8.1 Tracers are necessary for delineating metabolic pathways. In the scheme depicted, pyruvate, lactate, α-ketoglutarate (α-KG), malate and glutamic acid (Glu) participate in multiple pathways including glycolysis, the Krebs cycle, serinolysis, and glutaminolysis. Without labeled tracers, it is impractical to resolve the specific pathway(s) involved in their production. With the use of $^{13}C_6$-glucose, the synthesis of pyruvate, lactate, α-KG, malate, and Glu via glycolysis and the Krebs cycle can be distinguished from that via serinolysis or glutaminolysis by the ^{13}C labeling pattern of these metabolites. Further, the ^{13}C positional isotopomers of Glu (i.e. ^{13}C-2,3-Glu and ^{13}C-4,5-Glu) can be used to delineate respectively the Krebs cycle with or without pyruvate carboxylation (PC) input (Reproduced with permission from Ref. [69])

as ^2H, ^{13}C and ^{15}N can be incorporated as tracers and the formed metabolites can be detected through their isotopically enriched nuclei. Stable isotopes used in these studies are biocompatible and are easily distinguished from highly abundant natural isotopes using NMR spectroscopy; thus they facilitate distinguishing the same metabolite derived from multiple pathways, and identification and measurement of the contributions of specific pathways involved in the biosynthesis of specific metabolites.

To date, stable isotope based metabolomics methods are used for a number of applications including studies of cancer using cells, animal models and even human subjects [65–68, 77, 78]. These methods have also been applied to other higher organisms and plants [44, 79]. Cell studies are particularly important as they provide the means for understanding metabolic pathways under controlled conditions. On the other hand, studies using animal models enable understanding pathogenesis in more realistic tissues that are still under controlled conditions. Mouse models, for example, can provide reasonable (though not perfect) surrogates for human diseases. The knowledge gained from the studies of cells and animal models can then be translated to understanding diseases directly using humans [65, 67, 80, 81].

Mammalian cells depend on the catabolism of glucose and glutamine for their viability and growth [82, 83]. While it is well known that many cancer cell lines depend on high rates of glucose uptake and metabolism to maintain their viability, it has only recently been observed that high rates glutamine metabolism are also exhibited by some cancer cells. In 2007, it was shown that the high rates at which the glioma cells uptake and metabolize glutamine exceed the cells' use of glutamine for protein and nucleotide biosynthesis [74]. Subsequently, based on identification of the downstream products of ^{13}C isotope labeled glutamine using NMR, oncogenes known to contribute to the malignant transformation of the cells were tested for the ability to induce glutaminolysis [84]. Results of this study show that the transcriptional regulatory oncogene Myc is involved in the catabolism of glutamine leading to glutamine addiction of the tumor cells to provide needed energy via the TCA cycle. It is thought that Myc binds to promoters and induces expression of many key regulatory genes that are associated with the glutamine catabolism. These studies promise a number of enzymatic targets through which the growth of Myc-transformed tumor cells can be prevented.

Numerous investigations are focused on improving isotope labeled approaches for metabolomics applications. For example, many studies have used a combination of isotopes such as ^{2}H and ^{13}C to trace pathways and analyze fluxes in humans and animal models [85–89]. The advantage of such an approach is that it allows the use of multiple isotope tracers, which enable investigations of multiple pathways such as glycolysis, gluconeogenesis and tricarboxylic acid cycle, simultaneously. Investigations of pathways using non-invasive approaches, using urine samples, for example, are shown to be useful in metabolic studies of populations where sampling of blood and tissue is limited [90]. Dynamic nuclear polarization of isotope labeled nuclei is shown to provide an additional advantage of improved sensitivity of nearly four orders of magnitude and enable real-time assessment of metabolic pathways [91, 92].

For flux analysis using cancer or control cells, the cells are typically grown in media containing isotope labeled substrates such as [U-^{13}C]-glucose or [U-^{13}C]-glutamine that serve as source of tracers. For studies using mice, isotope labeled substrates are often injected through tail vein; and for studies using humans, stable isotope labeled tracers such as [U-^{13}C]-glucose, ^{2}H$_2$O or [U-^{13}C]-propionate, that are fully compatible with human experimentation are injected intravenously into recruited patients prior to surgical resection or ingested. Isotope enriched metabolites in the resulting cells or animal/human tissue are then extracted and subjected to quantitative analysis using NMR methods.

Numerous isotope edited 2D NMR experiments have been used to identify and quantify specific isotopomers. These include ^{1}H-^{13}C HSQC, HSQC-TOCSY, HCCH-TOCSY, HACACO, ^{1}H-^{31}P HSQC-TOCSY, ^{1}H-^{15}N HSQC-TOCSY and ^{1}H-^{1}H TOCSY [69, 74]. Of these, 2D HSQC experiment is the most often used and is a highly valuable method for identifying essentially all the metabolites containing isotope tracers with good resolution and sensitivity. To aid identification of metabolites, a database of over 1,000 ^{1}H and ^{13}C chemical shifts corresponding to nearly 150 metabolites obtained under identical conditions has been developed [44]. A large number of metabolites peaks in cell extracts have been identified using the combination of 2D NMR and metabolite databases (Fig. 8.2). More recently, a new pulse

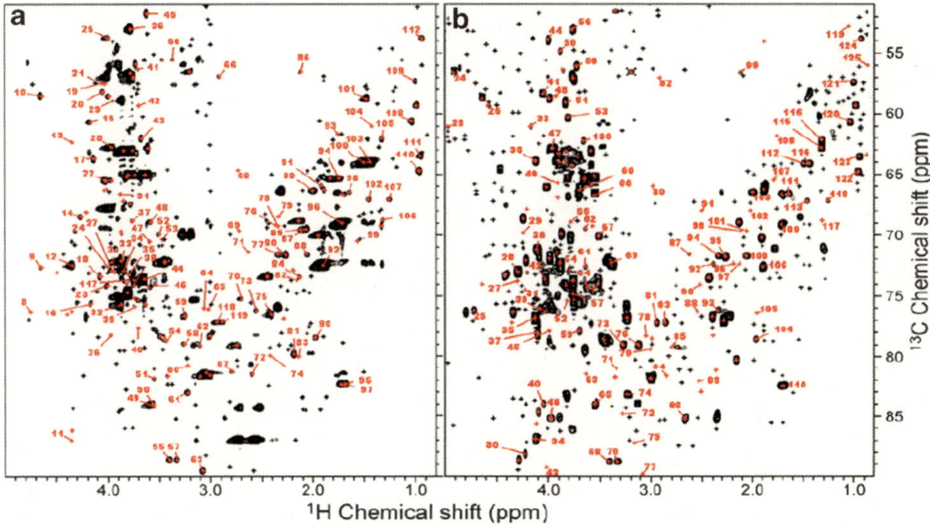

Fig. 8.2 Metabolite identification using chemical shift database. (**a**) A total of 453 peaks were detected in the *B. mori* HSQC spectrum, of which 174 had candidate matches in the database (*red*) and 119 were uniquely identified, and 279 had no candidate matches (*black*). (**b**) A total of 544 peaks were detected in extracts from T87 cultured cells, of which 192 had candidate matches in the database (*red*) and 124 were uniquely identified, and 353 had no candidate matches (*black*) (Reproduced from Ref. [44])

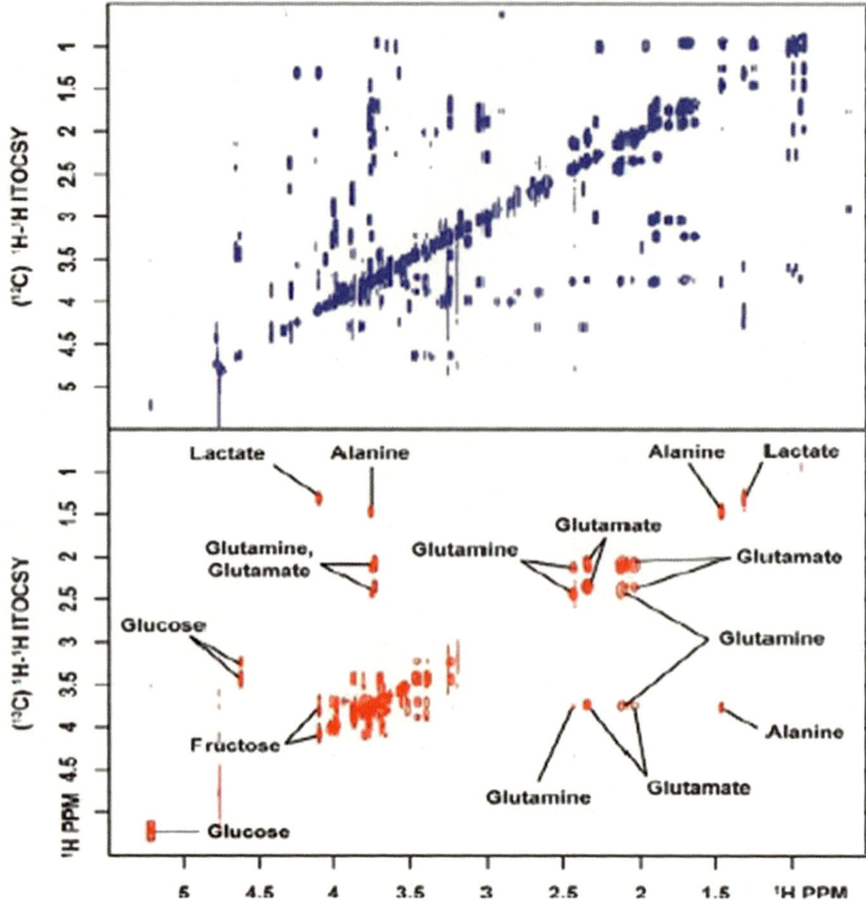

Fig. 8.3 Isotope-filtered (¹²C) and difference edited (¹³C) 2D ¹H-¹H isotope edited total correlation spectroscopy (*ITOCSY*) spectra of a synthetic mixture containing 30 unlabeled and six ¹³C-labeled metabolites. Resonance assignments for ¹³C-enriched compounds are shown (Reproduced with permission from Ref. [93])

sequence, isotope edited total correlation spectroscopy (ITOCSY), which filters two-dimensional ¹H-¹H NMR spectra from ¹²C- and ¹³C-containing molecules into separate and quantitatively equivalent spectra has been proposed for metabolic flux analysis [93]. The ITOCSY spectra of labeled and unlabeled molecules are directly comparable, and ITOCSY effectively separates signals from labeled and unlabeled molecules (Fig. 8.3).

8.2.2 *Ex Vivo Isotope Labeling for Metabolites Analysis*

The complexity and spectral overlap apparent in NMR spectra of biofluids remain significant obstacles to the identification and quantification of a large number of metabolites. Eukaryotic organisms possess more than 3,000 metabolites with very diversified molecular structures and physical/chemical properties [94]. The quantitative determination of this magnitude of compounds in a single analysis remains out of reach for current technologies. Only a small fraction of the metabolites can currently be accurately and precisely detected. The information derived from this small sampling of metabolites is insufficient

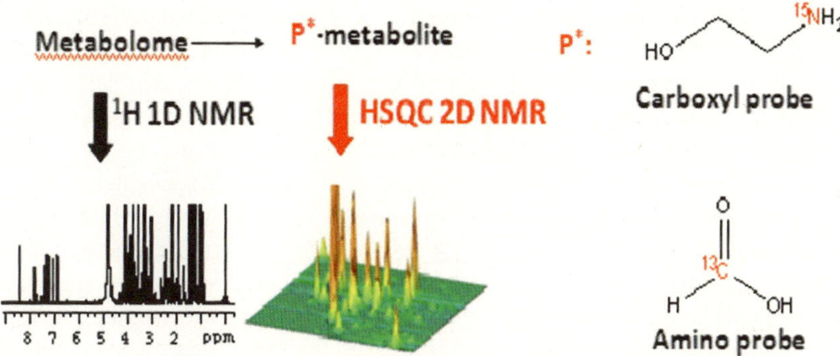

Fig. 8.4 Isotope tagging of metabolites in biological mixtures and their detection by 2D NMR enables access to low concentrated metabolites that are inaccessible to 1D NMR [95–97]

to reveal enough biochemical detail. To circumvent this problem, targeted profiling has shown to be a powerful approach for better understanding the metabolome in complex biological systems. As every metabolite contains at least one chemical functional group, the use of chemoselective tags can reduce the molecular complexity of the samples and potentially improve the detection of low-concentration metabolites by reducing the contribution of less interesting chemical signals.

Stable isotope tagging of metabolites in biofluids *ex vivo* using MS methods has found widespread applications for the identification as well as quantitation of metabolites [61–64]. The approach used for *ex vivo* isotope labeling for the NMR analysis of metabolites is shown in Fig. 8.4. To date, stable isotopes such as ^{13}C and ^{15}N and abundant heteronuclei such as ^{31}P have been used to tag metabolites with specific functional groups and thereby significantly enhance the resolution and sensitivity of the NMR experiments [95–99].

8.2.2.1 ^{13}C Isotope Labeling

^{13}C NMR can potentially serve as a useful alternative to ^{1}H NMR for identifying and quantifying metabolites [100]. The combination of the larger chemical shift range of ^{13}C and NMR experiments such as 2D ^{1}H-^{13}C HSQC significantly reduces spectral complexity by spreading the peaks in a two-dimensional plane. Recently constant time experiments have been developed to enable improved quantitation from HSQC type experiments [101, 102]. However, the low natural abundance of ^{13}C results in unacceptably long data acquisition times to compensate for the poor sensitivity. Hence, despite the improved resolution offered by ^{13}C NMR, it is impractical to use experiments involving ^{13}C at natural abundance for routine metabolomics applications unless concentrations are very high [11].

In contrast, isotope tagging of metabolites with the amine functional group using a ^{13}C isotope labeled tagging molecule significantly improves both the resolution and sensitivity of NMR methods for metabolite analysis [95, 96]. Tagging a specific functional group using ^{13}C enables selection of only a particular class of molecules, and using ^{1}H-^{13}C HSQC contributes to a significant simplification of the 2D spectrum. ^{13}C labeling using an acetylation reaction provides a useful approach for the analysis of the amine class of metabolites in complex mixtures [95]. The acetylation reaction is performed using 1,1'-^{13}C$_2$ acetic anhydride followed by 1D ^{13}C or ^{1}H-^{13}C HSQC experiments to obtain a spectrum enhanced in both resolution and sensitivity. Typically, the pH of aqueous solutions of the biofluids samples are set to 8.0, to which 1, 1'-^{13}C$_2$ acetic anhydride is added while stirring at room temperature. The pH of the reaction medium is maintained at 8.0 by addition of a 1 M NaOH solution at regular intervals. The resulting mixture containing ^{13}C labeled metabolites is then used for NMR analysis. Important aspects of this labeling approach are that it is facile, quantitative, and can be carried out directly in aqueous solution at ambient temperature. This is especially attractive for the analysis of complex mixtures such as urine, serum or other bio-fluids on a large set of samples.

More recently the performance of ^{13}C isotope enhanced amino metabolites profiling was improved by using ^{13}C-formic acid as the isotope tag [97, 99]. In the ^{13}C-formylated metabolite, the short, one-bond distance between the labeled ^{13}C and its closest ^1H produces a large J-coupling, which facilitates efficient transfer of polarization between ^1H and ^{13}C in HSQC experiments. In particular, the J-coupling of 200 Hz allows for a short 2.5 ms INEPT transfer delay compared to the 83 ms delay, due to the small J-coupling of 6 Hz, which was used for acetic anhydride tagging. For the ^{13}C-formic acid tagging, ^{13}C-formic acid and N-hydroxysuccinimide are dissolved in tetrahydrofuran; N, N-dicyclohexylcarbodiimide in tetrahydrofuran is then added to the mixture. After the reaction, the supernatant containing ^{13}C-N-formyloxysuccinimide is separated and added to the biofluid sample along with an aqueous solution of $NaHCO_3$. The mixture is then stirred at room temperature, dried under vacuum and dispersed in D_2O. This solution is then used for NMR analysis after adjusting the pH to 7.0. Alternatively, ^{13}C-N-formyloxysuccinimide can be purified by recrystallization in ethanol and used for the tagging reaction instead of *in situ* generation.

Amino group containing metabolites are an important class of molecules associated with biological processes. Amino acids, for example, are not only the building blocks for proteins but also precursors for nucleotides [103] and energy sources through transamination, urea cycle, citric acid cycle and gluconeogenesis [104–106]. Other common amino metabolites include derivatives of amino acids, taurine, dimethylamine, methylamine, and many neurotransmitters such as dopamine, serotonin and histamine. Drugs such as amphetamine, procaine, rimantadine and their metabolites also belong to this group of compounds. Selection of this class of metabolites using ^{13}C isotope enhanced NMR promises a number of metabolomics applications.

8.2.2.2 ^{15}N Isotope Labeling

Carboxyl functional group containing metabolites, which are associated with almost all the metabolic pathways, represent another major class of molecules in biological systems. Efforts focused on the detection of this class of metabolites using NMR spectroscopy have lead to the development of a ^{15}N isotope enhanced strategy, in which metabolites with carboxyl groups are chemically tagged with ^{15}N-ethanolamine and detected using a 2D heteronuclear correlation NMR experiment [96, 99]. This approach, which significantly improves the sensitivity and resolution of the NMR method, is capable of detecting metabolites at concentrations as low as a few micromolar in biological samples, both quantitatively and reproducibly. By this approach nearly 200 well-resolved signals corresponding to well over 100 carboxyl-containing metabolites can be routinely detected in serum and urine samples, which is unprecedented from the NMR point of view (Figs. 8.5 and 8.6). The resolution improvement, in this approach, is imparted from the dispersion of chemical shifts of the ^{15}N tag, and the high sensitivity is derived from the combination of factors such as isotope labeling and the strong, 90 Hz, J-coupling between the observed nuclei (^{15}N and ^1H). A single peak, devoid of multiplicity, for each tagged metabolite and effective filtering of nontagged metabolites significantly add to the sensitivity and background suppression. Further, the ^{15}N tag shows excellent reproducibility for complex biological samples. These characteristics are important for advanced metabolic profiling as well as for identifying unknown potential metabolite biomarkers.

Chemical derivatization reaction to tag ^{15}N isotope to carboxyl metabolites is performed by adding ^{15}N-ethanolamine and a catalyst, DMT-MM (4-(4,6-dimethoxy[1.3.5]triazin-2-yl)-4-methylmorpholinium chloride) to the biological mixture. In order to maintain the ^{15}N amide protonation in the tagged metabolites, the pH is adjusted to 5.0 by adding HCl or NaOH, and then the solution volume is adjusted to 600 μL by adding water prior to detection of isotope tagged carboxyl metabolites using 2D HSQC NMR experiment. While the samples such as urine are used with no pretreatment, samples

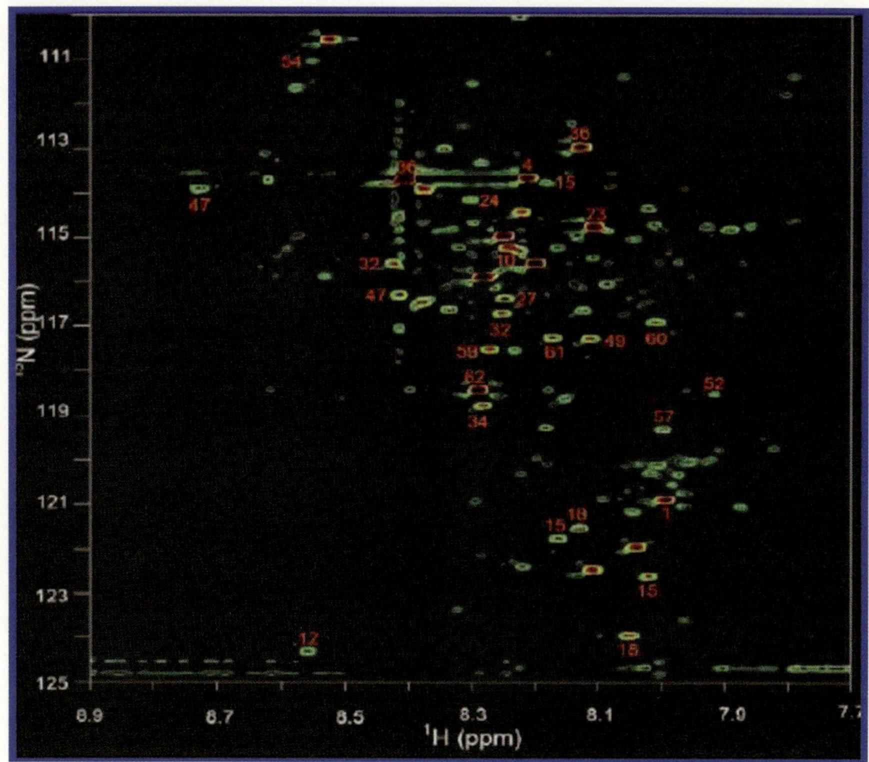

Fig. 8.5 A 2D ^{1}H-^{15}N HSQC spectrum of human serum obtained after tagging carboxyl-containing metabolites with ^{15}N-ethonalime. Nearly 180 metabolite peaks are detected, and the annotated peaks indicate identified metabolites (Reproduced with permission from Ref. [96])

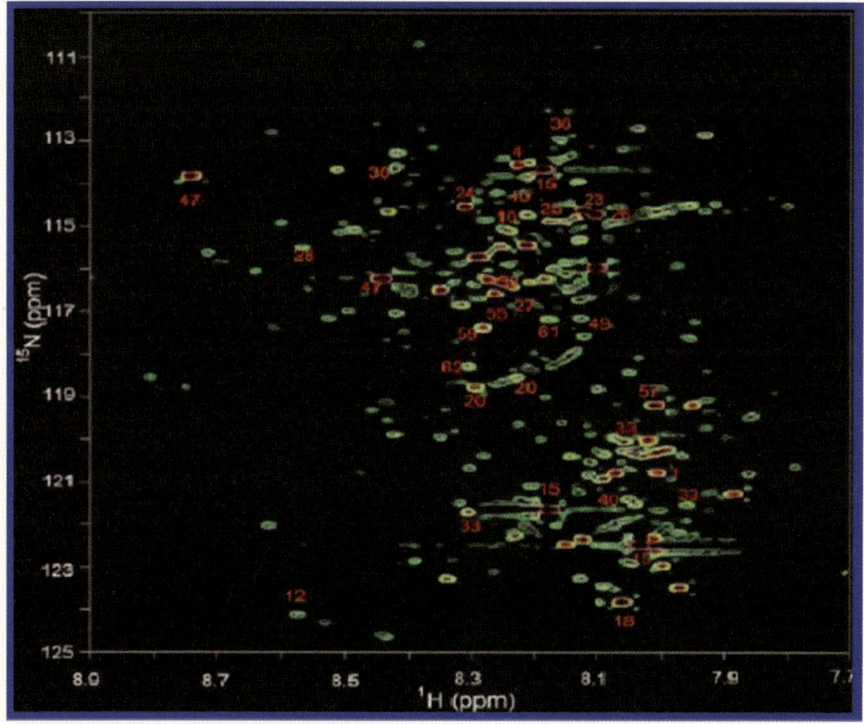

Fig. 8.6 A 2D ^{1}H-^{15}N HSQC spectrum of human urine obtained after tagging carboxyl-containing metabolites with ^{15}N-ethanolamine. Nearly 200 metabolites are detected, and the annotated peaks indicate identified metabolites (Reproduced with permission from Ref. [96])

such as blood serum/plasma are deproteinized before chemical derivatization, by precipitating out the proteins by mixing serum/plasma samples with methanol in a 1:2 (v/v) ratio.

An important aspect of the isotope labeling using ^{15}N-ethanolamine is that it easily and selectively combines with carboxyl-containing metabolites under aqueous conditions, which is suited for the study of biological mixtures without subjecting the samples to complex preprocessing. The use of the ^{15}N labeled tag ensures that other nitrogen containing metabolites are invisible in the spectrum because of the low natural abundance of ^{15}N (0.37%). The tagged metabolite retains the solubility since it contains at least one polar group (the hydroxyl in ethanolamine).

8.2.2.3 ^{31}P Labeling

The analysis of lipids in biological mixtures is a vital part of metabolomics studies due to the important role of lipids in human diseases and health. Using ^{1}H NMR, aliphatic fatty acid chain resonances can be assigned to sub-groups such as methyl, methylene, olefin, and allylic features but it is not possible to identify individual lipophilic components. Extending the scope of metabolite class selection using isotope labels, a method to detect members of the lipid class of metabolites with hydroxyl, carboxyl or aldehyde groups in complex biological fluids such as serum was recently developed [98]. This method employs the ^{31}P containing reagent, 2-chloro-4,4,5,5-tetramethyldioxaphospholane (CTMDP), to derivatize the lipid metabolites. Derivatized metabolites are then detected with enhanced resolution using one-dimensional ^{31}P NMR.

Typically, serum samples are lyophilized to dryness, reconstituted in stock solution (pyridine/chloroform), and subjected to derivatization with ^{31}P containing reagent, CTMDP. In the ^{31}P NMR spectra, newly formed P-O groups derived from hydroxyl groups appear in the 144.5–150 ppm region (Fig. 8.7), while the P-O groups derived from carboxylic acids appear at 134.5–135.5 ppm. ^{31}P labeled unesterified

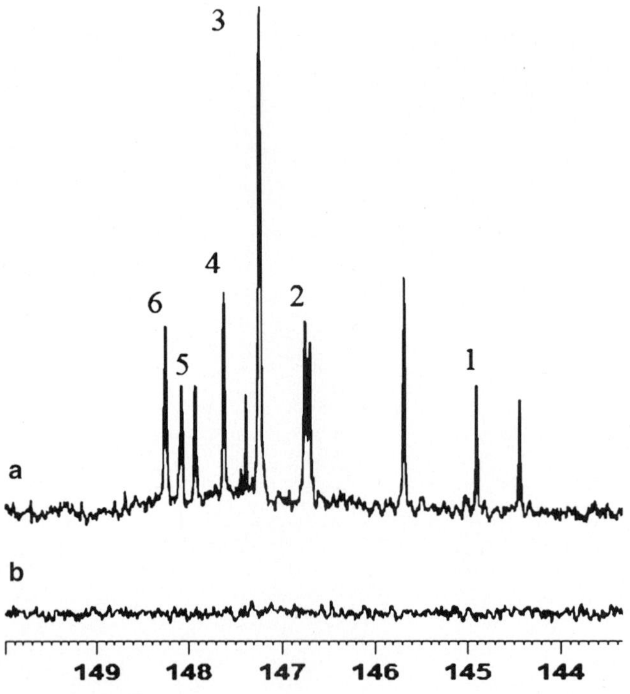

Fig. 8.7 A portion of the ^{31}P NMR spectrum of serum from a healthy individual: (**a**) ^{31}P-labeled and (**b**) Non-derivatized. ^{31}P labeled compounds include: *1* cholesterol, *2* lyso lipids, *3* possible fatty aldehydes, *4* phosphotidyl mono glycerides (primary), *5* free n-alkanol, *6* 1,2-diacylglycerol

cholesterol appears at 144.93 ppm. Mono and diglycerides can also be distinguished by this methodology, whereas derivatized 1, 2 -diglycerides in serum appear at 148.26 ppm, while for mono glycerides, derivatized primary and secondary hydroxyl groups appear at 147.6 and 146.54 ppm, respectively. Regardless of the length of the carbon chain, polar head group or unsaturation, lyso lipids produce two resonances at 146.68 and 146.74 ppm in serum. Free fatty acids appeared at 134.78 ppm.

Overall, this ^{31}P labeling strategy offers the possibility of identification of a number of lipophilic compounds that contain hydroxyl and carboxylic acid functionalities in different chemical environments. The method provides sufficient sensitivity and spectral resolution, and derivatized species have unique and well-resolved resonances in the ^{31}P NMR spectrum. This approach is consistent with the requirements for a fast screening method for lipid associated pathologies in human serum making possible the efficient use of ^{31}P NMR spectroscopy alongside ^{1}H NMR spectroscopy in metabolomics studies.

8.2.2.4 Applications of *Ex Vivo* Isotope Labeling

The reduced complexity of the spectra due to the absence of less interesting chemical signals is particularly important for the analysis of low-concentration metabolites. Chemoselective isotope tags produce a single peak for a metabolite with a single functional group. The resulting spectra from the isotope labeling strategies produce altogether new NMR chemical shifts for several hundreds of metabolites. Identification of these metabolites requires the knowledge of chemical shifts of isotope tags for authentic compounds. In view of this, based on the spectra of isotope labeled standard compounds, a chemical shifts library has been developed for nearly 150 metabolites. By matching with this library of chemical shifts, a number of metabolites in the ^{13}C, ^{15}N and ^{31}P isotope labeled spectra have so far been identified [96–98]. Subsequently, using a combination of isotope labeled methods, metabolites in human plasma procured from the National Institute of Standards and Technology (NIST, Gaithersburg, MD), were quantified after validating the experimental protocols using a mixture of synthetic compounds [99] (Fig. 8.8). It was demonstrated that the combination of isotope tagging approaches, along with conventional 1D and 2D NMR methods enables the quantitative analysis of a large number of metabolites in human biofluids on a routine basis. For example, the ^{13}C isotope labeling approach was used for the analysis of urine from patients with pre-diagnosed metabolic disorders such as tyrosinemia type II, argininosuccinicaciduria, homocystinuria and phenylketonuria [95].

The combination of improved sensitivity and resolution, and less time required when compared to natural abundance heteronuclear NMR methods, is attractive for the routine and accurate analysis of metabolites in complex biological samples. Although, the isotope tagging methods often use 2D NMR experiments, each 2D experiment requires only 30 min or less, and hence the approach can be useful for high throughput analysis of human plasma as well as other biological fluids.

8.3 Conclusions

While mass spectrometry is highly sensitive for metabolomics applications, the performance of NMR spectroscopy is significantly better in both reproducibility and quantitative ability and hence it promises a number of unique applications in the area of metabolomics. Despite vast advancements in the area, however, challenges remain for NMR for the detection, identification and quantification of a large number of metabolites in complex biological samples. Aimed at alleviating these limitations and gaining insights into *in vivo* biochemistry in cells, animal models and humans, the number of NMR investigations utilizing stable isotope labeling methods *in vivo* and *ex vivo* is growing rapidly. In *in vivo* isotope labeled studies, the targeted metabolites are always products of the metabolism of isotope labeled substrates and their detection, selectively, prevents the contributions from many confounding factors. This is extremely important for metabolomics studies since confounding factors arising from unrelated pathways or source are still a major challenge to the growth of metabolomics research area.

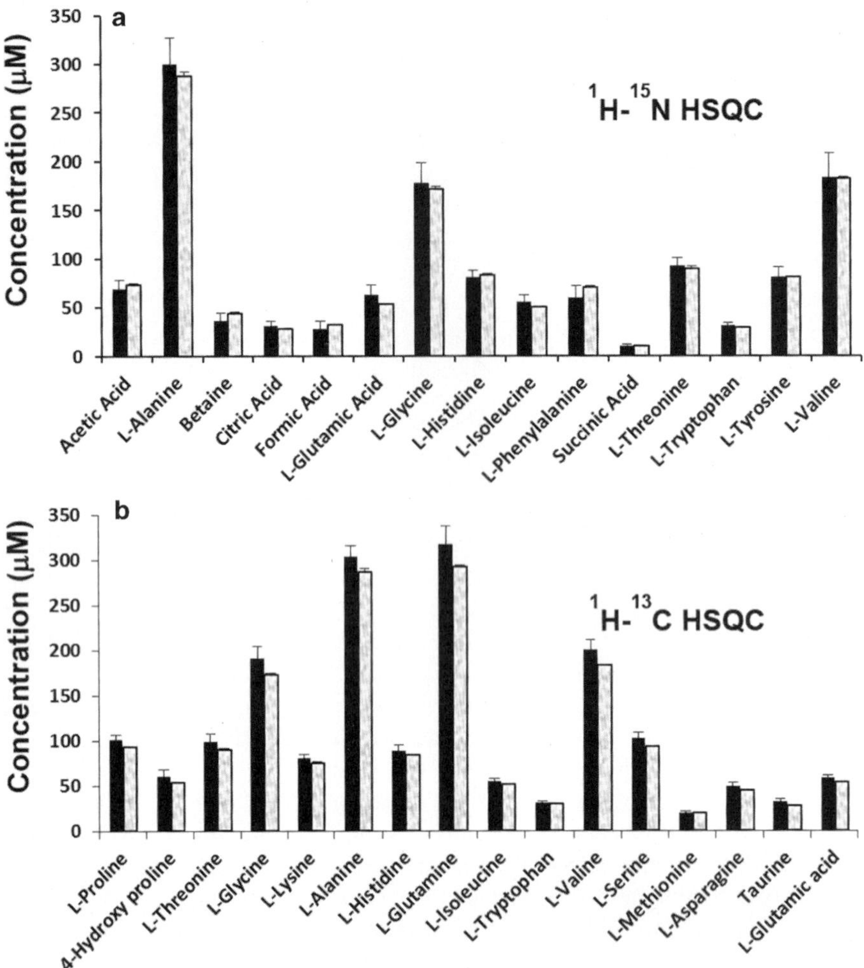

Fig. 8.8 Concentration of metabolites obtained with ^{15}N or ^{13}C tagging: (**a**) obtained from ^{1}H-^{15}N HSQC NMR after ^{15}N tagging; (**b**) obtained from ^{1}H-^{13}C HSQC NMR after ^{13}C tagging. The shaded bar on the right in each pair represents the actual concentration of the metabolite

Development of *ex vivo* isotope labeling methods has enabled the detection of hundreds of metabolites with improved sensitivity and resolution, from a single experiment, which is unprecedented for NMR of complex biological samples. Further, a combination of the isotope tagging approach with the latest advancements in NMR technology, such as detection using micro-coil probes and/or cryoprobes, for example, can significantly minimize the volume of biofluid samples and time required for routine analysis. Future advancements in this area promise far reaching implications for systems biology, early detection of diseases and therapy monitoring, and personalized medicine.

References

1. Nicholson JK, Lindon JC, Holmes E (1999) 'Metabonomics': understanding the metabolic responses of living systems to pathophysiological stimuli via multivariate statistical analysis of biological NMR spectroscopic data. Xenobiotica 29:1181–1189
2. Fiehn O (2002) Metabolomics-the link between genotype and phenotype. Plant Mol Biol 48:155–171

3. Saghatelian A, Cravatt BF (2005) Global strategies to integrate the proteome and metabolome. Curr Opin Chem Biol 9:62–68
4. Assfalg M, Bertini I, Colangiuli D et al (2008) Evidence of different metabolic phenotypes in humans. Proc Natl Acad Sci USA 105:1420–1424
5. Nicholson JK, Lindon JC (2008) Systems biology: metabonomics. Nature 455:1054–1056
6. van der Greef J, Smilde AK (2005) Symbiosis of chemometrics and metabolomics: past, present, and future. J Chemom 19:376–386
7. Nagana Gowda GA, Zhang S, Gu H et al (2008) Metabolomics based methods for early disease diagnostics: a review. Exp Rev Mol Diagn 8:627–633
8. NaganaGowda GA, Ijare OB, Shanaiah N et al (2009) Combining NMR spectroscopy and mass spectrometry in biomarker discovery. Biomark Med 3:307–322
9. Holmes E, Wilson ID, Nicholson JK (2008) Metabolic phenotyping in health and disease. Cell 134:714–717
10. Weljie AM, Newton J, Mercier P et al (2006) Targeted profiling: quantitative analysis of ^{1}H NMR metabolomics data. Anal Chem 78:4430–4442
11. Lewis IA, Schommer SC, Hodis B et al (2007) Method for determining molar concentrations of metabolites in complex solutions from two-dimensional ^{1}H-^{13}C NMR spectra. Anal Chem 79:9385–9390
12. Robinette SL, Zhang F, Brüschweiler-Li L et al (2008) Web server based complex mixture analysis by NMR. Anal Chem 80:3606–3611
13. Chikayama E, Sekiyama Y, Okamoto M et al (2010) Statistical indices for simultaneous large-scale metabolite detections for a single NMR spectrum. Anal Chem 82:1653–1658
14. Keun HC (2006) Metabonomic modeling of drug toxicity. Pharmacol Ther 109:92–106
15. Coen M, Holmes E, Lindon JC et al (2008) NMR-based metabolic profiling and metabonomic approaches to problems in molecular toxicology. Chem Res Toxicol 2:9–27
16. Bollard ME, Stanley EG, Lindon JC et al (2005) NMR-based metabonomic approaches for evaluating physiological influences on biofluid composition. NMR Biomed 18:143–162
17. Nagana Gowda GA, Ijare OB, Somashekar BS et al (2006) Single-step analysis of individual conjugated bile acids in human bile using ^{1}H NMR spectroscopy. Lipids 41:591–603
18. Nagana Gowda GA (2010) Human bile as a rich source of biomarkers for hepatopancreatobiliary cancers. Biomark Med 4:299–314
19. Nagana Gowda GA (2011) NMR spectroscopy for discovery and quantitation of biomarkers of disease in human bile. Bioanalysis 3:1877–1890
20. Bala L, Ghoshal UC, Ghoshal U et al (2006) Malabsorption syndrome with and without small intestinal bacterial overgrowth: a study on upper-gut aspirate using ^{1}H NMR spectroscopy. Magn Reson Med 56:738–744
21. Griffin JL, Kauppinen RA (2007) Tumour metabolomics in animal models of human cancer. J Proteome Res 6:498–505
22. Villas-Boas SG, Hojer-Pedersen J, Akesson M et al (2005) Global metabolite analysis of yeast: evaluation of sample preparation methods. Yeast 22:1155–1169
23. Bollard ME, Xu JS, Purcell W et al (2002) Metabolic profiling of the effects of D-galactosamine in liver spheroids using ^{1}H NMR and MAS-NMR spectroscopy. Chem Res Toxicol 15:1351–1359
24. Boroujerdi AF, Vizcaino MI, Meyers A et al (2009) NMR-based microbial metabolomics and the temperature-dependent coral pathogen Vibrio coralliilyticus. Environ Sci Technol 43:7658–7664
25. Lyng H, Sitter B, Bathen TF et al (2007) Metabolic mapping by use of high-resolution magic angle spinning ^{1}H MR spectroscopy for assessment of apoptosis in cervical carcinomas. BMC Cancer 7:11
26. Sitter B, Bathen T, Hagen B et al (2004) Cervical cancer tissue characterized by high-resolution magic angle spinning MR spectroscopy. Magn Reson Mater Phys Biol Med 16:174–181
27. Schenetti L, Mucci A, Parenti F et al (2006) HR-MAS NMR spectroscopy in the characterization of human tissues: application to healthy gastric mucosa. Concepts Magn Reson A 28A:430–443
28. Beckonert O, Keun HC, Ebbels TMD et al (2007) Metabolic profiling, metabolomic and metabonomic procedures for NMR spectroscopy of urine, plasma, serum and tissue extracts. Nat Protoc 2:2692–2703
29. Beckonert O, Coen M, Keun HC et al (2010) High-resolution magic-angle-spinning NMR spectroscopy for metabolic profiling of intact tissues. Nat Protoc 5:1019–1032
30. Nicholson JK, Foxall PJD, Spraul M et al (1995) 750 MHz ^{1}H and ^{1}H-^{13}C NMR spectroscopy of human blood-plasma. Anal Chem 67:793–811
31. Balayssac S, DelsucM-A GV et al (2009) Two-dimensional DOSY experiment with Excitation Sculpting water suppression for the analysis of natural and biological media. J Magn Reson 196:78–83
32. Van Lokeren L, Kerssebaum R, Willem R et al (2008) ERETIC implemented in diffusion-ordered NMR as a diffusion reference. Magn Reson Chem 46:S63–S71
33. Smith LM, Maher AD, Cloarec O et al (2007) Statistical correlation and projection methods for improved information recovery from diffusion-edited NMR spectra of biological samples. Anal Chem 79:5682–5689
34. Simpson AJ, Brown SA (2005) Purge NMR: effective and easy solvent suppression. J Magn Reson 175:340–346

35. Ogg RJ, Kingsley PB, Taylor JS (1994) Wet, a T-1-insensitive and B-1-insensitive water-suppression method for in-vivo localized ^{1}H NMR spectroscopy. J Magn Reson B 104:1–10
36. Mo H, Raftery D (2008) Improved residual water suppression: WET180. J Biomol NMR 41:105–111
37. Hoult DI (1976) Solvent peak saturation with single-phase and quadrature fourier transformation. J Magn Reson 21:337–347
38. Sandusky P, Raftery D (2005) Use of selective TOCSY NMR experiments for quantifying minor components in complex mixtures: application to the metabonomics of amino acids in honey. Anal Chem 77:2455–2463
39. Sandusky P, Raftery D (2005) Use of semiselective TOCSY and the pearson correlation for the metabonomic analysis of biofluid mixtures: application to urine. Anal Chem 77:7717–7723
40. Sandusky P, Appiah-Amponsah E, Raftery D (2011) Use of optimized 1D TOCSY NMR for improved quantitation and metabolomic analysis of biofluids. J Biomol NMR 49:281–290
41. Dumas ME, Canlet C, André F et al (2002) Metabonomic assessment of physiological disruptions using ^{1}H-^{13}C HMBC-NMR spectroscopy combined with pattern recognition procedures performed on filtered variables. Anal Chem 74:2261–2273
42. Tang HR, Wang Y, Nicholson JK et al (2004) Use of relaxation-edited one-dimensional and two dimensional nuclear magnetic resonance spectroscopy to improve detection of small metabolites in blood plasma. Anal Biochem 325:260–272
43. Xi Y, de Ropp JS, Viant MR et al (2006) Automated screening for metabolites in complex mixtures using 2D COSY NMR spectroscopy. Metabolomics 2:221–233
44. Chikayama E, Suto M, Nishihara T et al (2008) Systematic NMR analysis of stable isotope labeled metabolite mixtures in plant and animal systems: coarse grained views of metabolic pathways. PLoS One 3:e3805
45. Fan TWM, Bandura LL, Higashi RM et al (2005) Metabolomics-edited transcriptomics analysis of Se anticancer action in human lung cancer cells. Metabolomics 1:325–339
46. Ludwig C, Viant MR (2010) Two-dimensional J-resolved NMR spectroscopy: review of a key methodology in the metabolomics toolbox. Phytochem Anal 21:22–32
47. Parsons HM, Ludwig C, Gunther UL et al (2007) Improved classification accuracy in 1-and 2-dimensional NMR metabolomics data using the variance stabilising generalised logarithm transformation. BMC Bioinform 8:234
48. Hyberts SG, Heffron GJ, Tarragona NG et al (2007) Ultrahigh-resolution ^{1}H-^{13}C HSQC spectra of metabolite mixtures using nonlinear sampling and forward maximum entropy reconstruction. J Am Chem Soc 129:5108–5116
49. Viant MR (2003) Improved methods for the acquisition and interpretation of NMR metabolomic data. Biochem Biophys Res Commun 310:943–948
50. Lindon JC, Holmes E, Nicholson JK (2006) Metabonomics techniques and applications to pharmaceutical research & development. Pharm Res 23:1075–1088
51. Kind T, Tolstikov V, Fiehn O et al (2007) A comprehensive urinary metabolomic approach for identifying kidney cancer. Anal Biochem 363:185–195
52. Wilson ID, Nicholson JK, Castro-Perez J et al (2005) High resolution "ultra performance" liquid chromatography coupled to a-TOF mass spectrometry as a tool for differential metabolic pathway profiling in functional genomic studies. J Proteome Res 4:591–598
53. Spraul M, Freund AS, Nast RE et al (2003) Advancing NMR sensitivity for LC-NMR-MS using a cryo-flow probe: application to the analysis of acetaminophen metabolites in urine. Anal Chem 75:1536–1541
54. Lacey ME, Subramanian R, Olson DL et al (1999) High-resolution NMR spectroscopy of sample volumes from 1 nL to 10 μL. Chem Rev 99:3133–3152
55. Olson DL, Peck TL, Webb AG et al (1995) High-resolution microcoil ^{1}H-NMR for mass-limited, nanoliter-volume samples. Science 270:1967–1970
56. Webb AG (2005) Microcoil nuclear magnetic resonance spectroscopy. J Pharm Biomed Anal 38:892–903
57. Kc R, Henry ID, Park GHJ et al (2009) Design and construction of a versatile dual volume heteronuclear double resonance microcoil NMR probe. J Magn Reson 197:186–192
58. Henry D, Park GHJ, Kc R et al (2008) Design and construction of a microcoil NMR probe for the routine analysis of 20-mu L samples. Concepts Magn Reson B Magn Reson Eng 33B:1–8
59. Bergeron SJ, Henry ID, Santini RE et al (2008) Saturation transfer double-difference NMR spectroscopy using a dual solenoid microcoil difference probe. Magn Reson Chem 46:925–929
60. Kc R, Gowda YN, Djukovic D et al (2010) Susceptibility-matched plugs for microcoil NMR probes. J Magn Reson 205:63–68
61. Guo K, Bamforth F, Li L (2011) Qualitative metabolome analysis of human cerebrospinal fluid by 13C-/12C-isotope dansylation labeling combined with liquid chromatography fourier transform ion cyclotron resonance mass spectrometry. J Am Soc Mass Spectrom 22:339–347
62. Huang X, Regnier FE (2008) Differential metabolomics using stable isotope labeling and two-dimensional gas chromatography with time-of-flight mass spectrometry. Anal Chem 80:107–114
63. Yang WC, Adamec J, Regnier FE (2007) Enhancement of the LC/MS analysis of fatty acids through derivatization and stable isotope coding. Anal Chem 79:5150–5157

64. Yang WC, Regnier FE, Sliva D et al (2008) Stable isotope-coded quaternization for comparative quantification of estrogen metabolites by high-performance liquid chromatography-electrospray ionization mass spectrometry. J Chromatogr B Anal Technol Biomed Life Sci 870:233–240

65. Lane AN, Fan TWM, Bousamra M II et al (2011) Stable isotope-resolved metabolomics (SIRM) in cancer research with clinical application to nonsmall cell lung cancer. OMICS A J Integr Biol 15:173–182

66. Fan TW, Lane AN, Higashi RM et al (2011) Stable isotope resolved metabolomics of lung cancer in a SCID mouse model. Metabolomics 7:257–269

67. Fan TW, Lane AN, Higashi RM et al (2009) Altered regulation of metabolic pathways in human lung cancer discerned by (13)C stable isotope-resolved metabolomics (SIRM). Mol Cancer 8:41

68. Locasale JW, Grassian AR, Melman T et al (2011) Phosphoglycerate dehydrogenase diverts glycolytic flux and contributes to oncogenesis. Nat Genet 43:869–874

69. Fan TW, Lane AN (2011) NMR-based stable isotope resolved metabolomics in systems biochemistry. J Biomol NMR 49:267–280

70. Petch D, Butler M (1994) Profile of energy metabolism in a murine hybridoma: glucose and glutamine utilization. J Cell Physiol 161:71–76

71. Portais JC, Voisin P, Merle M et al (1996) Glucose and glutamine metabolism in C6 glioma cells studied by carbon-13 NMR. Biochimie 78:155–164

72. Mazurek S, Grimm H, Oehmke M et al (2000) Tumor M2-PK and glutaminolytic enzymes in the metabolic shift of tumor cells. Anticancer Res 20(6D):5151–5154

73. DeBerardinis RJ, Mancuso A, Daikhin E et al (2007) Beyond aerobic glycolysis: transformed cells can engage in glutamine metabolism that exceeds the requirement for protein and nucleotide synthesis. Proc Natl Acad Sci USA 104:19345–19350

74. Lane AN, Fan TW, Higashi RM et al (2009) Prospects for clinical cancer metabolomics using stable isotope tracers. Exp Mol Pathol 86:165–173

75. Lloyd SG, Zeng H, Wang P et al (2004) Lactate isotopomer analysis by ^{1}H NMR spectroscopy: consideration of long-range nuclear spin-spin interactions. Magn Reson Med 51:1279–1282

76. Lane AN, Fan TW (2007) Quantification and identification of isotopomer distributions of metabolites in crude cell extracts using ^{1}H TOCSY. Metabolomics 3:79–86

77. Burgess SC, Babcock EE, Jeffrey FM et al (2001) NMR indirect detection of glutamate to measure citric acid cycle flux in the isolated perfused mouse heart. FEBS Lett 505:163–167

78. Perdigoto R, Furtado AL, Porto A et al (2003) Sources of glucose production in cirrhosis by ^{2}H$_2$O ingestion and ^{2}H NMR analysis of plasma glucose. Biochim Biophys Acta 1637:156–163

79. Kikuchi J, Shinozaki K, Hirayama T (2004) Stable isotope labeling of Arabidopsis thaliana for an NMR-based metabolomics approach. Plant Cell Physiol 45:1099–1104

80. Lane AN, Fan TW, Higashi RM (2008) Isotopomer based metabolic analysis by NMR and mass spectrometry. Biophys Tools Biol 84:541–588

81. Lane AN, Fan TW, Xie Z et al (2009) Isotopomer analysis of lipid biosynthesis by high resolution mass spectrometry and NMR. Anal Chim Acta 651:201–208

82. Coles NW, Johnstone RM (1962) Glutamine metabolism in Ehrlich ascites-carcinoma cells. Biochem J J83:284–291

83. Eagle H (1955) Nutrition needs of mammalian cells in tissue culture. Science 122:501–514

84. Wise DR, DeBerardinis RJ, Mancuso A et al (2008) Myc regulates a transcriptional program that stimulates mitochondrial glutaminolysis and leads to glutamine addiction. Proc Natl Acad Sci USA 105:18782–18787

85. Weis BC, Margolis D, Burgess SC et al (2004) Glucose production pathways by ^{2}H and ^{13}C NMR in patients with HIV-associated lipoatrophy. Magn Reson Med 51:649–654

86. Jones JG, Solomon MA, Cole SM et al (2001) An integrated (2)H and (13)C NMR study of gluconeogenesis and TCA cycle flux in humans. Am J Physiol Endocrinol Metab 281:E848–E856

87. Hausler N, Browning J, Merritt M et al (2006) Effects of insulin and cytosolic redox state on glucose production pathways in the isolated perfused mouse liver measured by integrated ^{2}H and ^{13}C NMR. Biochem J 394(Pt 2): 465–473, Erratum in: Biochem J 2006; 395:663

88. Perdigoto R, Rodrigues TB, Furtado AL et al (2003) Integration of [U-13C]glucose and ^{2}H$_2$O for quantification of hepatic glucose production and gluconeogenesis. NMR Biomed 16:189–198

89. Jin ES, Jones JG, Merritt M et al (2004) Glucose production, gluconeogenesis, and hepatic tricarboxylic acid cycle fluxes measured by nuclear magnetic resonance analysis of a single glucose derivative. Anal Biochem 327:149–155

90. Burgess SC, Weis B, Jones JG et al (2003) Noninvasive evaluation of liver metabolism by ^{2}H and ^{13}C NMR isotopomer analysis of human urine. Anal Biochem 312:228–234

91. Merritt ME, Harrison C, Sherry AD et al (2011) Flux through hepatic pyruvate carboxylase and phosphoenolpyruvate-carboxykinase detected by hyperpolarized ^{13}C magnetic resonance. Proc Natl Acad Sci USA 108:19084–19089

92. Schroeder MA, Atherton HJ, Ball DR et al (2009) Real-time assessment of Krebs cycle metabolism using hyperpolarized ^{13}C magnetic resonance spectroscopy. FASEB J 23:2529–2538

93. Lewis IA, Karsten RH, Norton ME et al (2010) NMR method for measuring carbon-13 isotopic enrichment of metabolites in complex solutions. Anal Chem 82:4558–4563

94. Fernie AR, Trethewey RN, Krotzky AJ et al (2004) Metabolite profiling: from diagnostics to systems biology. Nat Rev Mol Cell Biol 5:763–769

95. Shanaiah N, Desilva A, Nagana Gowda GA et al (2007) Metabolite class selection of amino acids in biofluids using chemical derivatization and their enhanced ^{13}C NMR. Proc Natl Acad Sci USA 104:11540–11544

96. Ye T, Mo H, Shanaiah N et al (2009) Chemoselective ^{15}N tag for sensitive and high-resolution nuclear magnetic resonance profiling of the carboxyl-containing metabolome. Anal Chem 81:4882–4888

97. Ye T, Zhang S, Mo H et al (2010) ^{13}C-formylation for improved NMR profiling of amino metabolites in biofluids. Anal Chem 82:2303–2309

98. DeSilva MA, Shanaiah N, Nagana Gowda GA et al (2009) Application of ^{31}P NMR spectroscopy and chemical derivatization for metabolite profiling of lipophilic compounds in human serum. Magn Reson Chem 47:S74–S80

99. NaganaGowda GA, Tayyari F, Ye T et al (2010) Quantitative analysis of blood plasma metabolites using isotope enhanced NMR methods. Anal Chem 82:8983–8990

100. Fan TW (1996) Metabolite profiling by one- and two-dimensional NMR analysis of complex mixtures. Prog NMR Spectrosc 28:161–219

101. Hu K, Ellinger JJ, Chylla RA et al (2011) Measurement of absolute concentrations of individual compounds in metabolite mixtures by gradient-selective time-zero (1)H-(13)C HSQC with two concentration references and fast maximum likelihood reconstruction analysis. Anal Chem 83:9352–9360

102. Hu K, Wyche TP, Bugni TS et al (2011) Selective quantification by 2D HSQC(0) spectroscopy of thiocoraline in an extract from a sponge-derived Verrucosispora sp. J Nat Prod 74:2295–2298

103. Berg JM, Tymoczko JL, Stryer L (2002) Biochemistry. W. H. Freeman & Co, New York

104. Sakami W, Harrington H (1963) Amino acid metabolism. Annu Rev Biochem 32:355–398

105. Brosnan JT (2000) Glutamate, at the interface between amino acid and carbohydrate metabolism. J Nutr 130: 988S–990S

106. Young VR, Ajami AM (2001) Glutamine: the emperor or his clothes? J Nutr 131:2449S–2459S

Part IV
Expression Systems

Chapter 9
Cell-Free Protein Synthesis Using *E. coli* Cell Extract for NMR Studies

Mitsuhiro Takeda and Masatsune Kainosho

Abstract The use of cell-free protein production systems for producing isotope labeled proteins generates new opportunities to perform unprecedented NMR studies. As compared with conventional cellular expression systems, the scrambling and dilution of amino acids are highly suppressed in the cell-free reaction, allowing the production of proteins with a wide variety of residue and site-specific isotope labeling patterns. In this chapter, the procedure for cell-free protein synthesis for NMR studies, using an *E. coli* extract, is introduced.

9.1 Introduction

The production of isotopically labeled proteins is an initial step in the analysis of a protein by NMR. Cell-free protein synthesis is a practical method for protein production, along with cell-based expression systems. In the *in vitro* expression system, a target protein is produced in a vessel, where the protein synthesis machinery is reconstituted. As compared with conventional *in vivo* expression, amino acid metabolic conversion is suppressed in the *in vitro* expression system, thus providing an opportunity to produce proteins with a wide variety of residue- and site-specific isotope labeling patterns [1]. In addition, the open nature of the *in vitro* expression enables various modifications of the reaction conditions, such as supplementation with chaperones, inhibitors, and detergents/lipids.

Historically, the *in vitro* expression of proteins has utilized an extract from *E. coli* cells [2–7]. The *E. coli* extract contains a large number of proteins needed for protein synthesis (e.g., ribosomal proteins and initiation factors). When energy sources and template DNA are added to the extract, the encoded protein is synthesized in the vessel. However, the protein synthesis soon ceases, due to the depletion of energy and substrate. Over the past decade, many efforts have been directed toward improving its productivity. Now, the *E. coli* cell-free expression method has become a practical means

M. Takeda
Structural Biology Research Center, Graduate School of Science,
Nagoya University, Furo-cho, Chikusa-ku, Nagoya 464-8602, Japan

M. Kainosho (✉)
Structural Biology Research Center, Graduate School of Science,
Nagoya University, Furo-cho, Chikusa-ku, Nagoya 464-8602, Japan

Center for Priority Areas, Graduate School of Science and Technology,
Tokyo Metropolitan University, 1-1 Minami-ohsawa, Hachioji 192-0397, Japan
kainosho@tmu.ac.jp

H.S. Atreya (ed.), *Isotope Labeling in Biomolecular NMR*, Advances in Experimental Medicine and Biology 992,   167
DOI 10.1007/978-94-007-4954-2_9, © Springer Science+Business Media Dordrecht 2012

for producing proteins on a preparative scale. Furthermore, new types of cell-free synthesis systems, such as the wheat germ system [8, 9] and one exclusively composed of recombinant proteins [10, 11], have appeared as alternative cell-free methods. Especially, wheat germ cell-free protein synthesis has become a practical method for producing proteins for NMR studies, along with the *E. coli* cell-free expression system.

The advent of cell-free protein synthesis systems provides opportunities to perform NMR approaches that are not amenable to cellular-based expression system. For example, a wider range of isotope labeled amino acids can be incorporated into a target protein by using cell-free synthesis, due to the suppressed inter-conversion of amino acids [1, 12, 13]. In addition, the incorporation of unnatural amino acids can be accomplished using *E. coli* cell-free synthesis [14–16]. The cell-free protein synthesis system has also been used for the production of membrane proteins [17–19]. In general, membrane proteins need proper membrane-mimicking components (e.g., detergents and lipids) to exist in a folded state in an aqueous solution. The merit of cell-free protein synthesis is that the membrane protein can be expressed in the presence of a wide variety of detergents/lipids, owing to the open nature of the reaction, thereby increasing the chance of expressing the membrane protein in the properly folded state. Another important application of cell-free protein synthesis is the production of SAIL proteins [20, 21]. The SAIL protein is composed of chemically synthesized amino acids that are stereo- and region-selectively isotope labeled. Therefore, the SAIL amino acids must be incorporated into a target protein efficiently, without metabolic scrambling. The details of the method are described in another chapter in this book. In our laboratory, several SAIL proteins were produced by using an *E. coli* cell-free system [22–24].

This chapter provides an overview of the procedure for *E. coli* cell-free protein production, with a focus on NMR study. In the early version of the *E. coli* cell-free reaction system, the protein synthesis soon ceased, and this problem remained unsolved. By virtue of extensive efforts to improve the productivity over the past few decades, however, it has become possible to produce recombinant proteins in milli-gram quantities by using an *E. coli* cell-free system [3, 5, 25]. Another issue in the *E. coli* cell-free protein synthesis is the low amino acid labeling efficiency. The *E. coli* extract prepared by the previous protocol contained a large amount of endogenous, unlabeled amino acids, which dilute the added isotope labeled amino acid in the resulting proteins. Furthermore, in some cases, the processing of the N-terminal formyl methionine by a peptide deformylase does not occur completely in the *in vitro* expression system [26, 27]. The presence of the formyl moiety causes large chemical shift changes in nearby residues, giving rise to doubled peaks corresponding to the formylated and deformylated forms. It is also somewhat problematic that isotope scrambling and dilution still occur even in the *E. coli* cell-free system due to residual enzymatic activity, although they are relatively suppressed as compared to cellular expression.

In the following, the procedures for the preparation of the *E. coli* cell-free extract and the cell-free protein synthesis will be described.

9.2 Preparation of the *E. coli* S30 Extract for NMR Samples

The *E. coli* cell-free expression system is commonly used in a wide variety of fields, and thus *E. coli* extracts are commercially available. However, such extracts are not necessarily optimized for the production of NMR samples. One potential problem is that they may contain large amounts of endogenous amino acids, which dilute the added isotope labeled amino acids. Therefore, we developed an *E. coli* extract preparation method. The protocol for the preparation of the *E. coli* extract consists of culturing the *E. coli* cells, washing the cells, a run-off reaction, dialysis, gel-filtration and condensation. This procedure was established by modifying pre-existing protocols [2, 5, 27–32]. The preparation of the *E. coli* extract takes about 2 days for one person.

9.2.1   Protocol for Amino Acid-Free S30 Extract Preparation

1. Inoculate *E. coli* cells (e.g., A19, BL21 Star (DE3) strains) into 10 ml of LB medium and grow the cells overnight at 37°C with shaking.
2. Use the 10 ml culture to inoculate 1 l of incomplete rich medium (5.6 g/L KH_2PO_4, 28.9 g/L K_2HPO_4, 1 g/L Bacto yeast extract, 1.5 mg/L thiamine, 20 g/L D-glucose and 1 mM $Mg(OAc)_2$) in a 2-l flask.
3. Culture the cells at 37°C with shaking, until an OD_{650} of 0.7 is attained.
4. Centrifuge the cells (5,000 g, 4°C, 10 min) and wash them three times with 200 ml of ice-cold S30 buffer (10 mM Tris-acetate (pH 8.2), 14 mM $Mg(OAc)_2$, 60 mM KOAc, 1 mM DTT) containing 0.05% 2-mercaptoethanol.
5. Gently resuspend the cell pellet in 200 ml of ice-cold S30 buffer containing 0.05% (v/v) 2-mercaptoethanol. Centrifuge the suspension (5,000 g, 4°C, 10 min) and weigh the *E. coli* pellet. Resuspend the pellet in 1.27 ml of S30 buffer per gram of *E. coli*.
6. Disrupt the cells with a French Press at 20,000 psi (1,400 kg cm^{-2}). Add 30 μl of 1 M DTT to the lysate immediately after the disruption of the cells. Centrifuge the lysate (30,000 g, 4°C, 30 min) using RNase-free centrifuge tubes. Carefully remove approximately 1.4 ml of the supernatant per gram of *E. coli*, without mixing with the precipitate.
7. Transfer the supernatant to RNase-free tubes. Centrifuge them (30,000 g, 4°C, 30 min) and remove approximately 1.0 ml of the supernatant per gram of *E. coli* into a 50 ml tube.
8. Shake the tube at 37°C for 80 min.
9. Dialyze the solution at 4°C for 45 min against 2 l of S30 buffer, using a dialysis tube with a molecular weight cut off of 6,000–8,000. Repeat the dialysis twice, and then centrifuge the solution (15,000 g, 10 min, 4°C) and collect the supernatant.
10. Uniformly fill an open column (Econo-column chromatography column, 2.5 × 20 cm) with Sephadex G25 resin (GE Healthcare), and place the column vertically in a cold space (4°C). Attach an Econo-column funnel to the top end of the column. Pour 500 ml of S30 buffer through the funnel into the column.
11. Apply the supernatant from step 9 to the column that was pre-equilibrated at 4°C in step 10. After loading the supernatant, continue to supply the funnel with S30 buffer, to maintain the flow in the column. When the first fraction reaches the bottom, start to collect 1.4 times the volume of the applied extract. The first fraction is determined by its color (yellow) and turbidity.
12. Dialyze the eluate at 4°C against 700 ml of an equal weight mixture of polyethyelene glycol (PEG)-8000 and S30 buffer. Before use, the PEG-S30 buffer (at 4°C) should always be stirred to avoid PEG deposition. Adjust the dialysis time so as to concentrate the extract up to 0.86 times the volume. Dialyze the concentrated extract at 4°C for 60 min against 2 l of S30 buffer.
13. Transfer the extract to 1.5 ml tubes. Freeze the tubes in liquid nitrogen. Store them at 80°C.

9.2.2   Considerations About the Preparation Protocol

The *E. coli* cells are grown in either complex (LB) medium or glucose minimal medium. The type of growth medium affects both the activity and composition of the resulting extract [33]. In addition, the type of *E. coli* cells affects the expression level and the residual enzymatic activity to some extent. In our laboratory, *E. coli* A19 and BL21 Star (DE3) (Invitrogen, Carlsbad, CA) strains are preferably used, for their high expression levels. When culturing *E. coli* cells, the growth of the cells is an important factor that significantly influences the activity of the extract [34]. Therefore, the growth rate of the cells should be monitored. If the growth is slower than usual, then it is likely that the resulting extract will have lower activity.

Cells within the exponential growth phase are harvested by centrifugation, and then washed four times with buffer. After the final wash, the cells are disrupted. This step also affects the activity of the resulting extract. The cells should be extensively disrupted. However, the protein synthesis machinery, such as ribosomes, should not be damaged. The solution containing the disrupted cells is centrifuged at 30,000 g to eliminate the insoluble materials. It may be noteworthy that one preparation protocol performs this centrifugation at 12,000 g [35]. According to a report by Pederson and coworkers, the use of this extract (termed S12 extract) leads to a 30% increase in the expression level [36].

A run-off reaction (or called pre-incubation) is then performed to release ribosomes/polysomes from the endogenous mRNA. In the original procedure, a solution containing amino acids and polymerases without template DNA was added and incubated for 40–80 min. However, the addition of amino acids to the extract can cause the dilution of the added isotope labeled amino acids. Therefore, we perform the run-off reaction without adding the solution. Fortunately, the run-off reaction reportedly proceeds even in the absence of the solution containing amino acids and polymerases [30].

After the run-off reaction, the solution is dialyzed, using a semi-permeable membrane with a molecular weight cut off of 6–8,000. By doing so, small molecules, (e.g., ADP, inorganic phosphate and amino acids) are eliminated from the extract solution. The elimination of the small molecules is important, as they disturb the protein synthesis and cause the dilution of isotope labeled amino acids.

In our laboratory, a gel filtration step is performed after the dialysis to minimize the amount of endogenous amino acids present in the resulting extract [27, 31, 32]. Our tests revealed that some endogenous amino acids remained even after the dialysis. The gel-filtration step separates the extract from the endogenous amino acids. For example, when [U-^{2}H, ^{13}C, ^{15}N]-calmodulin (CaM) protein was produced by using an *E. coli* S30 extract prepared by the original protocol, the labeling efficiency was only ~90%, but it was improved up to 96% by introducing the gel filtration step [27] (Fig. 9.1).

After the dialysis, the resulting extract is condensed by using polyethylene glycol (PEG). The condensation of extract reportedly increases the productivity of the extract [4, 5]. In our experience, however, excess condensation leads to the precipitation of the extract, and thus the duration of the PEG condensation should be adjusted such that the highest activity is obtained.

Finally, the extracts are retrieved and stored at −80°C. To evaluate the presence of endogenous amino acids in the prepared extract, an isotope labeled protein is expressed using the prepared sample, and then the labeling efficiency is monitored by either mass spectroscopy or a ^{13}C-filtered ^{1}H1D experiment [27].

9.3 *E. coli* Cell-Free Protein Synthesis

9.3.1 *Configuration of the E. coli Cell-Free Protein Synthesis*

The cell-free systems are classified into two types: batch-type and continuous exchange cell-free (CECF) configurations. In the batch-type configuration, the experiment is performed on small-scales (~a few μl). In the early days, the protein synthesis in the batch-configuration ceased soon after the initiation of protein synthesis, probably due to the depletion of the energy sources and the accumulation of inorganic phosphate. Therefore, the batch-configuration mode is mainly used for checking the expression level, and is not used for preparative scale production. Recently, however, the productivity of the batch-mode has been dramatically increased, by improving the ATP (adenosine triphosphate) regeneration system. In the case of the CECF configuration, the reaction solution is connected to the dialysis solution via a semi-permeable membrane, such that the energy sources are continuously supplied from the dialysis solution, and the by-products generated in the reaction solution are diluted [4]. One representative of CECF is a dialysis system, where the reaction solution and the dialysis solution are partitioned by using a semi-permeable membrane (Fig. 9.2). The protein synthesis occurs in the

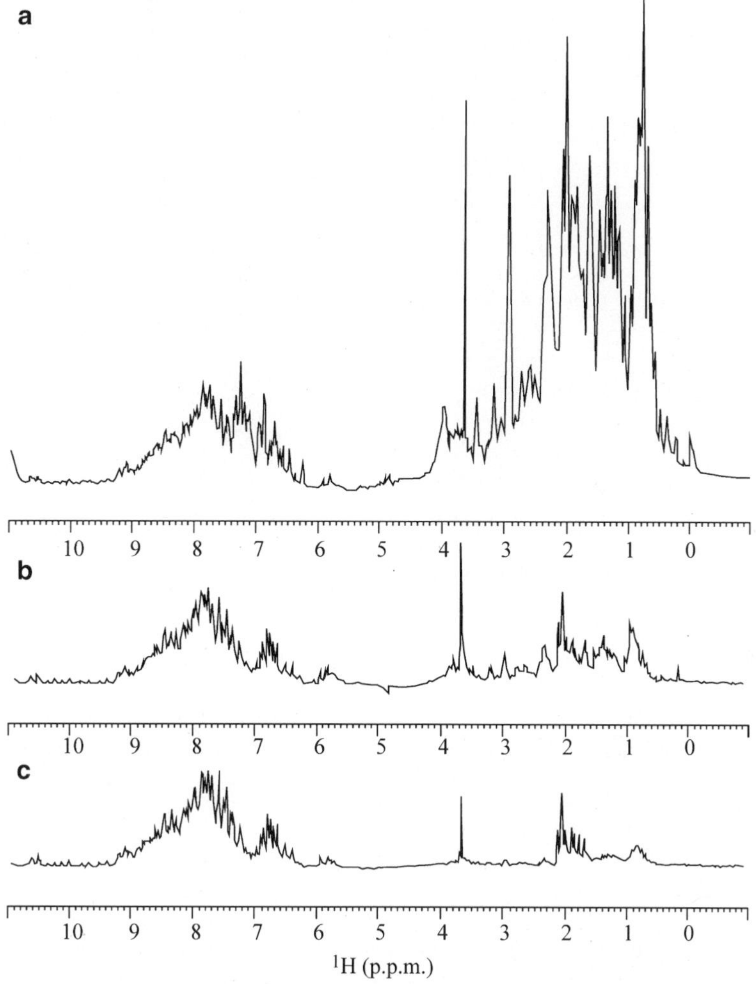

Fig. 9.1 ^{13}C-filtered ^{1}H-NMR spectra of CaM. (**a**) ^{15}N-labeled CaM synthesized with the conventional extract. (**b**) ^{2}H, ^{13}C, ^{15}N-labeled CaM synthesized with the conventional S30 extract. (**c**) ^{2}H, ^{13}C, ^{15}N-labeled CaM synthesized with the improved, dialyzed S30 extract. All of the spectra are adjusted for the intensities of the amide region. Since the peaks of the protons attached to ^{13}C are filtered, only the protons attached to ^{12}C give rise to resonances in the aliphatic region (Reproduced from Ref. [27]. With permission from Elsevier Science)

reaction solution, and the dialysis solution supplies energy sources and dilutes harmful by-products, such as adenosine diphosphate (ADP) and inorganic phosphate. In the following section, the cell-free synthesis using the dialysis system is introduced.

9.3.2 Composition of the E. coli Cell-Free Reaction Solution

The typical compositions of the solutions in the dialysis system are shown in Table 9.1. For the dialysis system, the reaction and dialysis solutions are prepared separately. The reaction solution contains components required for producing the proteins encoded by the added DNA template. On the other hand, the dialysis solution provides energy sources for the reaction solution and dilutes the generated by-products.

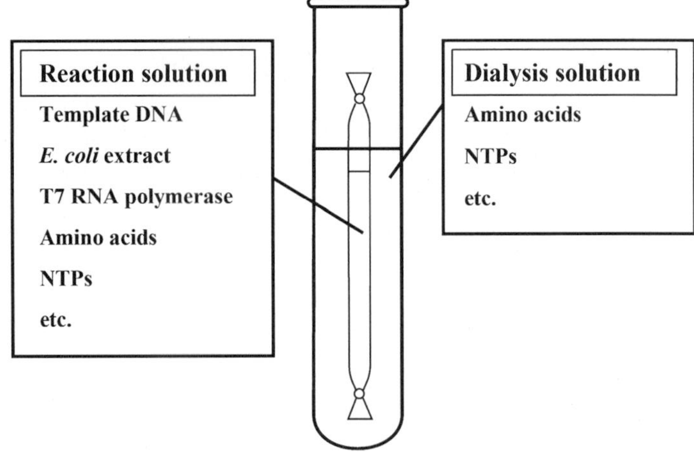

Fig. 9.2 The dialysis system for cell-free expression

Table 9.1 Composition of reaction and dialysis solutions in cell-free reactions

Stock solution	Reaction solution	Dialysis solution
RNase-free water	112 µl	1160.8 µl
1.4 M NH$_4$OAc	9.8 µl	39.2 µl
0.5 M Mg(OAc)$_2$	15 µl	60 µl
Amino acid mixture (25 mM each)	20 µl	80 µl
0.645 M creatine phosphate	40 µl	160 µl
Low molecular weight mixture[a]	125 µl	500 µl
1 mg/ml template DNA	10 µl	–
11 mg/ml T7 RNA polymerase	4.5 µl	–
40 units/µl RNase inhibitor	1.25 µl	–
10 mg/ml creatine kinase	12.5 µl	–
E. coli S30 extract	150 µl	–
Total volume	0.5 ml	2 ml

[a]The low molecular weight (LM) mixture is prepared by combining the following components: 22 ml of 2 *M* HEPES-KOH (pH 7.5), 33.4 ml of 6 *M* K(OAc), 210 mg of DTT, 530 mg of ATP (Sigma), 338 mg of CTP (Sigma), 335 mg of GTP (Sigma), 310 mg of UTP (Sigma), 172 mg of cAMP (Wako, Osaka, Japan), 28 mg of folinic acid (Sigma) and 140 mg of tRNA (Roche, Basel, Switzerland), and, if needed, 64 ml 50% (w/v) PEG-8000. RNase-free water is then added, to bring the volume up to 200 ml

In *E. coli* cell-free expression, the transcription and translation occur in a vessel in a coupled manner. Most of the currently employed *E. coli* cell-free synthesis systems are designed such that the mRNA is transcribed from the template DNA by T7 RNA polymerase [37]. Therefore, a plasmid encoding a DNA sequence under the control of the T7 promoter, such as the pET vectors (Novagen), can be used as-is. An advantage of the cell-free system over cellular expression is that PCR products can also be used as templates, thus accelerating the functional and structural analyses of proteins [38].

The continuous supply of ATP to the reaction system is essential for high productivity. For this purpose, a compound carrying a high-energy phosphate bond and the corresponding kinase (e.g., phospho(enol)pyruvate (PEP)/pyruvate kinase and creatine phosphate (CP)/creatine kinase) are included in the reaction solution. One problem with the energy source regeneration system is that the high-energy compound is degraded by the phosphatases present in the extract, which leads to the production of inorganic phosphate [39]. The accumulation of inorganic phosphate leads to the cession of protein synthesis. To mitigate this problem, CP is used as an energy source rather than PEP, as CP does not exist in *E. coli* cells and thus there are fewer phosphatases able to degrade CP in the *E. coli* extract [5].

The 20 amino acids are supplemented as precursors in the cell-free system. Therefore, residue-selective labeling can readily be achieved by using an amino acid bearing an intended isotope labeling pattern. For the production of uniformly ^{13}C and/or ^{15}N labeled proteins, an algal hydrolysate containing a mixture of ^{13}C and/or ^{15}N labeled amino acids is useful. However, it should be noted that the composition of each amino acid in the algal hydrolysate varies, depending on the lot and vendor. Especially, most algal hydrolysates lack Asn, Gln, Trp and Cys residues. In this regard, Asn and Gln are synthesized from Asp and Glu by transaminases present in the *E. coli* extract, and the side-chain amino groups of Asn and Gln are transferred from ammonium acetate. Therefore, by using ^{15}N ammonium acetate, the side-chain amino groups are enriched by ^{15}N. On the other hand, Trp and Cys are not synthesized, and thus they must be supplemented if the target protein contains them in its amino acid sequence.

9.3.3   Protocol for Cell-Free Expression

1. Wear sanitary gloves and thaw the LM mixture and the S30 extract.
2. Prepare the reaction and dialysis solutions, as shown in Table 9.1. In the case of a small-scale reaction for evaluating expression levels, the typical volumes of the reaction and dialysis solutions are 0.5 and 2.0 ml, respectively. For large production quantities, each volume is scaled up while maintaining the volume ratio of the reaction solution to the dialysis solution.
3. Transfer the dialysis solution to an RNase-free vessel. Tie one end of the dialysis tube firmly. Transfer the reaction solution into the dialysis tube, and tie off the other end. Fold the tubing four to six times, and place the dialysis tube into the vessel, such that the dialysis tube is completely submerged in the dialysis solution.
4. Incubate the vessel with shaking for 4–8 h at 37°C.
5. Retrieve the reaction and dialysis solutions. If the produced expressed protein has a molecular weight smaller than molecular weight cut-off of the membrane, then the produced protein can be found in both the reaction and dialysis solutions.

9.3.4   Optimization of Conditions for Small-Scale Reactions

The optimal conditions for cell-free reactions vary, depending on the kind of protein, and thus the reaction conditions should be optimized to some extent before proceeding to the expression of isotopically labeled proteins. The optimization is performed based on pilot experiments with small-scale reactions.

As the expression level of the target protein in an *E. coli* cell-free expression system is affected by the sequence of the template DNA, it is worth considering the introduction of a silent mutation to the DNA sequence. Even if they encode the same amino acid, distinct DNA sequences lead to different expression levels, and thus the introduction of a mutation is worth trying, unless the amino acid is altered. Empirically, the DNA sequence following the translation initiation site greatly affects the expression level. Unfortunately, however, there is no rational strategy to design silent mutations, and thus the optimized sequence must be found by trial and error. To systematically optimize the DNA sequence of the N-terminus, we transfer the target DNA sequence into a line of vectors with different silent mutations [27]. These vectors were constructed by introducing the DNA sequence encoding the N-terminal His-tag sequence of the pET15 vector (Novagen) into the multi-cloning site of the pIVEX2.3d vector (Roche). The DNA sequence encoding a target protein is introduced into the sites following the His-tag sequence in the line of vectors. While all of the vectors encode the target protein fused with the N-terminal His-tag, their expression levels differ, and thus a vector with high productivity can be identified (Fig. 9.3).

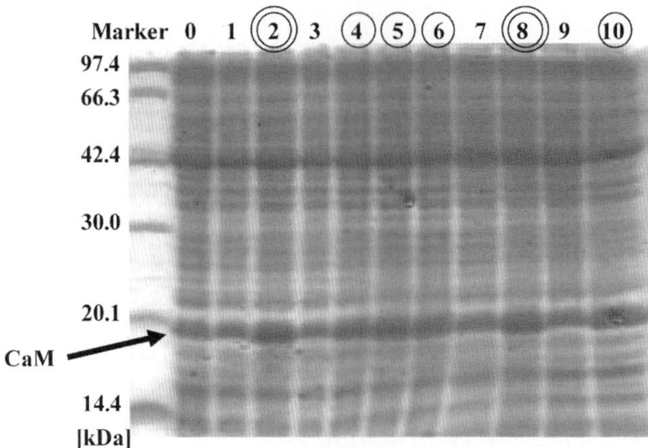

Fig. 9.3 SDS–PAGE of CaM, synthesized using plasmids without or with silent mutations predicted to enhance production. Lane 0: protein from the construct without the silent mutations. Lanes 1–10: proteins from constructs with ten different silent mutations. In the comparison of the intensities of lanes 1–10 to lane 0, the candidates with higher intensity are shown by a *circle* and those with much higher intensity by a *doubled circle* (Reproduced from Ref. [27]. With permission from Elsevier Science)

The concentration of magnesium is also an important parameter. During a cell-free reaction, phosphate accumulates due to the presence of endogenous phosphatases, and magnesium precipitates upon binding to the phosphate. This problem could be mitigated by adding a large amount of magnesium. In practical applications, however, the optimal concentration of magnesium differs depending on the target protein, and thus the best magnesium concentration should be determined in small-scale reactions.

The volume ratio of the dialysis and reaction solutions is an important parameter. By increasing the ratio, the productivity is expected to increase, due to the increased supply of energy and efficient dilution of harmful by-products. However, the isotope labeled amino acids are membrane-permeable, and thus the efficiency of amino acid usage becomes worse with a higher volume ratio, in terms of the cost of isotope labeled amino acids. In our laboratory, the ratio of the dialysis solution to the reaction solution ranges from 4 to 10.

9.3.5 Checking the NMR Spectral Quality of Proteins Produced by the Cell-Free Synthesis System

Once the conditions for achieving an acceptable expression level are established in small-scale reactions, we recommend scaling-up the reaction volume to produce a uniformly ^{15}N-labeled protein under the established conditions. A comparison of its [^{1}H, ^{15}N]-HSQC spectrum with that produced by cellular expression reveals the differences between them. As the enzyme activity is reduced in the *in vitro* expression, the property of the resulting protein may be altered. Special attention should be paid to the incomplete deformylation of the N-terminus of proteins in the cell-free reaction [26, 27]. To overcome this problem, the use of the aforementioned engineered N-terminal cleavable tag is helpful [27]. The inhomogeneous N-terminus is cleaved by a protease, and thus the homogeneous N-terminus is obtained (Fig. 9.4).

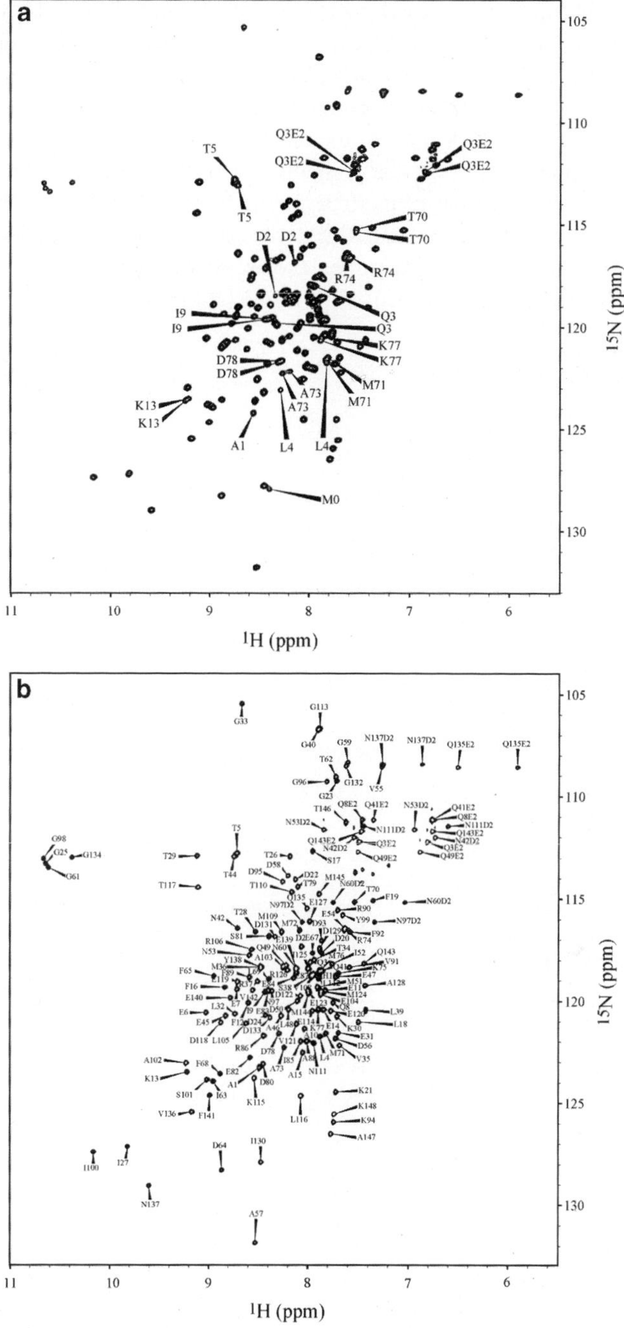

Fig. 9.4 Incompete deformylation leads to the doubling of NMR peaks. ¹H-¹⁵N HSQC spectra of [U-¹³C, ¹⁵N] CaM synthesized by an *E. coli* cell-free reaction gives rise to doubled peaks for residues spatially proximal to its N-terminus (**a**) This peak doubling was eliminated by producing CaM as a fusion with an N-terminal cleavable tag and digestion by the protease (**b**) The extra peaks in (**a**) and all peaks in (**b**) are labeled with their assignments (Reproduced from Ref. [27]. With permission from Elsevier Science)

It should be noted that metabolic conversion is not completely suppressed even in the *E. coli* cell-free system. For example, the exchange of the α- and, to a lesser extent, β-proton(s) of an amino acid with the solvent water can occur in the *E. coli* cell-free system [40, 41]. Therefore, if the expression of a deuterated protein is intended, the reaction should be performed in a D_2O solution; otherwise, the α-protons and, to a lesser extent, β-protons of the amino acids will exchange with the solvent water. To suppress the activity of the enzyme, the use of inhibitors has been proposed [26, 41–44]. However, an excess amount of inhibitor can exert harmful effects on the expression of the target proteins. Therefore, the optimal concentration should be evaluated in terms of a compromise between the inhibition of scrambling and the expression level. In the future, the use of the PURE system, in which the protein synthesis machinery is reconstituted exclusively with recombinant proteins, will ultimately resolve this problem.

9.4 Conclusions

In this chapter, we described the procedure for *E. coli* cell-free protein synthesis. Along with *E. coli* cell-free expression, wheat-germ cell-free expression has now become a practical method for producing NMR samples. In addition, the application of the PURE system to NMR sample production has high potential, due to the ability to completely control the enzymatic activity. The continuous improvement of cell-free protein synthesis and the increasing number of NMR applications will pave the way toward unprecedented NMR studies.

References

1. Kigawa T, Muto Y, Yokoyama S (1995) Cell-free synthesis and amino acid-selective stable isotope labeling of proteins for NMR analysis. J Biomol NMR 6:129–134
2. Zubay G (1973) In vitro synthesis of protein in microbial systems. Ann Rev Genet 7:267–287
3. Spirin AS, Baranov VI, Ryabova LA, Ovodov SY, Alakhov YB (1988) A continuous cell-free translation system capable of producing polypeptides in high yield. Science 242:1162–1164
4. Kim DM, Kigawa T, Choi CY, Yokoyama S (1996) A highly efficient cell-free protein synthesis system from *Escherichia coli*. Eur J Biochem 239:881–886
5. Kigawa T, Yabuki T, Yoshida Y, Tsutsui M, Ito Y, Shibata T, Yokoyama S (1999) Cell-free production and stable-isotope labeling of milligram quantities of proteins. FEBS Lett 442:15–19
6. Kramer G, Kudlicki W, Hardesty B (1999) Cell-free coupled transcription-translation systems from *Escherichia coli*. Oxford University Press, New York, pp 129–165
7. Kim DM, Swartz JR (2000) Prolonging cell-free protein synthesis by selective reagent additions. Biotechnol Prog 16:385–390
8. Madin K, Sawasaki T, Ogasawara T, Endo Y (2000) A highly efficient and robust cell-free protein synthesis system prepared from wheat embryos: plants apparently contain a suicide system directed at ribosomes. Proc Natl Acad Sci USA 97:559–564
9. Endo Y, Sawasaki T (2003) High-throughput, genome-scale protein production method based on the wheat germ cell-free expression system. Biotechnol Adv 21:695–713
10. Shimizu Y, Inoue A, Tomari Y, Suzuki T, Yokogawa T, Nishikawa K, Ueda T (2001) Cell-free translation reconstituted with purified components. Nat Biotechnol 19:751–755
11. Shimizu Y, Ueda T (2010) PURE technology. Methods Mol Biol 607:11–21
12. Parker MJ, Aulton-Jones M, Hounslow AM, Craven CJ (2004) A combinatorial selective labeling method for the assignment of backbone amide NMR resonances. J Am Chem Soc 126:5020–5021
13. Wu PS, Ozawa K, Jergic S, Su XC, Dixon NE, Otting G (2006) Amino-acid type identification in ^{15}N-HSQC spectra by combinatorial selective ^{15}N-labelling. J Biomol NMR 34:13–21
14. Ozawa K, Headlam MJ, Mouradov D, Watt SJ, Beck JL, Rodgers KJ, Dean RT, Huber T, Otting G, Dixon NE (2005) Translational incorporation of L-3,4-dihydroxyphenylalanine into proteins. FEBS J 272:3162–3171
15. Ugwumba IN, Ozawa K, de la Cruz L, Xu ZQ, Herlt AJ, Hadler KS, Coppin C, Brown SE, Schenk G, Oakeshott JG, Otting G (2011) Using a genetically encoded fluorescent amino acid as a site-specific probe to detect binding of low-molecular-weight compounds. Assay Drug Dev Technol 9:50–57

16. Loscha KV, Herlt AJ, Qi R, Huber T, Ozawa K, Otting G (2012) Multiple-site labeling of proteins with unnatural amino acids. Angew Chem Int Ed 51:1–5
17. Sobhanifar S, Reckel S, Junge F, Schwarz D, Kai L, Karbyshev M, Löhr F, Bernhard F, Dötsch V (2010) Cell-free expression and stable isotope labelling strategies for membrane proteins. J Biomol NMR 46:33–43
18. Junge F, Haberstock S, Roos C, Stefer S, Proverbio D, Dötsch V, Bernhard F (2011) Advances in cell-free protein synthesis for the functional and structural analysis of membrane proteins. N Biotechnol 28:262–271
19. Reckel S, Gottstein D, Stehle J, Löhr F, Verhoefen MK, Takeda M, Silvers R, Kainosho M, Glaubitz C, Wachtveitl J, Bernhard F, Schwalbe H, Güntert P, Dötsch V (2011) Solution NMR structure of proteorhodopsin. Angew Chem Int Ed Engl 50:11942–11946
20. Kainosho M, Torizawa T, Iwashita Y, Terauchi T, Ono AM, Güntert P (2006) Optimal isotope labelling for NMR protein structure determinations. Nature 440:52–57
21. Kainosho M, Güntert P (2009) SAIL-Stereo-array isotope labeling. Q Rev Biophys 7:1–54
22. Takeda M, Chang CK, Ikeya T, Güntert P, Chang YH, Hsu YL, Huang TH, Kainosho M (2008) Solution structure of the C-terminal dimerization domain of SARS coronavirus nucleocapsid protein solved by the SAIL-NMR method. J Mol Biol 380:608–622
23. Takeda M, Sugimori N, Torizawa T, Terauchi T, Ono AM, Yagi H, Yamaguchi Y, Kato K, Ikeya T, Jee J, Güntert P, Aceti DJ, Markley JL, Kainosho M (2008) Structure of the putative 32 kDa myrosinase binding protein from Arabidopsis (At3g16450.1) determined by SAIL-NMR. FEBS J 275:5873–5884
24. Takeda M, Jee J, Ono AM, Terauchi T, Kainosho M (2011) Hydrogen exchange study on the hydroxyl groups of serine and threonine residues in proteins and structure refinement using NOE restraints with polar side-chain groups. J Am Chem Soc 133:17420–17427
25. Kim DM, Choi CY (1996) A semicontinuous prokaryotic coupled transcription/translation system using a dialysis membrane. Biotechnol Prog 12:645–649
26. Ozawa K, Headlam MJ, Schaeffer PM, Henderson BR, Dixon NE, Otting G (2004) Optimization of an Escherichia coli system for cell-free synthesis of selectively N-labelled proteins for rapid analysis by NMR spectroscopy. Eur J Biochem 271:4084–4093
27. Torizawa T, Shimizu M, Taoka M, Miyano H, Kainosho M (2004) Efficient production of isotopically labeled proteins by cell-free synthesis: a practical protocol. J Biomol NMR 30:311–325
28. Nirenberg M (1963) Cell-free protein synthesis directed by messenger RNA. Methods Enzymol 6:17–23
29. Pratt JM (1984) Coupled transcription-translation in prokaryotic cell-free systems. In: Hames BD, Higgins SJ (eds) Transcription and translation: a practical approach. IRL Press, Oxford, pp 179–209
30. Liu DV, Zawada JF, Swartz JR (2005) Streamlining *Escherichia coli* S30 extract preparation for economical cell-free protein synthesis. Biotechnol Prog 21:460–465
31. Takeda M, Ikeya T, Güntert P, Kainosho M (2007) Automated structure determination of proteins with the SAIL-FLYA NMR method. Nat Protoc 2:2896–2902
32. Takeda M, Kainosho M (2012) Cell-free protein production for NMR studies. Methods Mol Biol 831:71–84
33. Schindler PT, Baumann S, Reuss M, Siemann M (2000) In vitro coupled transcription translation: effects of modification in lysate preparation on protein composition and biosynthesis activity. Electrophoresis 21:2606–2609
34. Zawada J, Swartz JR (2006) Effects of growth rate on cell extract performance in cell-free protein synthesis. Biotechnol Bioeng 94:618–624
35. Kim TW, Keum JW, Oh IS, Choi CY, Park CG, Kim DM (2006) Simple procedures for the construction of a robust and cost-effective cell-free protein synthesis system. J Biotechnol 126:554–561
36. Pedersen A, Hellberg K, Enberg J, Karlsson BG (2011) Rational improvement of cell-free protein synthesis. N Biotechnol 30:218–224
37. Studier FW, Rosenberg AH, Dunn JJ, Dubendorff JW (1990) Use of T7 RNA polymerase to direct expression of cloned genes. Methods Enzymol 185:60–89
38. Wu PS, Ozawa K, Lim SP, Vasudevan SG, Dixon NE, Otting G (2007) Cell-free transcription/translation from PCR-amplified DNA for high-throughput NMR studies. Angew Chem Int Ed Engl 46:3356–3358
39. Kawarasaki Y, Nakano H, Yamane T (1998) Phosphatase-immunodepleted cell-free protein synthesis system. J Biotechnol 61:199–208
40. Etezady-Esfarjani T, Hiller S, Villalba C, Wüthrich K (2007) Cell-free protein synthesis of perdeuterated proteins for NMR studies. J Biomol NMR 39:229–238
41. Tonelli M, Singarapu KK, Makino S, Sahu SC, Matsubara Y, Endo Y, Kainosho M, Markley JL (2011) Hydrogen exchange during cell-free incorporation of deuterated amino acids and an approach to its inhibition. J Biomol NMR 51:467–476
42. Jia X, Ozawa K, Loscha K, Otting G (2009) Glutarate and N-acetyl-L-glutamate buffers for cell-free synthesis of selectively 15N-labelled proteins. J Biomol NMR 44:59–67
43. Yokoyama J, Matsuda T, Koshiba S, Tochio N, Kigawa T (2011) A practical method for cell-free protein synthesis to avoid stable isotope scrambling and dilution. Anal Biochem 411:223–229
44. Su XC, Loh CT, Qi R, Otting G (2011) Suppression of isotope scrambling in cell-free protein synthesis by broadband inhibition of PLP enymes for selective ^{15}N-labelling and production of perdeuterated proteins in H_2O. J Biomol NMR 50:35–42

Chapter 10
Isotope Labeling in Insect Cells

Alvar D. Gossert and Wolfgang Jahnke

Abstract Target proteins in contemporary NMR are becoming increasingly complex. The interest lies on membrane proteins, multi-domain proteins, secreted proteins with secondary modifications etc., which can often only be expressed in eukaryotic expression hosts. Although *E. coli* is the organism of choice for expression of proteins in isotope labeled form for NMR studies, bacterial cells have limitations concerning their ability of producing soluble and well-folded proteins of human origin. Insect cells are eukaryotic cells and therefore evolutionary closer to human cells than bacterial cells. Therefore a larger share of human proteins can be functionally expressed in them, for example multi-domain proteins, protein complexes, membrane proteins and proteins requiring post-translational modifications. In order to study these proteins by NMR, they are ideally prepared in isotope labeled form. In this chapter different strategies for isotope labeling in insect cells are described – uniform and amino acid specific. A general introduction to expression with baculovirus infected insect cells is given followed by a detailed descriptions of labeling approaches. The chapter is concluded with case studies, describing successful application of isotope labeling in insect cells for NMR studies including solid-state experiments, ligand binding studies and protein dynamics.

10.1   Introduction

Target proteins in contemporary NMR are becoming increasingly complex. The interest lies on membrane proteins, multi-domain proteins, secreted proteins with secondary modifications etc., which can often only be expressed in eukaryotic expression hosts. Although *E. coli* is the organism of choice for expression of proteins in isotope labeled form for NMR studies, bacterial cells have limitations concerning their ability of producing soluble and well-folded proteins of human origin. Insect cells are eukaryotic cells and therefore evolutionary closer to human cells than *i.e.* bacterial cells. Therefore a larger share of human proteins can be functionally expressed in them, especially multi-domain proteins and protein complexes are correctly folded and active. Furthermore, post-translational modifications are introduced into the polypeptides, like phosphorylation. Very similar secretion pathways as in human cells enable expression of membrane proteins and secretory proteins with correct disulfide formation and glycosylation patterns. These post-translational modifications are often essential

A.D. Gossert (✉) • W. Jahnke
Novartis Institutes for BioMedical Research, Novartis AG,
4002 Basel, Switzerland
e-mail: alvar.gossert@novartis.com

H.S. Atreya (ed.), *Isotope Labeling in Biomolecular NMR*, Advances in Experimental Medicine and Biology 992,   179
DOI 10.1007/978-94-007-4954-2_10, © Springer Science+Business Media Dordrecht 2012

for protein folding, stability and activity. Another advantage of insect cells is that small molecules can pass their delicate plasma membranes. Proteins can therefore be expressed in presence of stabilizing compounds or co-factors. All-in-all, a larger fraction of the human proteome is accessible for expression in insect cells compared to bacteria.

With more complex cells, however, come more complex protocols for heterologuous protein expression. This can be appreciated in a comparison of the general workflow with that of bacterial expression. Culturing of insect cells in shake flasks is practically as simple as growing bacteria, just with the added complication that all manipulations need to be carried out in a sterile environment. However, most other aspects of the expression workflow require higher efforts and more dedication. Bacteria have (i) very fast growth rates and high protein synthesis capacities, (ii) foreign genes can be introduced in matter of minutes and – important for isotope labeling – (iii) they are able to synthesize all molecules important for life from basic building blocks. On the insect cell side, this looks quite different: (i) The slow growth rate of insect cells leads to expression times of typically 2–3 days, up to 7 days, and protein yields can be expected to be in the range of a few mg/l. Continuous maintenance of cell stocks and lengthy building up of expression sized pre-cultures is also laborious and time consuming. (ii) For high levels of expression, the gene of interest needs to be delivered in an infectious particle, the baculovirus. Therefore, a second line of work is required for producing highly infectious recombinant baculovirus, which takes weeks. (iii) Growth media are more complex – and more expensive. They easily contain more than 60 ingredients, as nearly all nutrients like amino acids and vitamins need to be given to insect cells on a silver plate, due to their poor metabolic capacities.

However, despite these disadvantages, insect cells represent a very powerful tool for expression of difficult proteins for structural studies, making accessible a larger fraction of the proteome for NMR studies.

In order to give an overview of the applicability of insect cells for NMR studies, the further text is divided into three sections: A description of the general culturing process (Sect. 10.2), the adaptations needed for isotope labeling (Sect. 10.3) and a number of case studies of successful application of isotope labeling in insect cells (Sect. 10.4).

10.2 Insect Cells as Expression System

In this section a brief overview over the general procedures is given, which are necessary for successful over-expression of proteins in insect cells. Mainly the points are touched, which are important for understanding the special considerations required for isotope labeling. These are described in the next section. For actual comprehensive protocols the reader is referred to the standard works [10–12], where detailed references can be found.

10.2.1 Insect Cells

In the late 1980s to early 1990s, insect cells emerged as an alternative eukaryotic expression system for recombinant proteins. During the last decade insect cells have established their firm place as an expression platform for structural biology. The main insect cell lines for the purpose of expression are derived from two organisms. In 1977, ovary cells of the fall army worm, *spodoptera frugiperda*, were isolated and cultured resulting in the cell lines sf21 and sf9, where sf9 is a monoclonal derivative of sf21 [17]. In 1983 the Hi5 line was established from cells of the cabbage looper, *Trichopulsia ni* [6]. In conjunction with a virus vehicle, the baculovirus, these became the most popular eukaryotic cell lines for protein over-expression for structural biology.

10.2.2   The Baculovirus Life-Cycle and Its Use for Heterologuous Protein Expression

Baculoviruses are a family of arthropode specific viruses. They are rod shaped particles with a genome of 80–200 kbp of double stranded, circular DNA, the so-called bacmid. A prototypical member of this family is the *Autographa californica* multiple nuclear polyhedrosis virus (AcNPV), which is used as the main vehicle for heterologuous protein production in insect cells. The life-cycle of this virus starts with the primary infection of a host cell, where it is replicated. After 10–20 h, new virions emerge from the infected cell and infect other cells in the same organism, where the replication is re-initiated in a secondary infection phase, which ultimately leads to death of the host. The free virion form, however, is not a stable entity outside of the insect organism. Therefore, in the very late stage of infection, *i.e.* after more than 30 h, the virus induces production of copious amounts (up to 50% of total protein) of the proteins polyhedrin and p10 that form a cube-shaped protein matrix, in which virus particles are embedded. These are the so-called occlusion bodies. Occluded viruses are long-term stable outside of the host cells and are, eventually, ingested by another insect larva. In the alkaline conditions of the gut lumen, the polyhedrin matrix dissolves and releases virions that infect the gut cells. With this event, a new cycle in the baculoviral life is initiated.

For the purpose of over-expression of heterologuous proteins, the naturally highly expressed polyhedrin or p10 genes are replaced by the gene of interest. In culture conditions, the polyhedrin/p10 occlusion bodies are not vital for the viral life-cycle, as free virions are stable in cell culture media and conserve their infectivity. Therefore, baculoviruses carrying a heterologuous protein are as infective and proliferative as unaltered ones in cell-culture conditions and can be used to infect insect cells. The result is, that a foreign gene under the control of the strong polyhedrin or p10 promotor is expressed in high amounts by the insect cells 1–3 days post infection.

10.2.3   Lab Equipment

Insect cells and baculovirus can be cultured and handled in biosafety level 1 facilities. (Caveat: In Europe, Hi5 cells need to be handled in BL2 facilities, as they contain viral DNA potentially harmful to human beings). The basic large equipment needed for culturing of insect cells consists of a sterile bench for carrying out all manipulations, and incubators for culturing the cells in petri dishes and shake flasks of small and large volume. It is important to have enough autoclaving capacity for sterilizing the culture flask and consumables. Further essential equipment comprises an inverse light microscope for assessing cell parameters like viability, cell size and number. A waterbath for pre-warming media to 27°C is also important. Finally a centrifuge is needed for harvesting the cells. A state of the art facility for insect cell expression may have an automatic cell counter that fully replaces the microscope, and more sophisticated devices replacing shake flasks for large scale expression, like wave-bags or fermenters.

In general there are only small differences to a typical *E. coli* lab, except for the sterile bench and the microscope. However, it is strongly advisable not to share equipment with an *E. coli* expression lab, because of the high risk of cross-contamination.

10.2.4   Culturing of Insect Cells

Insect cells (*Spodoptera frugiperda, Trichopulsia ni*) are usually grown in erlenmeyer shake flasks at 27°C and gentle mixing at 70–90 rpm. Basic incubators can be used, as growth takes place in ambient atmosphere. Under these conditions doubling times are about 20–24 h in commercial media.

During culturing, cells are regularly monitored with light microscopy to assess their viability, proliferation and density. Viability is assessed with a dye that penetrates only the non-intact membranes of dead cells, coloring them blue. In this way the fraction of living and dead cells can be counted. The cell diameter, typically 14–18 μm, is a second parameter which allows evaluating viability, the smaller the better. Also, it allows characterizing the efficiency of viral infection, as it is manifested by massively enlarged cells with diameters of up to 24 μm. Proliferation can be monitored by growth rates but also by looking for cell doublets, which indicate cells in the process of doubling and are a sign of a well-growing culture. However, cell lumps of many cells indicate declining viability of the culture and call for measures to improve growth conditions.

The practical process to introduce and establish a new cell line in the lab is the following: Insect cells are obtained in frozen aliquots, *e.g.* from ATCC or Invitrogen, and need to be adapted very gently to growth in suspension in a chosen medium. To this end, the insect cell medium is supplemented with 10% fetal calf serum (FCS). Cells are grown in 50 ml erlenmeyer flasks and every 2–3 days, cells are taken up in fresh medium. At the beginning of an adaptation phase, 50% of the grown culture is mixed with 50% v/v of fresh medium. Progressively larger dilutions up to 10–90% are carried out and after 10–20 such passages a stable stock-culture is established. FCS can slowly be removed over a few passages, and a stock adapted to "serum free" medium is obtained. During this adaptation phase, the cells need to be carefully monitored as described above, and at first signs of cell lumps or reduced viability, adaptation should be slowed down or even reverted one or two steps. Once a stable stock is obtained, it is maintained by regular passaging into fresh medium twice a week. This can be carried on for up to 200 passages, however, generally after 50 passages a backup stock is started.

Most labs keep different stocks, as expression levels can vary largely (up to three fold) depending on combinations of cell strains and media. Adapting strains to different media requires gradual progressive change of the medium, often with the need of re-introducing FCS as a supplement.

For a large scale expression experiment, an expression culture is built up starting from a well-growing individual stock of a density of 4–6×10^6 cells/ml. Over the course of days, cells are grown in large volume erlenmeyer flasks, in order to build up the required number cells. Typically, the target density of an expression culture is 1.5×10^6 cells/ml. Alternatives to large erlenmeyer flasks are wave-bag devices or fermenters. For expression of a target protein, the culture is infected with baculovirus carrying the gene of interest at a given ratio of virus to cell, termed multiplicity of infection (MOI, typically 1–10). Roughly 20 h after infection, production of the target protein can be detected. Depending on the protein, expression is continued for 24–72 h, or even longer for secreted proteins. Harvesting takes place by gentle centrifugation at 400 g, as insect cells tend to rupture at more than 1,000 g.

10.2.5 *Generation of Virus*

Introduction of the gene of interest into insect cells is a scientific discipline on itself. There are several ways of delivering DNA to be transcribed to the cells: (1) transient transfection with pure DNA, (2) generation of stable cell lines carrying the gene of interest integrated in the genome and (3) introducing the gene into the host cells by means of a virus, the baculovirus. For structural biology, where high yields are needed (fulfilled by 2 and 3) and a certain flexibility for testing several constructs is often required (fulfilled by 1 and 3), strategy number 3 is the most promising one, and in fact, most often used. Therefore, the rest of the discussion will focus on the baculovirus system, which is used in conjunction with sf9/sf21 and Hi5 cells.

As described in Sect. 10.2.2, the strategy for over-expression of a given target protein consists in inserting the gene of interest in place of the polyhedrin or p10 gene in the baculovirus genome. The molecular biology required for gene insertion into a bacmid is not trivial, as the bacmid is roughly

100 kbp in size, compared to 4–6 kbp for a plasmid used in *E. coli* expression. A plethora of ever refined strategies have been put in place – and have been marketed – for creating recombinant virus. Generally, in a first step the gene of interest is inserted into a bacterial plasmid by conventional techniques. This plasmid contains elements for selection and propagation in *E. coli* and, most importantly, the gene of interest is flanked by sequences that are homologuous to the bacmid sequences flanking the poly-hedrin gene. Subsequently the gene on the plasmid needs to be inserted into the bacmid in a sequence specific manner. Homologuous recombination is a process, in which pieces of DNA flanked with identical sequences are exchanged by DNA-repair enzymes, leading in this case to introduction of recombinant DNA into a bacmid. Nowadays, the transfer of the gene of interest from the plasmid into the bacmid is usually accomplished in two major ways: (1) by using *E. coli* as an intermediate host or (2) directly in insect cells. (1) In the first strategy, the plasmid is introduced into special *E. coli* cells, which contain bacmid DNA and a transposition enzyme that will transfer the recombinant gene into the bacmid. The recombinant bacmid can then be isolated and introduced into insect cells (Bac-to-Bac, Invitrogen). (2) The gene of interest can also be introduced in a single experimental step, where insect cells are transfected simultaneously with an "empty" bacmid and the plasmid carrying the gene of interest. In the cell the recombination takes place and elaborate selection mechanisms allow for select-ing infectious, recombinant viruses. (*flash*Bac, OET; BacMagic Novagen). A few milliliters containing recombinant virus and infected cells are obtained by these procedures [8, 9, 11].

For large scale expression huge numbers of such viruses are needed, as all cells in the culture should be infected by at least one virus particle. In order to obtain such large numbers of viruses, cells are infected with virus at a certain virus-to-cell ratio (multiplicity of infection, MOI) and after prolif-eration of the virus the supernatant of the culture containing the new generation of viruses is kept. In order to further increase the concentration of the virus stock solution, it is in turn used to infect a much larger volume of cells. Usually three such amplification steps are required to obtain 50 ml of highly infectious virus stock, containing typically 10^7–10^8 plaque forming units.

The number of viruses is measured in plaque assays, where plaque forming units (pfu) are counted. In a plaque assay, a monolayer of adherent insect cell grown in a petri dish is infected with a given amount of virus solution. After 4–5 days of incubation, plaques become visible. Plaques are circular regions of killed cells in the monolayer, indicating infection with virus. Each plaque therefore represents an original infection event by a single virus. Consequently, the concentration of infectious virus – plaque forming units – can be estimated by plaque assays using appropriate dilutions of virus solution.

The plaque assay can also be used for isolating individual, monoclonal baculovirus in a process called plaque cloning. This will be discussed in the following section.

10.2.6   Optimizing Yields of Expressed Protein

As for labeling experiments in *E. coli*, due to cost reasons it is essential to ensure highest yields of expressed protein per liter of – expensive – medium. Two main factors in the culturing process affect the yields in a labeling experiment (1) the characteristics of the expression medium and pre-culture media and (2) the productivity of the baculovirus.

10.2.6.1   Expression Medium Change Leads to Reduced Yields

Insect cell media should promote highest viability of cells in pre-cultures on the one hand, but also ensure efficient infection by the baculovirus in the expression culture. Generally, stocks are main-tained in commercial and unlabeled media. For labeling therefore a medium change from pre-culture to labeled expression culture is needed. In this step, attention needs to be paid to the fact, that insect cells don't tolerate large changes in the medium very well. Changing from a commercial medium to a medium prepared in house usually results in lowered protein yields [5]. Therefore, pre-cultures should be maintained in a very similar medium as the main culture. This can be ensured by preparing all media following the same recipe or by relying on commercial media, as done in the case of BioExpress2000 (Cambridge Isotope Labs), where no change in protein production could be observed for labeled expression and expression in the corresponding unlabeled medium.

Most insect cell laboratories maintain a variety of stocks in different media of different brands that are tested for highest expression. That's because different formulations of media can lead to differences in expression levels of factors of 2–3. Some vendors offer amino acid free media upon request, which can be supplemented by labeled amino acids and yeast extract. This strategy ensures highest yields, as it gives the flexibility of using different media with their proprietary additive formulations, which are crucial for highly efficient infection by baculovirus and concomitant high level protein expression. As mentioned above, with this strategy the cells also don't experience a significant change of medium when changed form pre-culture to main culture.

10.2.6.2 Plaque Cloning of Virus

In our laboratory, for expensive labeling cultures, special attention is given to the baculovirus preparation. In a process called plaque cloning, single virus clones are selected and individual expression levels are determined. It is always astonishing, how different the expression levels of individual clones – all carrying the gene of interest – are. Plaque cloning takes an additional 1–2 weeks, but since expression levels can be more than doubled by this procedure it is often more economical to go through the process. For a labeling experiment it is therefore recommended to use plaque-cloned virus for highest expression levels.

10.3 Strategies for Isotope Labeling in Insect Cells

10.3.1 General Considerations for Isotope Labeling

The much reduced amino acid metabolism of insect cells represents the major difference important for isotope labeling compared to *E. coli*. Ten amino acids are essential and therefore need to be present in the cell culture medium. This offers an advantage for amino acid specific isotope labeling, as label dilution and metabolic scrambling to other amino acids is much reduced for these amino acids. Uniform isotope labeling, on the other hand, suffers from this fact. Insect cells are – by far – not able to synthesize a protein from basic carbon and nitrogen sources. Therefore most amino acids need to be included in the medium in labeled form, rendering uniform labeling a very costly endeavor.

10.3.2 Culture Media for Insect Cells

The typical composition of insect cell media is given in Table 10.1 [11, 20]. The ingredients can be grouped in different classes: Minerals, carbohydrates, amino acids, lipids and yeast extract. Additionally, additives like FCS or antibiotics are added. In order to produce isotope labeled proteins, obviously the carbon, nitrogen and – if deuterium labeling is needed – hydrogen sources need to be replaced by labeled ones. The most important sources for these elements that are incorporated into proteins are carbohydrates, amino acids and yeast extract.

Table 10.1 Composition of IPL41 insect cell medium

Group	Ingredient	Amount [mg/l]
	KCl	1,200
Salts	$MgSO_4$	920
	$MgCl_2 \times 6\ H_2O$	1,000
	NaCl	1,000
	$NaHCO_3$	350
	$NaH_2PO_4 \times H_2O$	1,160
	$CaCl_2$	500
	$FeSO_4 \times 7\ H_2O$	0.55
	$NH_4Mo_6O_{24}$	0.042
	$CoCl_2 \times 6\ H_2O$	0.05
	$CuSO_4$	0.4
	$MnCl_2$	0.02
	$ZnCl_2$	0.04
Carbohydrates	Glucose	2,500–10,000
	Maltose	1,000
	Sucrose	1,650
Amino acids (pure)	L-Alanine (−)	300
	L-Arginine (−)	800
	L-Asparagine (−)	1,300
	L-Aspartic acid (−)	1,300
	L-Glutamic acid (−)	1,500
	L-Glutamine (−)	2,000
	Glycine (−)	200
	L-Histidine (−)	200
	L-Isoleucine (+)	750
	L-Leucine (+)	250
	L-Lysin HCl (+)	700
	L-Methionine (+)	200–1,000
	L-Phenylalanine (+)	1,000
	L-Proline (−)	500
	L-Serine (−)	500
	L-Threonine (+)	200
	L-Tryptophan (+)	200
	L-Tyrosine 2 Na (+ Phe)	360
	L-Valine (+)	120
	L-Cystine 2 HCl (+ Met)	200
	L-Hydroxyproline	800
Vitamins	Inositol	0.4
	Nicotinic acid	0.16
	Pyridoxine HCl	0.4
	Thiamine HCl	0.08
	Ca Panthothenic acid	0.008
	p-Aminobenzenic acid	0.32
	Vitamin B12	0.24
	Biotine	0.16
	Choline chloride	20
	Folic acid	0.08
	Riboflavine	0.08
Organic acids	Succinic acid	4.8
	L-Malic acid	53.6
	Fumaric acid	4.4
	Ketoglutaric acid	30

(continued)

Table 10.1 (continued)

Group	Ingredient	Amount [mg/l]
	Cholesterol	2.81
Lipids	Tween 80	15.62
	Tocopherolacetate	1.25
	Cod liver oil*	6.25
	Pluronic	1,000
Yeast or algal extract	Yeast extract/Yeastolate*	4,000–6,000

The ingredients are divided into major groups (left)

The pH of insect cell media is adjusted to 6.2, and the osmolarity is typically 320 mOsm/kg

(+) and (−) indicate whether an amino acid is essential or not, respectively. Tyrosine is not essential if Phenylalanine is present, Cysteine if Methionine is present

*asterisks indicate undefined mixtures of animal origin

Firstly, as apparent from Table 10.1, labeling in insect cells is therefore very expensive. Secondly, another obstacle is that insect cell media contain components like cod liver oil, FCS, pluronic and yeast extract or yeastolate, which are poorly defined mixtures and are not easily available in isotope labeled form. To our knowledge, it is not possible to formulate a "minimal medium" with pure nitrogen and carbon sources for insect cells as it is done for *E. coli*. While cells can grow well without FCS and cod liver oil, yeast extract and pluronic are essential. Therefore, these two ingredients need to be discussed in greater detail:

Yeast extract or the ultra-filtrated aqueous solution of it, yeastolate, is present in concentrations of up to 6 g/l in standard insect cell media. It represents a major obstacle for isotope labeling because of its high amino acid content. It has been replaced by pure components like nucleic acids, amino acids, lipids etc. in numerous – unsuccessful – attempts. The viability of cells in yeast extract-free media is inevitably reduced, and it is not possible to maintain cells for several passages in such media. However, cells can survive for a few days without yeast extract, which is enough time for an expression experiment. Still, reproducibility of experiments is not satisfactory and infection with baculovirus seems to be less effective without yeast extract as witnessed by three to five fold lowered expression amounts.

Lipids and the lipid mixture pluronic are also vital components of insect cell media and cannot be replaced. They are crucial as a protectant against cell rupture in shaking suspension cultures and play an important role in ensuring efficient infection by the baculovirus.

Therefore, for reasons of cost and/or feasibility, only parts of the ingredients are exchanged for isotope labeled ones and some unlabeled components are always present in the medium. It is therefore important to assess, which ingredients are metabolized into amino acids, as these will need to be labeled. Fortunately, lipids don't seem to be significantly metabolized into amino acids. The challenges represented by the yeast extract are much greater.

10.3.3 Isotope Incorporation Ratios

With the published protocols, isotope incorporation of maximally 90–94% can be obtained [5, 13]. Unlabeled components, which are present in the labeling medium as described above, are not the main reason for lowered incorporation. There are three more important sources of unlabeled amino acids and carbohydrates, which need to be considered: (1) The endogenous pool of unlabeled amino acids and carbohydrates of insect cells from carry over from the pre-culture and (2) from the suspension of BV, which is unlabeled. Additionally, (3) metabolic scrambling of the isotope labeled atoms in the labeled compounds, e.g. amino acids or sugars, can occur.

10.3.3.1 Pre-culture

Stocks and pre-cultures are usually grown in unlabeled media or commercial media due to cost reasons. Unlike *E. coli*, insect cell pre-cultures cannot be diluted strongly, e.g. 1:1,000, for inoculating the main expression culture. A maximum of 1:4 is usually attainable. Therefore, cells are gently centrifuged (3–10 min at 400 g) and only the cell pellet is re-suspended in the labeling medium. Optionally, cell pellets can be washed in PBS to further dilute the unlabeled pre-culture medium. By this method the carry over of medium is minimized. However, the endogenous pool of unlabeled amino acids remains. Fortunately, production of the target protein starts typically only 8 h post-infection. During this time the endogenous unlabeled amino acids are strongly diluted.

If cost permit, the pre-culture may be run in labeled medium. However, self made labeling media are often not as proliferative as commercial ones. This may lead to reduced viability in the main culture and lower expression yields.

10.3.3.2 BV Suspension

The second source of unlabeled components is the suspension of baculovirus. Baculovirus is produced in unlabeled media, usually supplemented with 10% FCS. Depending on the multiplicity of infection needed, an important volume of this suspension needs to be added to the expression culture, leading to significant label dilution. It is therefore essential, to produce a highly concentrated virus solution with more than 10^8 pfu. This allows limiting the virus solution to 1–2% of the final culture volume. Such high virus titers can be achieved by extra rounds of virus amplification, or simply by gently concentrating the virus solution. The first of the two methods is the preferred one, as virus may loose infectivity by the concentration process.

Again, here an alternative would be to carry out the last amplification round in labeled medium.

10.3.3.3 Metabolic Scrambling of Isotopes

^{15}N label dilution and scrambling only appears for the few amino acids biosynthetically directly related to the citric acid cycle, *i.e.* Asn, Asp, Gln, Glu, and Ala as well as Gly. For ^{13}C, label dilution from carbohydrates in the medium (foremost glucose) occurs in varying degrees. In an experiment with only carbohydrate labeled BioExpress2000 medium, 77% of carbons of Ala were labeled. This occured also for Gln/Glu (28%) and Asn/Asp (16%) [14]. For ^{2}H we found no examples in literature, but we expect that the same rules apply for it as for carbon labeling.

In general, as mentioned above, less scrambling is happening in insect cells than in bacterial cells. In Fig. 10.1, spectra from a ligand binding experiment based on ^{15}N-Phe labeling in insect cells are shown, the advantage of reduced cross-labeling in insect cells compared to *E. coli* was evident. While in *E. coli* pronounced cross-labeling of Phe to Tyr, and to a lesser degree to Asp and Glu, is usually observed, in insect cells only the expected number of signals arising from Phe was detected.

10.3.4 Amino Acid Specific Labeling

Insect cells lend themselves very well for amino acid specific labeling, as their amino acid metabolism is much reduced compared to *E. coli*. Ten amino acids are essential and most others are not readily metabolized (Table 10.1). Therefore, for most amino acids no special attention needs to be paid to scrambling or label dilution (Fig. 10.1). In practice, in the published protocols the amino acid of

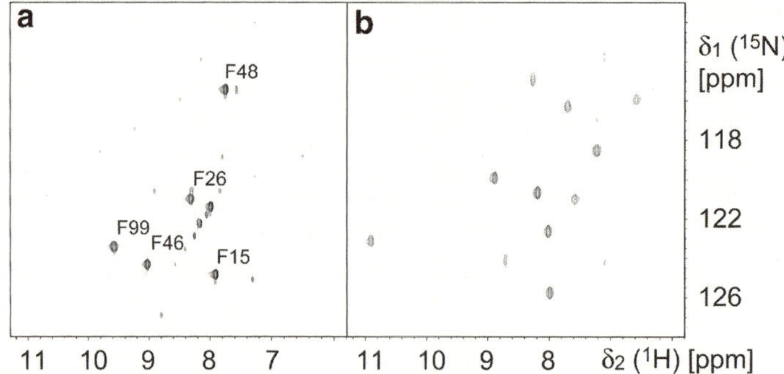

Fig. 10.1 Reduced metabolic scrambling of isotopes in insect cells: example of ¹⁵N-Phe labeling. (a) [¹⁵N,¹H]-HSQC spectrum of FKBP labeled with ¹⁵N-Phe in *E. coli*, using a optimized protocol based on Muchmore et al. No specific enzymatic inhibitors or auxotroph strains are used. The assignments of the five expected signals are shown. Scrambling of the ¹⁵N label to Tyr, and to a lesser degree to Asp and and Glu, leads to additional signals. In [¹⁵N,¹H]-HSQC spectra of a kinase produced in insect cells (**b**) only the expected number of 11 signals for ¹⁵N-Phe are visible

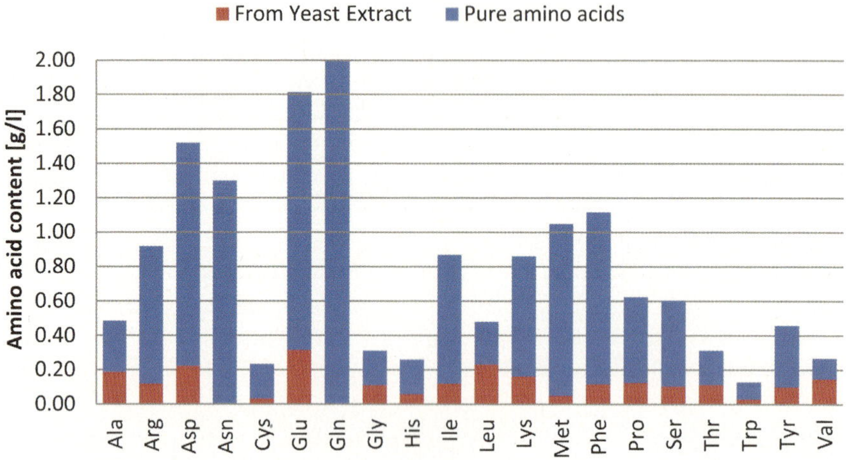

Fig. 10.2 Relative amounts of amino acids in insect cell media originating from pure sources (*blue*) or from yeast extract (*red*). The amino acid composition of a typical insect cell medium is given, containing 6 g/l of yeast extract

interest is just replaced by the isotope labeled version, when preparing the expression medium. Two approaches have been followed: Preparing the media from scratch [1] or ordering amino acid-free versions of commercial media, *i.e.* without amino acids and yeast extract [2, 5, 13, 18].

In this way ²H, ¹⁵N and ¹³C nuclei can be introduced fairly easily in an amino acid specific manner into the protein. With this simple procedure, already in 1998 and 1999, ¹⁵N₂-Lysine and ring-²H₄-Tyrosine labeled samples of rhodopsin were produced with isotope incorporation ratios of 60 and 70%, respectively [2, 4]. In 2003, Brüggert et al. brought forward a very similar protocol, where they presented a complete medium formulation for preparing media from scratch. Incorporation ratios for this medium were probably similar as above, but exact numbers were not stated in the publication [1].

In amino acid specific labeling, isotope incorporation ratios represent the main challenge. Due to the presence of significant amounts of amino acids from yeast extract in the medium, incorporation ratios can be low (Fig. 10.2). For certain applications, notably dual amino acid labeling [19], higher incorporation ratios are needed and strategies have been devised to increase them.

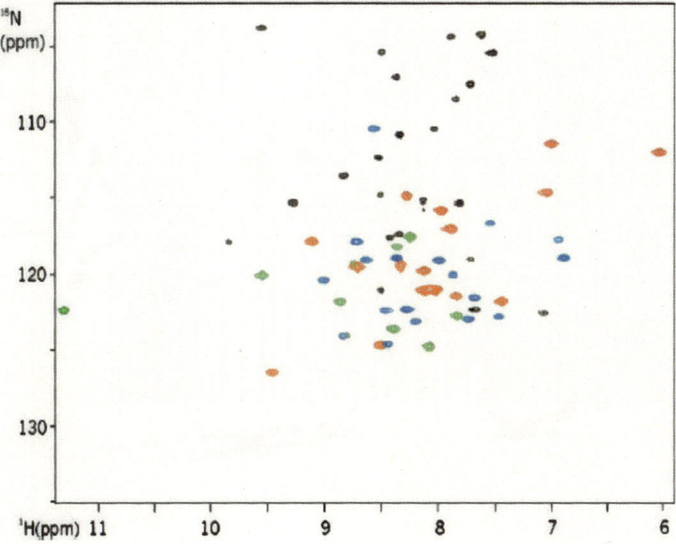

Fig. 10.3 Amino acid type selective labeling of Abl-kinase in BV-infected insect cells. 2D [^{15}N,^{1}H]-HSQC spectra of Abl kinase labeled with ^{15}N-Phe (*green*), ^{15}N-Tyr (*red*), ^{15}N-Val (*blue*) or with ^{15}N-Gly (*black*). Note secondary signals in the Gly spectrum, which arise from cross-labeling to Ser. Gly is one of the few amino acids, where such scrambling is observed in insect cells. (From Strauss et al. [13])

Strauss et al. opted for completely leaving away the yeast extract from the medium. Incorporation ratios of 91–94% are obtainable by this approach for non-metabolized amino acids, and yielded high quality spectra (Fig. 10.3, [13]). This, however, comes at the cost of three to five fold lowered protein yields and less reproducibility. Probably this is due to factors in the yeast extract that facilitate virus entry into the cell. The cells generally have a 1–3 µm smaller diameter at the end of an expression experiment in media without yeast extract than in full media, which may indicate less efficient infection. Nevertheless, this seems to be the most often and most broadly applied strategy for amino acid selective isotope labeling. A number of successful studies are based on this protocol.

In a recent publication based on the above protocol, we have shown that adding 10% of the usual amount of yeast extract can rescue yields and reproducibility to similar levels as in full media. At the same time incorporation ratios are only reduced by 2% compared to the yeast extract-free approach. Knowledge of the amino acid composition of yeast extract enabled adjusting its amount in such a way that no amino acid was present in the medium at more than 5% in unlabeled form [5].

The knowledge of the amino acid composition of yeast extract also makes an approach possible without much work required and high protein yields. The total amount of a given amino acid present in a commercial medium can be estimated. Labeled amino acid is added in such an excess to yield a targeted final incorporation ratio. We showed an incorporation ratio of 80% for ^{15}N-Leu labeling of Abl Kinase with this approach [5].

Feasibility of the approach depends on the solubility of the amino acid of interest, its toxicity to insect cells at a high concentration, and the kind of money available for the experiment. The advantage is that no special medium needs to be mixed from many components, the amino acid of interest is just added to a commercial medium. Expression levels are usually highest with this approach, as commercial media are most productive and the cells don't need to adapt to a different medium for expression.

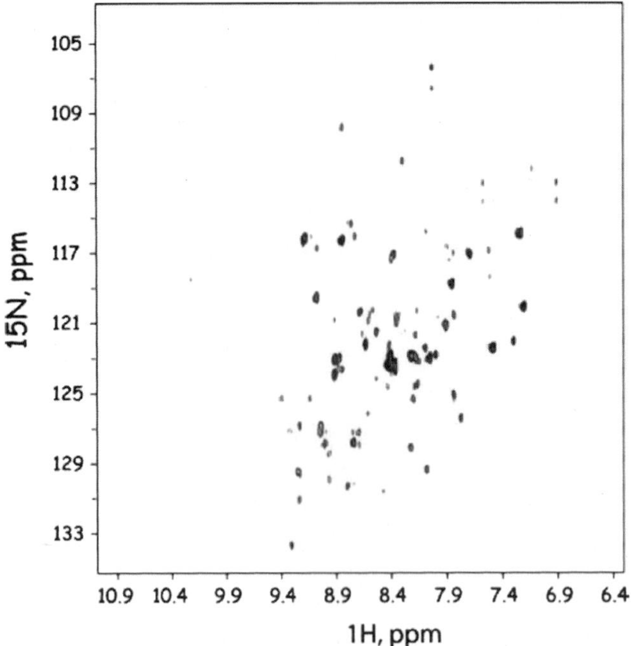

Fig. 10.4 [^{15}N,^{1}H]-HSQC spectrum of uniformly ^{15}N-labeled glycoprotein Thy-1. Additionally, the glycosyl moieties of the protein were ^{13}C-labeled by using ^{13}C-glucose in the medium. (From Walton et al. [18])

10.3.5   Uniform Labeling

Very similar considerations are applicable to uniform labeling. Here however, not only the amino acids present in pure form in the medium need to be replaced by labeled ones – which in itself is already a very costly procedure. On top of this, the yeast extract as well as the carbohydrates need to be isotope labeled. There are two records of uniform labeling in insect cells, one using self-made medium [18] and one relying on a commercial medium [14]. In the first approach, a ^{15}N labeled algal extract was used as the main amino acid source (1.5 g/l). ^{15}N-Cys and Trp were added, as they were not present in the algal extract. Additional ^{15}N sources were included in the medium, ^{15}N$_2$-Gln, ^{15}N-Glu and ^{15}N-NH$_4$Cl, yielding a relatively inexpensive recipe for uniform ^{15}N labeling (Fig. 10.4). Unfortunately, ^{15}NH$_4$Cl seemed to have a negative impact in reducing protein yields by a factor of four. At the same time, yeastolate was added to the medium, thereby deliberately introducing unlabeled amino acids. The levels of incorporation of ^{15}N was not stated in the publication. However, the main focus of the work lied on labeling carbohydrates, which was achieved.

In the other study, BioExpress2000 medium for ^{15}N and ^{15}N/^{13}C labeling was used. Isotope incorporation ratios were above 90% for all amino acids except for Gly (81%), Arg (81%) and His (73%) (Figs. 10.3 and 10.8). Judging from published spectra, this method is superior to the one of Walton et al., but it is also more expensive.

We can only speculate about the formulation of this commercial medium. Probably, a yeast or algal extract is used as the main amino acid source and a few selected amino acids are added, which would be underrepresented or not present at all in the medium if only the extract was used. Additionally, glucose as the main carbohydrate component should be present in labeled form, as this is important for ^{13}C labeling of Ala, Gln/Glu and Asn/Asp, and it is not expensive [14]. Other carbohydrates are probably not labeled, since they are not readily available and are not metabolized into amino acids by insect cells. In this way, theoretically all combinations of uniform ^{2}H, ^{13}C and ^{15}N labeling patterns can be achieved.

10.4 Case Studies

Isotope labeling of proteins in insect cell media have enabled a number of diverse applications with proteins otherwise not amenable to NMR studies. Examples range from simple detection of ligand binding to studies of different conformational states and dynamics of proteins. Even complete backbone assignments were achieved for the 31 kDa protein Abl-kinase based on uniform labeling in insect cells.

10.4.1 First Applications with Rhodopsin

Interestingly, isotope labeling of proteins in insect cells was not used for NMR experiments in the first place, but for fourier transform infrared spectroscopy (FTIR). In a publication of 1998, ring-^{2}H$_4$-Tyrosine was introduced into bovine rhodopsin and the involvement of Tyrosine in photo-activation was studied [4]. However, it did not take long until NMR spectroscopists exploited the labeling method developed for FTIR. One year later, 1D solid state spectra of ^{15}N$_2$-Lysine labeled rhodopsin were published (Fig. 10.5, [2]). In the spectra obtained under magic angle spinning conditions, isotropic chemical shifts of Lysine residues in rhodopsin could be determined. A strong signal of side chain ε-nitrogens was visible as well as a broader signal of backbone amides from Lysines. Probably, natural abundance ^{15}N from other amino acids also contributed to that signal. Finally, a well-separated resonance was observed, which was attributed to the nitrogen involved in the Schiff-base with retinal. The characteristic chemical shift of that nitrogen confirmed a protonated state of the Schiff-base. Further, calculations were carried out that suggested stabilization of the Schiff-base by a counter-ion in close proximity.

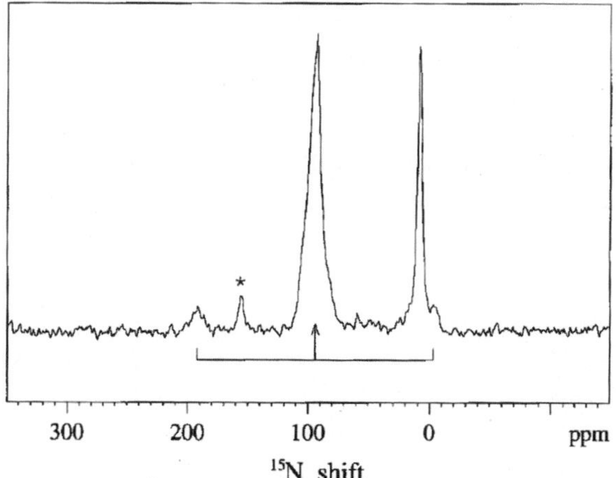

Fig. 10.5 1D ^{15}N solid state spectrum of ^{15}N$_2$-Lysine labeled rhodopsin. The *asterisk marks* the charachteristic signal of the ^{15}N^ε of the lysine involved in the Schiff-base. The *sharp signal* at 10 ppm arised from free ^{15}N^ε-groups of other Lysines. The *broad central signal* indicated by the *arrow* represents backbone amides, most likely not only from Lysines but also form other amino acids which where labeled by metabolic scrambling of the ^{15}N. Finally, two spinning side bands at Δδ 100 ppm from the major signal are indicated (From Creemers et al. [2]. Figure provided by Dr. HJM de Groot, Leiden University, The Netherlands)

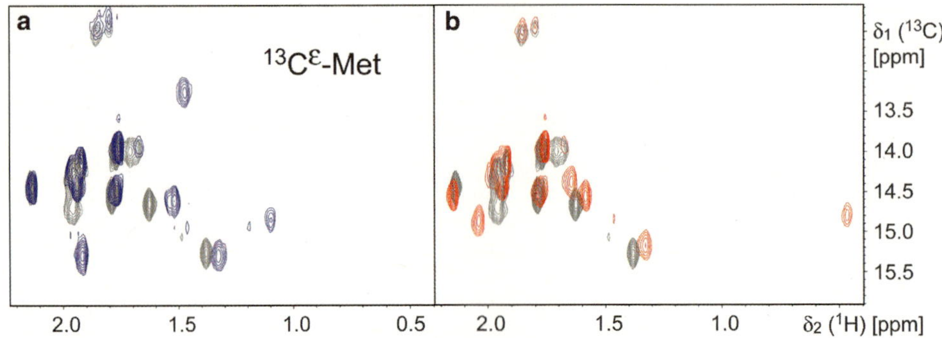

Fig. 10.6 Identification of ligands using a ¹³Cᵉ-Met labeled kinase (33 kDa). 2D [¹³C,¹H]-HMQC spectra of the protein with two different ligands ((**a**) *blue*, (**b**) *red*) are shown, individually superimposed on the spectrum of apo protein (*grey*). Specific binding can be proven by chemical shift perturbations. Spectra were recorded with 50 μM protein samples with 200 μM compound in 30 min (From Gossert et al. [5])

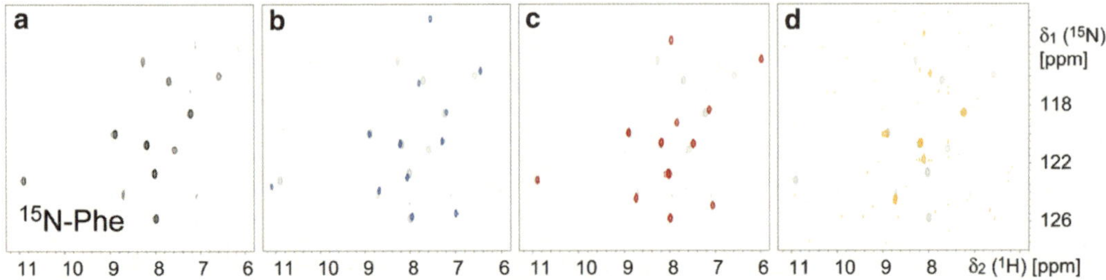

Fig. 10.7 Identification of specific ligand binding using a ¹⁵N-Phe labeled kinase. The 2D [¹⁵N,¹H]-HSQC spectrum of the apo form of the protein is shown in *panel* **a**. The exact number of expected signals is visible in the spectra of the inactive form of the kinase (**a**). Specific binding of different compounds is evidenced by chemical shift perturbations of the resonances, compared to the apo spectrum in *grey* (**b**, **c**). The compound Dasatinib binds to the active state of this kinase [3], leading to a qualitatively different spectrum due to internal dynamics (**d**). Spectra were recorded with 50 μM protein samples with 200 μM compound in 4 h (From Gossert et al. [5])

10.4.2 Amino Acid Selective Labeling for Binding Determination of Ligands

There are a number of NMR methods for detection of ligand binding to a protein without the need to isotope labeled protein. However, these approaches suffer from the risk of false positives. The gold standard in ligand binding detection are protein-observation spectra relying on chemical shift perturbation. In crowded 1D ¹H spectra, it is very often difficult to assess such chemical shift perturbations induced by true binding of a ligand. Here, simplified 2D HSQC spectra of amino acid selective labeled proteins represent an ideal alternative.

In a recent study, we used a selectively ¹³Cᵉ-Met labeled kinase produced in insect cells to study ligand binding. Ligand binding or non-binding was clearly demonstrated in the simplified [¹³C ¹H]-HSQC spectra by presence or absence of chemical shift distortions upon addition of ligand. The same can be very well appreciated in [¹⁵N ¹H]-HSQC spectra. The first approach, however, has the advantage of much reduced measurement times. (Figs. 10.6 and 10.7, [5])

10.4.3 Abl-Kinase

As the following examples all deal with kinases and their conformational states, a brief introduction is given here. Kinases are proteins with considerable inherent flexibility. Foremost two polypeptide regions, the phosphate binding loop (P-loop) and the activation loop, can adopt different conformations and dynamics, which are associated with activation or inactivation of the kinase. These non-rigid regions are important for the fine-tuning of kinase activity in the living cell.

Many forms of cancer are linked to mutations of kinases. In the case of Abl-kinase, constitutively activated protein leads to life-threatening chronic myelogenous leukemia. Therefore, attempts were made to find inhibitors that bind Abl-kinase and keep it in an inactive state.

10.4.4 Uniform, Dual and Single Amino Acid Labeling for Complete Assignments of the 31 kDa Protein Abl-Kinase

In a tour-de-force Vajpai and co-workers managed to obtain 96% of the assignments of backbone $^1H^N$, ^{15}N, $^{13}C^\alpha$ and $^{13}C'$resonances of the 277 residue protein Abl-kinase (Fig. 10.8). Protein samples with different labeling patterns were produced in insect cells [14]. Notably, uniformly $^{13}C,^{15}N$ labeled protein enabled backbone assignments based on HNCO, HNCA, HN(CO)CA spectra and an ^{15}N-edited $^1H,^1H$-NOESY [16]. Due to the lack of deuteration, experiments encompassing $^{13}C^\beta$ were not sensitive enough. Therefore, the assignment was aided and verified by spectra of a total of 15 selectively labeled samples that had ^{15}N, $^{13}C^\alpha$ and $^{13}C'$ nuclei introduced specifically for certain amino acid types or combinations thereof. These assignments enabled functional studies of Abl-kinase.

10.4.5 Quadruple Amino Acid Labeling for Understanding Activation States

With NMR studies based on chemical shift perturbations, residual dipolar couplings, and ^{15}N relaxation, different ATP-site binders of Abl-kinase were characterized [15]. As uniform labeling in insect cells is very expensive, the experiments for functionally characterizing Abl-kinase were focussed on the P-loop and the activation loop. By labeling only four amino acid types, FGMY, most residues of the two loops were labeled. As a consequence of metabolic scrambling from Glycine also Serine residues were labeled. Besides the amino acid-type information, this selective labeling scheme had the advantage that all key resonances were completely free of overlap in the simplified spectra, making possible precise measurements of residual dipolar couplings and ^{15}N-relaxation parameters. These data were in agreement with the binding sites of the ligands that were previously defined in crystal structures. More importantly, the solution NMR data showed that the inactive and active states of the kinase induced by the ligands were present in solution and were not just consequences of crystal packing or preferred conformations in crystals.

10.4.6 A Conformational Assay

A different type of ligand binding assay was brought forward by [7]. In a drug discovery effort, allosteric ligands were found, which induced the activated or the inactivated state of the kinase, depending

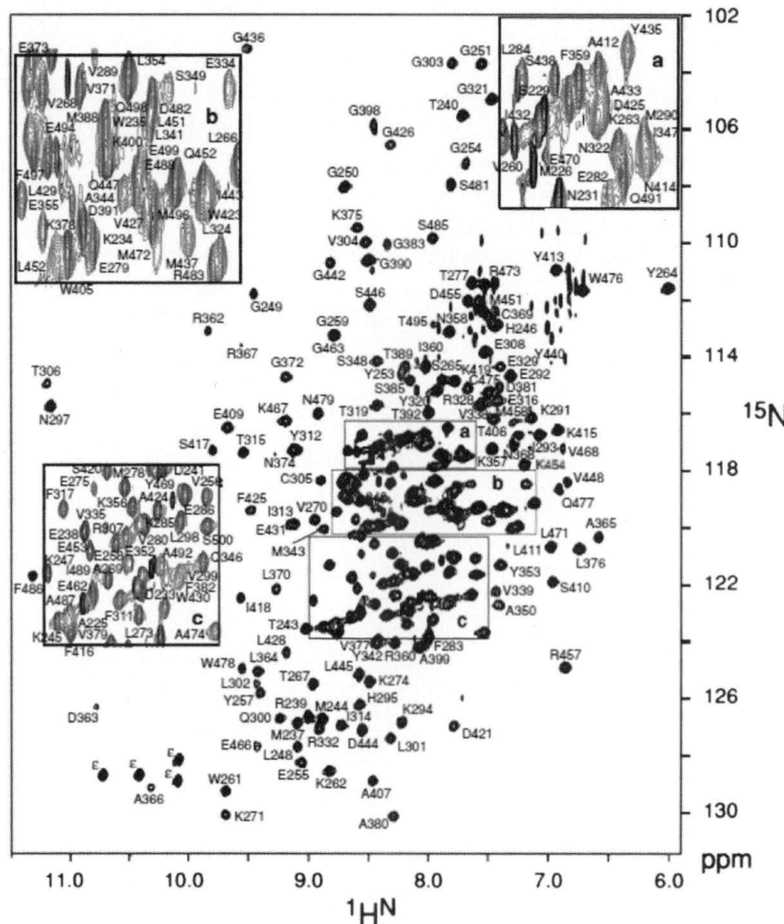

Fig. 10.8 Backbone assignments of Abl-kinase obtained from uniformly and amino acid specifically labeled samples produced in insect cells. A 2D [^{15}N, ^{1}H]-TROSY correlation spectrum is shown, with the assignments indicated (From Vajpai et al. [16])

on the ligand. Only the ligands leading to the inactive state of the kinase were true inhibitors and had a potential as anti-cancer drugs.

Here, not the question of binding and non-binding was central, but the conformations of Abl-kinase that the different ligands induced. When going from the inactive conformation to the active one, the C-terminal alpha helix of Abl kinase would unfold. In the [^{15}N, ^{1}H]-HSQC spectrum of ^{15}N-Val labeled Abl kinase, a single signal would at the same time become much more intense. This signal arised from the Valine in the C-terminal helix, and upon unfolding of the helix the signal became much sharper, due to the fast dynamics of the unfolded state. In summary, a fast assay was implemented based on simplified [^{15}N, ^{1}H]-HSQC spectra, which allowed quickly characterizing the conformation of Abl kinase in presence of different ligands (Fig. 10.9). This was a much faster approach than crystallization of each complex and as measurements were carried out in solution, crystal packing could not influence conformations. Ultimately, this study was enabled by expression of selectively ^{15}N-Valine labeled Abl-Kinase in insect cells by the method of [13].

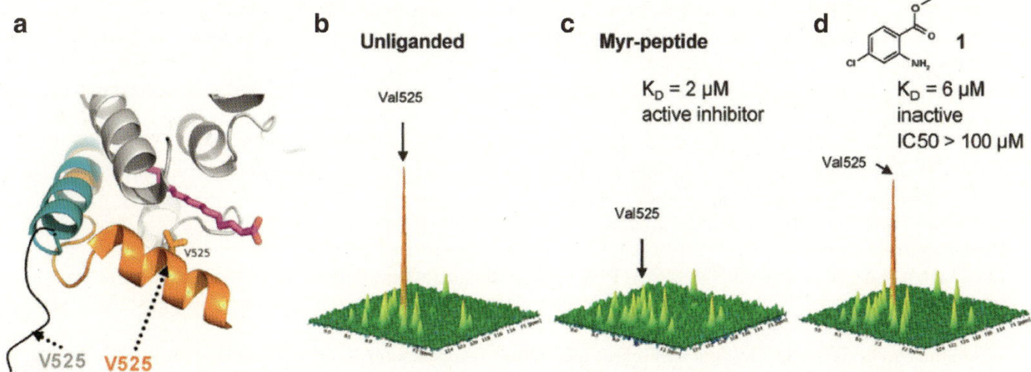

Fig. 10.9 A conformational assay for testing the activation state of Abl kinase upon binding of different compounds. In (**a**) the conformations of the C-terminal helix of Abl kinase domain in the inactive state (*orange*, helix formed) and the active state (*black,* flexibly disordered) are shown in a ribbon model. The position of V525 is indicated. In (**b–d**), [^{15}N, ^{1}H]-HSQC spectra of ^{15}N-Valine labeled Abl-kinase are shown, with the resonance of Val525 indicated. The spectra were recorded without a ligand present (**b**), with a myristoylated peptide (**c**) and with a compound (**d**). The intensitiy of the signal of Val525 indicates the flexibly disordered state of the C-terminus (**b** and **d**) or the well structured helical state (**c**), reflecting the active state and inactive state of the kinase. Therefore, the compound shown in (**d**) doesn't yield an inactive conformation of Abl kinase and is therefore not an inhibitor, although it binds to the myr-pocket (From Jahnke et al. [7])

10.4.7   Conclusion

The above examples show that with expression in insect cells difficult-to-express proteins can be studied by NMR. While uniform labeling is possible, selective labeling is in most cases sufficient to play out all the strengths of NMR – characterizing interactions, conformations and dynamics of proteins in solution.

Acknowledgements We would like to thank André Strauss for his work in establishing isotope labeling in insect cells in our lab and for sharing his knowledge. Further we thank Sébastien Rieffel and Binesh Shrestha for discussions on general insect cell expression techniques.

References

1. Brüggert M, Rehm T, Shanker S, Georgescu J, Holak TA (2003) A novel medium for expression of proteins selectively labeled with N-15-amino acids in Spodoptera frugiperda (Sf9) insect cells. J Biomol NMR 25:335–348
2. Creemers AF, Klaassen CH, Bovee-Geurts PH, Kelle R, Kragl U, Raap J, de Grip WJ, Lugtenburg J, de Groot HJ (1999) Solid state 15N NMR evidence for a complex Schiff base counterion in the visual G-protein-coupled receptor rhodopsin. Biochemistry 38:7195–7199
3. Das J, Chen P, Norris D, Padmanabha R, Lin J, Moquin RV, Shen Z, Cook LS, Doweyko AM, Pitt S, Pang S, Shen DR, Fang Q, de Fex HF, McIntyre KW, Shuster DJ, Gilooly KM, Behnia K, Schieven GL, Wityak J, Barrish JC (2006) 2-aminothiazole as a novel kinase inhibitor template. Structure-activity relationship studies toward the discovery of N-(2-chloro-6-methylphenyl)-2-[[6-[4-(2-hydroxyethyl)-1- piperazinyl)]-2-methyl-4-pyrimidinyl]amino)]-1,3-thiazole-5-carboxamide (dasatinib, BMS-354825) as a potent pan-Src kinase inhibitor. J Med Chem 49:6819–6832
4. DeLange F, Klaassen CH, Wallace-Williams SE, Bovee-Geurts PH, Liu XM, DeGrip WJ, Rothschild KJ (1998) Tyrosine structural changes detected during the photoactivation of rhodopsin. J Biol Chem 273:23735–23739

5. Gossert AD, Hinniger A, Gutmann S, Jahnke W, Strauss A, Fernandez C (2011) A simple protocol for amino acid type selective isotope labeling in insect cells with improved yields and high reproducibility. J Biomol NMR 51:449–456
6. Granados RR, Derksen AC, Dwyer KG (1986) Replication of the Trichoplusia ni granulosis and nuclear polyhedrosis viruses in cell cultures. Virology 152:472–476
7. Jahnke W, Grotzfeld RM, Pelle X, Strauss A, Fendrich G, Cowan-Jacob SW, Cotesta S, Fabbro D, Furet P, Mestan J, Marzinzik AL (2010) Binding or bending: distinction of allosteric Abl kinase agonists from antagonists by an NMR-based conformational assay. J Am Chem Soc 132:7043–7048
8. Kitts PA, Possee RD (1993) A method for producing recombinant baculovirus expression vectors at high frequency. Biotechniques 14:810–817
9. Luckow VA, Lee SC, Barry GF, Olins PO (1993) Efficient generation of infectious recombinant baculoviruses by site-specific transposon-mediated insertion of foreign genes into a baculovirus genome propagated in *Escherichia coli*. J Virol 67:4566–4579
10. Murhammer D (2007) Baculovirus and insect cell expression protocols, vol 388, Methods in molecular biology. Humana Press, Totowa
11. O'Reilly DR, Miller LK, Luckow VA (1994) Baculovirus expression vectors: a laboratory manual. Oxford University Press, Oxford
12. Richardson C (1995) Baculovirus expression protocols, vol 39, Methods in molecular biology. Humana Press, Totowa
13. Strauss A, Bitsch F, Cutting B, Fendrich G, Graff P, Liebetanz J, Zurini M, Jahnke W (2003) Amino-acid-type selective isotope labeling of proteins expressed in Baculovirus-infected insect cells useful for NMR studies. J Biomol NMR 26:367–372
14. Strauss A, Bitsch F, Fendrich G, Graff P, Knecht R, Meyhack B, Jahnke W (2005) Efficient uniform isotope labeling of Abl kinase expressed in Baculovirus-infected insect cells. J Biomol NMR 31:343–349
15. Vajpai N, Strauss A, Fendrich G, Cowan-Jacob SW, Manley PW, Grzesiek S, Jahnke W (2008) Solution conformations and dynamics of ABL kinase-inhibitor complexes determined by NMR substantiate the different binding modes of imatinib/nilotinib and dasatinib. J Biol Chem 283:18292–18302
16. Vajpai N, Strauss A, Fendrich G, Cowan-Jacob SW, Manley PW, Jahnke W, Grzesiek S (2008) Backbone NMR resonance assignment of the Abelson kinase domain in complex with imatinib. Biomol NMR Assign 2:41–42
17. Vaughn JL, Goodwin RH, Tompkins GJ, McCawley P (1977) The establishment of two cell lines from the insect Spodoptera frugiperda (Lepidoptera; Noctuidae). In Vitro 13:213–217
18. Walton WJ, Kasprzak AJ, Hare JT, Logan TM (2006) An economic approach to isotopic enrichment of glycoproteins expressed from Sf9 insect cells. J Biomol NMR 36:225–233
19. Weigelt J, van Dongen M, Uppenberg J, Schultz J, Wikström M (2002) Site-selective screening by NMR spectroscopy with labeled amino acid pairs. J Am Chem Soc 124:2446–2447
20. Weiss SA, Smith GC, Kalter SS, Vaughn JL (1981) Improved method for the production of insect cell-cultures in large volume. In Vitro J Tissue Cult Assoc 17:495–502

Chapter 11
Mammalian Expression of Isotopically Labeled Proteins for NMR Spectroscopy

Mallika Sastry, Carole A. Bewley, and Peter D. Kwong

Abstract NMR spectroscopic characterization of biologically interesting proteins generally requires the incorporation of $^{15}N/^{13}C$ and/or ^{2}H stable isotopes. While prokaryotic incorporation systems are regularly used, mammalian ones are not: of the nearly 9,000 NMR macromolecular structures currently deposited in the Protein Data Bank, only a handful (<0.5%) were solved with proteins expressed in mammalian systems. This low number of structures is largely a reflection of the difficulty in producing uniformly labeled, mammalian-expressed proteins. This is unfortunate, as many interesting proteins require mammalian cofactors, chaperons, or post-translational modifications such as N-linked glycosylation, and mammalian cells have the necessary machinery to produce them correctly. Here we describe recent advances in mammalian expression, including an efficient adenoviral vector-based system, for the production of isotopically enriched proteins. This system allows for the expression of mammalian proteins and their complexes, including proteins that require post-translational modifications. We describe how this system can produce isotopically labeled ^{15}N and ^{13}C post-translationally modified proteins, such as the outer domain of HIV-1 gp120, which has 15 sites of N-linked glycosylation. Selective amino-acid labeling is also described. These developments should reduce barriers to the determination of NMR structures with isotopically labeled proteins from mammalian expression systems.

Abbreviations

CHO	Chinese Hamster Ovary cells
HEK	Human Embryonic Kidney cells
BHK21	Baby Hamster Kidney cells
HIV-1	Human Immunodeficiency Virus Type 1
RSV	Rous Sarcoma Virus
CMV	Cytomegalovirus
PEI	Polyethyleneimine
ITR	Inverted terminal repeat

M. Sastry (✉) • P.D. Kwong
Vaccine Research Center, National Institute of Allergy and Infectious Diseases, National Institutes of Health, 40 Convent Drive; Building 40, Room 2613B, Bethesda, MD 20892-3027, USA
e-mail: sastrym@mail.nih.gov

C.A. Bewley
Laboratory of Bioorganic Chemistry, National Institute of Diabetes and Digestive and Kidney Diseases, National Institutes of Health, Bethesda, MD 20892, USA

H.S. Atreya (ed.), *Isotope Labeling in Biomolecular NMR*, Advances in Experimental Medicine and Biology 992, DOI 10.1007/978-94-007-4954-2_11, © Springer Science+Business Media Dordrecht 2012

BGHpA   Bovine growth hormone polyadenylation signal
SPR   Surface plasmon resonance
PBS   Phosphate buffered saline
ATCC   American Type Culture Collection
DMEM   Dubelco's modified eagle media
FBS   Dialyzed fetal bovine serum
CAR   Coxsackie-and Adenovirus Receptor.

11.1 Introduction

NMR spectroscopic characterization of proteins generally requires the incorporation of ^{15}N, ^{13}C, or ^{2}H stable isotopes. In most cases, full length proteins or their individual domains are expressed and purified from prokaryotic expression systems, and a number of these methods are discussed in Chap. 1 of this book. An analysis of the macromolecular structures deposited in the PDB (December 2011 release) reveals that over 99% of proteins studied by NMR spectroscopy are expressed prokaryotically (Table 11.1). A large number of biologically interesting proteins, however, require eukaryotic cofactors, chaperons or post-translational modifications such as glycosylation for proper folding and activity, and bacterial expression systems may lack the necessary cellular machinery to produce correctly folded functional proteins that are suitable for NMR spectroscopy. Why not express and purify post-translationally modified proteins from eukaryotic sources? Traditionally, eukaryotic systems such as insect, lower eukaryotes, and mammalian cells produce low yields of the desired protein, are time consuming, and/or are prohibitively expensive. Recent advances in expression techniques, such as multi-host vectors and the use of codon-optimized genes, have however reduced barriers to the evaluation and optimization of gene expression in different hosts. Furthermore, use of bioreactors and advances in cell culture techniques (such as the development of large scale mammalian transient expression systems) have allowed milligram- to gram-scale production of proteins suitable for structural analysis. Indeed, recombinant proteins produced from mammalian cells such as Chinese Hamster Ovary (CHO) and Human Embryonic Kidney (HEK) cells are routinely used in cryo-EM and X-ray crystallographic studies. An analysis of the PDB reveals that ~98% of the entries that used proteins obtained from mammalian expression systems are for structures solved by X-ray crystallography.

Despite these advances the production of isotopically enriched proteins suitable for NMR spectroscopy has been stymied by the absence of a suitable expression system as well as the cost of labeled media. We recently reported the adaptation of an efficient adenoviral vector-based mammalian expression system for the production of isotopically enriched proteins [1]. The high yields obtained from the adenoviral coupled mammalian expression system balanced the cost of labeled media and allowed for the production of post-translationally modified proteins suitable for NMR.

This chapter provides a summary of different systems used for mammalian expression and explicit details for the adenoviral vector-based expression system. We compare the adenoviral vector-based expression system to other methods – through a case study of a variant of the HIV-1 gp120 outer domain comprising 230 amino acids, 15 potential sites of *N*-linked glycosylation, and four disulfide bonds – and extend our previous characterization of this expression system to include selective amino-acid labeling.

11.2 Overview of Mammalian Expression

Recent development of eukaryotic systems including yeast, insect, and mammalian cells and associated vectors – which allows for the transient, inducible or constitutive expression of a target gene – has allowed researchers to investigate the possibility of overexpressing proteins in eukaryotic systems [2–4].

Table 11.1 Protein Data Bank[a] entries solved by solution NMR spectroscopy

Experimental method	Total	Protein	Mammalian expression	Isotopically enriched mammalian expression
Solution NMR spectroscopy	9,210	8,052	8	4

[a]December 2011 release

Mammalian cells have the necessary co-factors, chaperons and cellular machinery to produce correctly folded post-translationally modified functional proteins. These systems are currently used to produce proteins for therapeutic purposes and use primarily Chinese Hamster Ovary (CHO), Human Embryonic Kidney (HEK), Baby Hamster Kidney (BHK21), human fibrosarcoma (HT1080) and human lymphoma (Namalwa) cells. We describe expression methods in the following sections and provide details for production of isotopically enriched proteins using the HIV-1 gp120 outer domain as an example from the adenoviral vector-based mammalian expression system.

11.2.1   Transient Expression in Mammalian Cells

Transient expression of a target gene is defined as the temporary production of a protein. Transfection is achieved by well-established techniques wherein plasmid DNA containing the target gene is mixed with calcium phosphate, a cationic lipid, or other reagents [5] such as Lipofectamine (Fig. 11.1a). DNA of interest is introduced into the cells either by transiently opening pores or by fusion of liposomes with the cell membrane. Transfection can also be achieved by electroporation [6]. The factors governing gene expression in eukaryotes are similar to those in the prokaryotic systems discussed in preceding chapters. A strong promoter such as the SV40 early promoter, the Rous Sarcoma Virus (RSV) promoter or the cytomegalovirus (CMV) very early promoter can be used to increase protein expression [7]. Transient transfection of HEK 293E cells with polyethyleneimine (PEI) mediated transfections has been used to obtain milligram to gram quantities of secreted proteins [8, 9]. Transient transfection can be achieved in adherent as well as suspension cells, with the possibility of scale up [10, 11]. These proteins are functionally active and have been used in biophysical and crystallographic studies [12–14]. This method is simple to implement and requires very little infrastructure. In principle, protein expression from adherent cells using ^{15}N, ^{15}N/^{13}C labeled media is feasible although potentially prohibitively expensive. Lack of suitable labeling media for suspension cells complicates production of labeled proteins from transient expression. As a result a number of investigators have turned to constitutive expression in mammalian cells.

11.2.2   Constitutive Expression in Mammalian Cells

Plasmid DNA containing the target gene and a strong promoter (SV40, RSV or CMV) is co-transfected with another plasmid containing a selectable marker (e.g. hygromycin, tetracycline, neomycin or G418) [15, 16]. The mammalian cells expressing the selectable marker are screened for drug resistance and a clonal cell line that expresses the protein of interest is established (Fig. 11.1b). Protein expression is dependent on the toxicity of the heterologous protein, the promoter used and the position of gene integration. A number of research groups have attempted to produce isotopically labeled proteins from mouse or Chinese Hamster Ovary (CHO) cell lines using either a mixture of algal/bacterial hydrolysates [17] or a mixture of labeled amino acids [18–21]. Unfortunately these methods result in low yields and can be prohibitively expensive. There are only four reports in the protein data bank

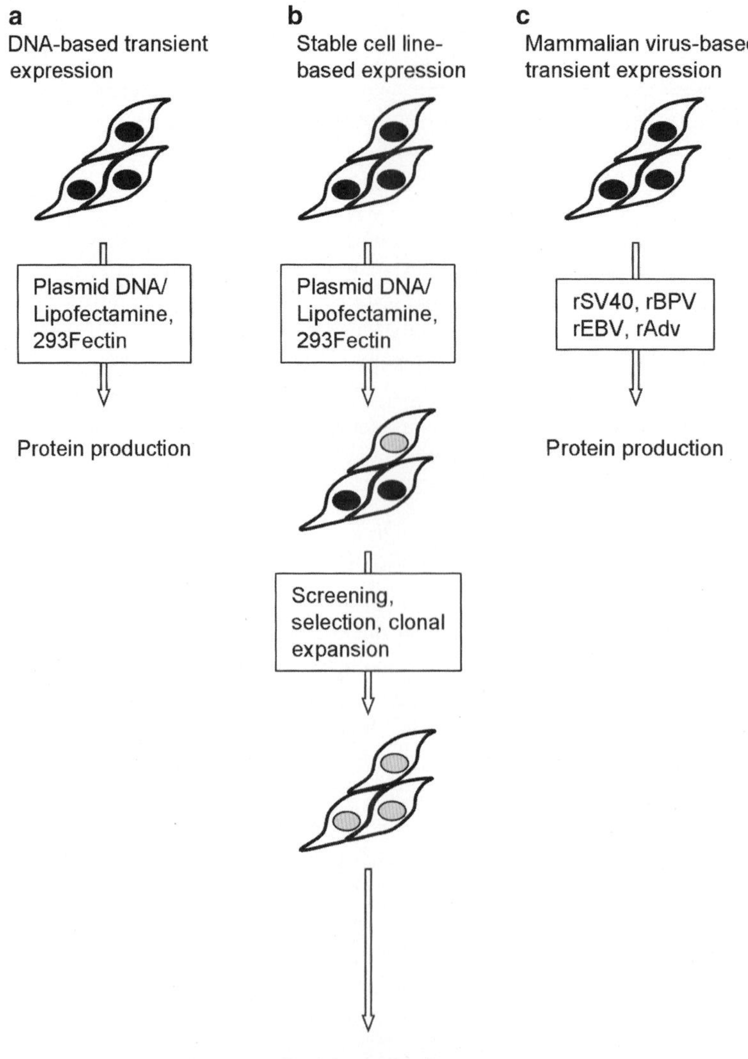

Fig. 11.1 Mammalian expression systems used for protein production. An overview of expression systems used to obtain proteins from (**a**) transient, (**b**) stable cell line based mammalian expression and (**c**) mammalian viral vectors based transient expression

(1URK, 1GYA, 1KLA, 1AH1) for structures obtained using heteronuclear NMR experiments from mammalian cell lines (Table 11.1) [22–25]. Researchers have thus focused on partial [26] or amino-acid type-specific labeling [27–29] of proteins from mammalian expression systems.

11.2.3 *Transient Expression Using Mammalian Viruses*

SV40, poxviruses, herpesviruses, papillomaviruses and adenoviruses have been used as expression vectors [15, 30–33]. Viral vectors wherein nonessential genes are substituted for a foreign gene [34] as well as vectors resulting in defective viral genomes can be used for transient protein expression by infecting a wide range of mammalian cells (Fig. 11.1c). Protein expression is driven by viral promoters and, at the expense of production of cellular proteins, can increase the overall yield of the heterologous protein. The following section focuses on the adenoviral expression system that we have adapted to produce isotopically enriched glycoproteins.

11.3 Recombinant Adenoviruses as a Tool to Obtain Isotopically Labeled Proteins

Adenoviruses are double-stranded DNA viruses. Their genome comprises ~36 kb of linear double-stranded DNA [35]. Adenoviral genome that includes the E1-E4 region is transcribed early in the life cycle and is involved in the replication of the virus. Deletion of the E1 region renders the virus replication incompetent and thus safe to use as a gene delivery vector. A further deletion of the E3 region allows for insertion of up to 8-kilobases of recombinant transgenes into the E1 region [32, 36, 37]. An E1/E3 deleted virus can thus be used to deliver target genes with very high efficiency to many different cell types. A very high efficiency of transfection and specific translational discrimination between viral and cellular mRNA together facilitate the exceptional expression of adenovirus-vectored proteins from mammalian cells [38, 39].

11.3.1 Design and Composition of Recombinant Adenoviral Vectors

A schematic outline of the design and composition of an adenoviral vector is shown in Fig. 11.2. Specifically, the adenoviral cosmid (pVRC1194) consists of 9.2–100 m.u (map units) of the adenoviral genome, along with a deletion in the E3 region and a loxP site at 9.2 m.u. The shuttle vector (pVRC1290) into which the target gene is cloned, contains the adenoviral 5' inverted terminal repeat (ITR), a 0–1 m.u packaging signal followed by the target gene, the bovine growth hormone polyadenylation signal (BGHpA), a loxP site and 9.2–16.1 m.u of the adenoviral genome [40]. The shuttle plasmid and the adenoviral cosmid are linearized and recombined *in vitro* using Cre recombinase (Novagen-EMD Biosciences, Madison, Wisconsin) to obtain recombinant adenoviral genome. Recombinant adenoviral DNA can be purified by standard methods [41, 42]. The recombinant adenoviral DNA obtained from the Cre-Lox recombination reaction contains the target gene flanked by adenoviral sequences. These flanking sequences consist of DNA packaging signal as well as the adenoviral genome to reconstitute defective adenoviruses. Since generation of recombinant adenovirus can be a lengthy process, prior to embarking on recombinant adenoviral production, construct design and functional properties of the target protein were tested using transient expression in either HEK293 adherent or suspension cells.

11.3.2 Generation of Adenoviruses Containing Gene of Interest

To produce recombinant adenovirus containing the target gene, recombinant Ad5 DNA obtained from the Cre-Lox reaction is transfected into helper 293 cells (Fig. 11.2) using calcium phosphate transfection methodology. Since the E1 region of the adenoviral DNA is deleted, viral production can only take place in mammalian cells such as HEK293 with endogenous E1 proteins that can complement the defect in the recombinant adenovirus [44]. HEK293 cells transfected with recombinant adenoviral

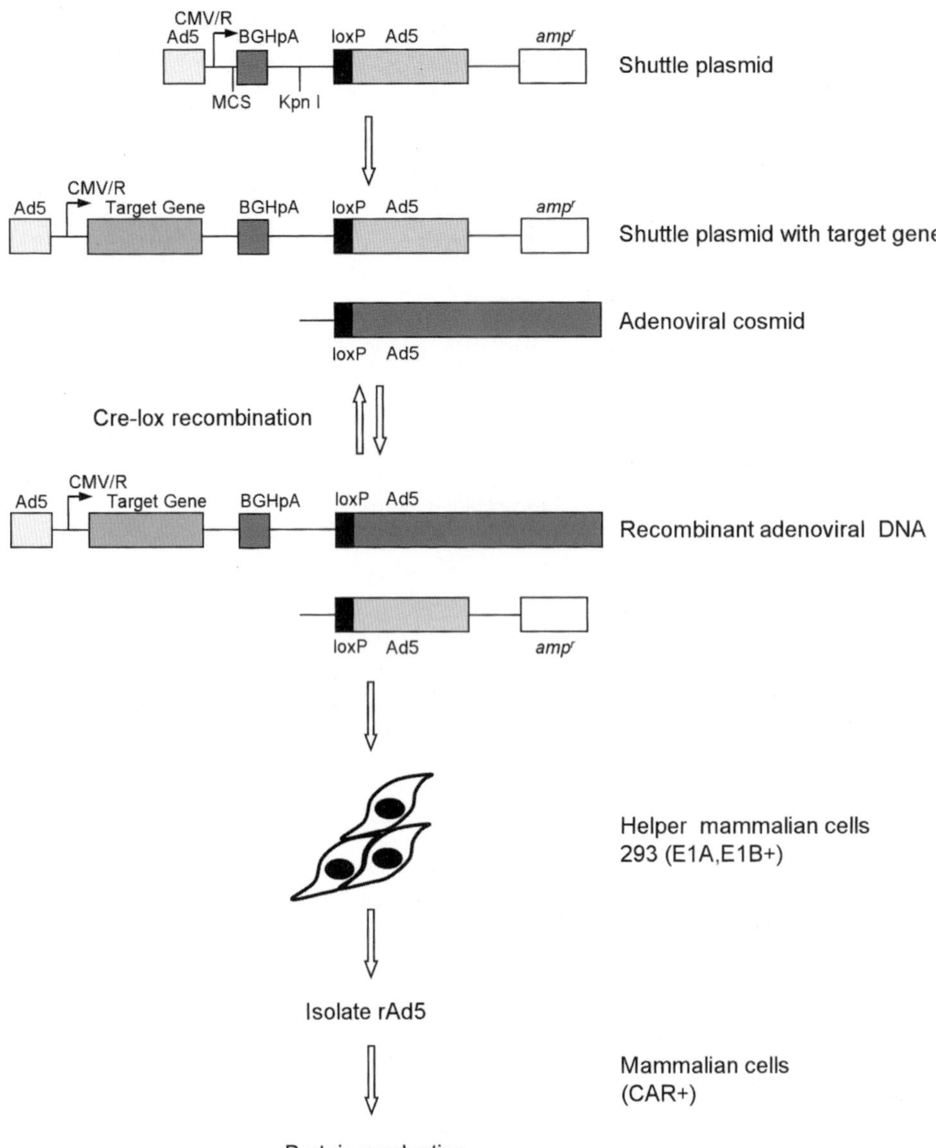

Fig. 11.2 Schematic overview of the adenoviral based mammalian expression system. The target gene is cloned into a shuttle plasmid (pVRC1194) using the restriction sites present in the multiple cloning site (*MCS*); linearized adenoviral cosmid DNA (pVRC1290) and shuttle plasmid are then recombined *in vitro* with Cre-Lox recombinase to obtain recombinant adenoviral genome (Aoki and Nabel [40]). The recombined adenoviral type 5 DNA is transfected into helper mammalian cells and recombinant adenovirus is isolated using well established methods [41, 42]. Target protein production is achieved by infecting CAR[+] mammalian cells. There are reports in the literature that Ad3 entry may be mediated by CD46 positive cells [43]

DNA are monitored for plaque formation for up to 10 days. Upon plaque formation, cells are harvested, the recombinant virus is isolated and protein expression of the desired gene tested by infecting A549 lung carcinoma cells with the crude virus. Protein expression of the target gene can be monitored and confirmed by western blots. Once protein expression is confirmed, the crude supernatant or cell lysate can be further evaluated for the presence of a correctly folded protein of interest by surface plasmon resonance (SPR) analysis [45, 46] or equivalent technique. Once protein expression is confirmed, the crude virus preparation is used to infect low passage HEK293 cells to produce recombinant adenovirus. Specifically, HEK293 cells are seeded at 2.0×10^7 cells per 15 cm plate 24 h prior to infection. The cells should be ~90% confluent at the time of infection. Cells are harvested ~30 h post infection with the crude virus. Harvested cells are spun down at $311 \times g$ for 10 min, the cell pellet is washed twice with phosphate buffered saline (PBS) and subsequently resuspended in 10 mM Tris–HCl pH 8.0.

11.3.3   Isolation and Purification of Recombinant Adenoviruses

Recombinant virus is released from the cells by three cycles of freezing in a dry ice/ethanol bath and thawing at 37°C. The cell lysate is spun down at $552 \times g$ for 10 min. The Ad5 virus present in the supernatant is further purified twice using CsCl gradient centrifugation, followed by a desalting column to remove the CsCl [41, 47–49]. The purified adenovirus is quantitated and aliquoted aseptically; these aliquots can be stored at −20°C and used as needed for protein expression. The purified adenovirus is an infectious agent and each researcher should consult their institution's guidelines for handling biohazardous material.

11.3.4   Growth Characteristics of Cell Lines

HEK293 and A549 cells can be obtained from American Type Culture Collection (ATCC). Primary HEK cells are the best hosts for the replication of human adenoviruses [50]. A549 cells are lung carcinoma cells; these cells grow in monolayers and can grow in Dubelco's modified eagle media (DMEM) supplemented with 10% heat inactivated, dialyzed fetal bovine serum (FBS) and 1% penicillin/streptomycin. Recombinant adenoviruses can infect a large number of cell types that express the Coxsackie-and Adenovirus Receptor (CAR) protein. Because, adenoviruses also increase glycolysis in continuous cell lines and thereby induce the cells to produce large quantities of acid [51] the DMEM is buffered with 25 mM Hepes to maintain a neutral pH during cell culture. As in the case of all eukaryotic cells, A549 cells require a sterile environment and must be kept from overgrowing and losing viability.

Depending on the post-translational modifications present in a given target protein, a researcher may need to investigate protein expression in different cell lines; each of these may have specific requirements for growth media. As a result, understanding cell growth in the expression media to be used is one of the first steps in obtaining isotopically labeled proteins. For our case study A549 lung carcinoma cells are brought up in DMEM, ^{15}N-CGM6000, or ^{15}N/^{13}C-CGM6000 media supplemented with 10% heat inactivated, dialyzed FBS and 1% penicillin/streptomycin and seeded in six well plates at 0.2×10^6 cells/well and incubated at 37°C for 6 days. Aliquots are taken every 24 h and cells counted for each media type. Perdeuteration or fractional deuteration of amino acid side chains is essential for the study of large proteins [52]. This can be achieved in principal by the growth of mammalian cells in deuterated media. A549 cells can be readily adapted to 20% D_2O-containing DMEM, 10% heat inactivated FBS and 1% penicillin/streptomycin [53]. The adapted cells are seeded at 1×10^6 cells/10 cm plate in DMEM containing 20, 45 or 70% D_2O. Viable cells are counted every 24 h for 6 days to obtain growth curves. Once it has been established that the cells are viable in the chosen growth media for the necessary time period, the researcher may test protein expression in the chosen cell line.

11.3.5 Protein Expression

Once the growth characteristics and media requirements of the favored cell line have been established, small scale protein expression can be initiated in six well plates. This is necessary prior to embarking on large scale protein production to assess the expression conditions as well as proper folding and activity of the desired protein. Expression of glycoproteins from mammalian cells can result in heterogeneous glycosylation due to variation in the occupancy as well as the nature of glycan. The glycosylation pattern obtained from mammalian cells can be high-mannose, hybrid or complex in nature. In our case study, we used a combination of kifunensine, a potent inhibitor of α-mannosidase I [54], as well as swainsonine, a potent inhibitor of α-mannosidase II [55], to obtain Endoglycosidase H-sensitive high mannose glycans [56, 57].

For small scale protein expression using the adenoviral expression system in six-well plates, A549 adherent cells are routinely maintained in DMEM. Typically, cells are seeded at 0.8×10^6 cell/well on Day 1 in fresh DMEM containing 10% heat inactivated dialyzed FBS, 1% penicillin/streptomycin and allowed to grow overnight at 37 °C and 5% CO_2. The following day (Day 2), media is replaced with labeled ^{15}N, ^{15}N/^{13}C CGM6000 or fresh DMEM containing 10% heat inactivated dialyzed FBS and 1% penicillin/ streptomycin. A549 cells are then infected with recombinant adenovirus (rAd) to a final concentration of 2,500 particles/cell. In general, a high multiplicity of infection is used so that all cells in the culture are synchronously infected. Protein expression of a cytoplasmic or secreted protein is monitored 72–96 h (Days 4–5) post infection either by harvesting cells or the culture media, respectively, and testing by a method of choice such as surface plasmon resonance, immunoprecipitation or western blotting.

For large scale production of unlabeled or isotopically enriched glycoprotein, A549 cells are seeded at $12–15 \times 10^6$ cells/15 cm plate on Day 1 in fresh DMEM containing 10% heat inactivated dialyzed FBS, 1% penicillin-streptomycin and allowed to grow overnight at 37°C and 5% CO_2. The following day (Day 2), media is replaced with ^{15}N, ^{15}N/^{13}C CGM6000 or fresh DMEM containing 10% heat inactivated dialyzed FBS, 1% penicillin/streptomycin as well as the glycosidic inhibitors kifunensine (12.5 mg/L) and swainsonine (5 mg/L). A549 cells are infected with recombinant adenovirus containing the gene for HIV-1 gp120 outer domain (rAdOD) 1–2 h later to a final concentration of 2,500 particles/cell. For secreted proteins, the culture supernatant is harvested 96–108 h post infection; cell debris is spun down at $365 \times g$. The culture supernatant is filtered aseptically through a 0.2 µm filter and, if necessary, the cell free supernatant can be concentrated five to tenfold by tangential flow filtration [58] prior to protein purification. The expressed protein (HIV-1 gp120 outer domain) is purified by immobilized nickel- and antibody b12-affinity chromatography. Fractions containing the outer domain are pooled, concentrated, dialyzed against PBS and deglycosylated using EndoH$_f$ followed by size-exclusion chromatography to obtain the deglycosylated protein [1]. For cytoplasmic proteins; culture supernatant is removed and adherent A549 cells are harvested following treatment with Trypsin-EDTA. Cells are pelleted at $300 \times g$, washed twice with PBS and subsequently lysed with cell lysis buffer (Cell Signaling) and the protein of interest purified by the method of choice.

11.3.6 Selective Labeling of Specific Amino Acids

Mammalian expression systems as well as insect cells are currently used to obtain amino-acid type specific labeling of proteins [27, 59, 60]. A similar approach is possible in the adenoviral mediated expression system. The methodology to produce selectively labeled protein is similar to that described in Sect. 11.3.5. CGM-6750 containing ^{15}N labeled glycine with all other amino acids and media components unlabeled can be used to produce protein specifically labeled with ^{15}N glycine (Fig. 11.5a). We did observe minor peaks in the glycine enriched sample possibly due to scrambling to serine or cysteine residues. CGM-6750 containing ^{15}N labeled valine with all other media components unlabeled can be used to produce protein selectively labeled with ^{15}N valine (Fig. 11.5b) We did observe minor peaks in valine enriched spectra possibly due to scrambling to isoleucine or leucine residues. We estimate yields of 60 and 49 mg/l of pure glycosylated ^{15}N glycine and ^{15}N valine enriched outer domain proteins, respectively. These yields balance out the high cost of the labeled media and allow for the production of selectively labeled proteins.

11.3.7   Quantification of Isotope Incorporation

Typically quantification of isotopic incorporation is obtained from MALDI TOF spectra of an enriched protein. However for glycoproteins such as the HIV-1 gp120 outer domain the inherent heterogeneity in glycosylation makes such an analysis very complicated. Thus, a glycoprotein is subjected to proteolytic digestion to identify peptide fragments devoid of N/O linked glycans followed by mass spectrometric analysis to obtain the isotope incorporation levels [1, 61].

11.4   NMR Characterization of Expressed Protein

Initially, the unlabeled full length or a target protein domain expressed either transiently or from the adenoviral expression system can be characterized using 1D or 2D NMR spectroscopy. Thus, in our example the HIV-1 gp120 outer domain obtained from adenovirus expression system exhibits a one dimensional ^{1}H NMR spectrum (Fig. 11.3a) that is well dispersed with resolved, upfield-shifted methyl protons as well as a relatively well dispersed amide region, indicative of a well folded protein. The ^{1}H-^{15}N HSQC spectra of the outer domain is also of very high quality and exhibits resolved backbone and side-chain amides (Fig. 11.3b). The ^{1}H-^{13}C HSQC (Fig. 11.3c) exhibits very good chemical shift dispersion along with upfield shifted methyl resonances, indicative of a structured protein.

To assess uniform ^{15}N/^{13}C incorporation necessary for assignment of backbone and side chain atoms, two data sets were recorded. In Fig. 11.4a, a high quality ^{1}H-^{13}C plane of a 3D HNCACB spectrum acquired at 900 MHz and 25°C shows correlations from the amide protons to the C_α and C_β carbons and demonstrates the uniform enrichment of C_α and C_β, a critical experiment for structural characterization of a protein by NMR spectroscopy. The ^{1}H-^{13}C projection of a 3D HNCO spectrum (Fig. 11.4b) showing correlations between the amide NH protons and the backbone CO resonances demonstrates conclusively that the adenoviral expression system provides sufficient isotope enrichment to allow acquisition of triple resonance experiments to obtain full backbone and side chain resonance assignments. In many instances, especially for large proteins, protein-protein complexes or membrane proteins, selective labeling of specific amino acids along with perdeuteration or random fractional deuteration becomes necessary to study specific interactions or complement assignments in extremely crowded regions of 3D spectra. Selectively labeled outer domain spectra enriched with ^{15}N glycine and valine (Fig. 11.5a, b) provide further proof that the adenoviral expression system is suitable for obtaining isotopically enriched proteins for structural characterization of proteins and protein complexes by NMR spectroscopy.

11.5   Conclusions

Characterization of the structural and dynamic properties of post-translationally modified proteins by NMR spectroscopy has been hindered by the absence of suitable expression systems to obtain isotopically enriched proteins. Recent advances in molecular biology and cell culture techniques

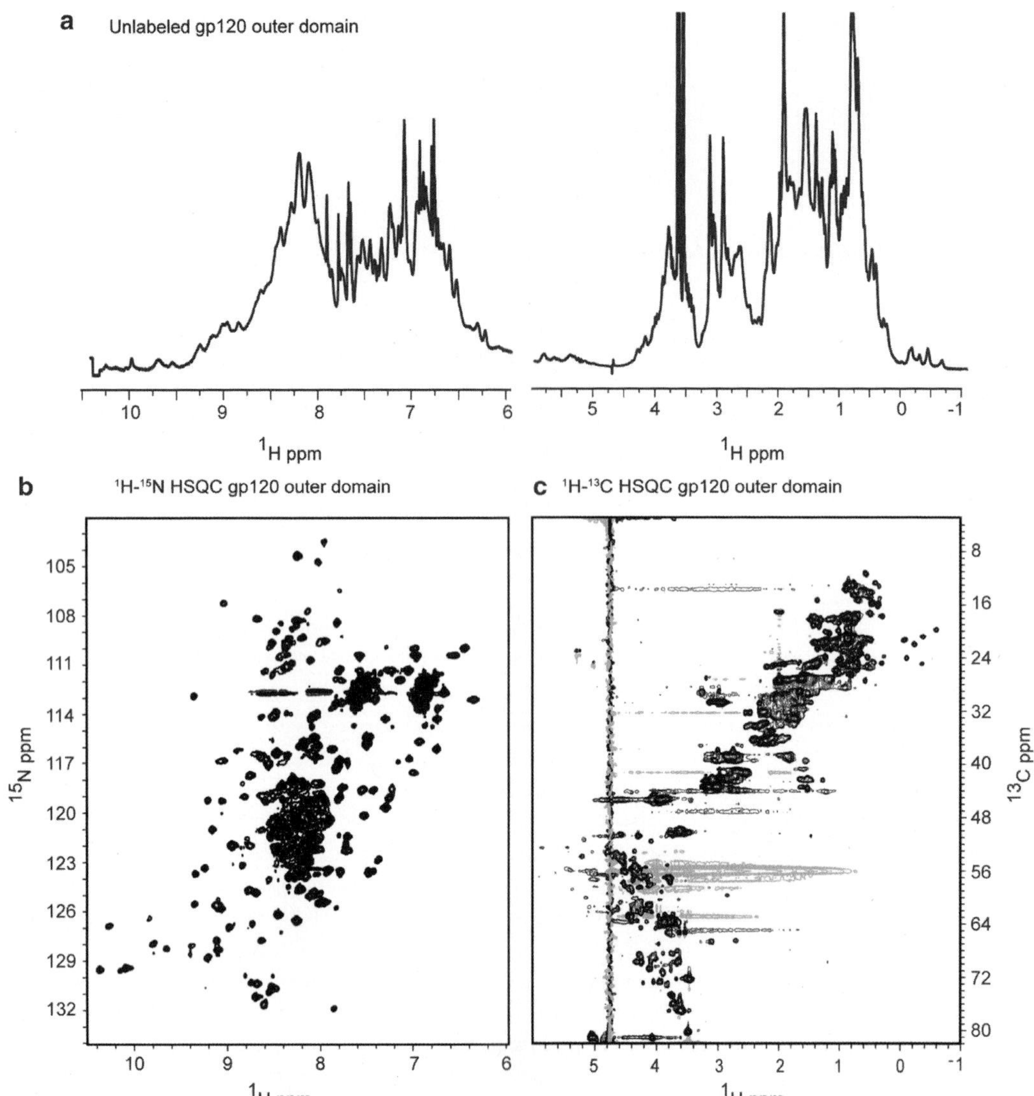

Fig. 11.3 **HIV-1 gp120 outer domain expressed and purified from the adenoviral expression system is a functionally active and well folded protein.** One dimensional proton spectra of unlabeled deglycosylated HIV-1 gp120 outer domain acquired at 600 MHz and 25°C is shown in panel A. ¹H-¹⁵N and ¹H-¹³C HSQC of ¹⁵N/¹³C labeled outer domain acquired at 900 MHz and 25°C are shown in (**b**) and (**c**), respectively (Reproduced from Sastry et al. [1]. With permission from Springer)

have revolutionized recombinant protein production from mammalian expression systems. In this chapter we have described an adenoviral based mammalian expression system that exploits the high level of protein expression obtained from an adenoviral vector when coupled to lung carcinoma cells. This system was developed for the expression of transgenes in the context of vaccines and gene therapy [62] The methodology described in this chapter to obtain uniform as well as selective amino-acid labeled proteins should reduce the barrier to structure determination of post-translationally modified proteins and their complexes by NMR spectroscopy.

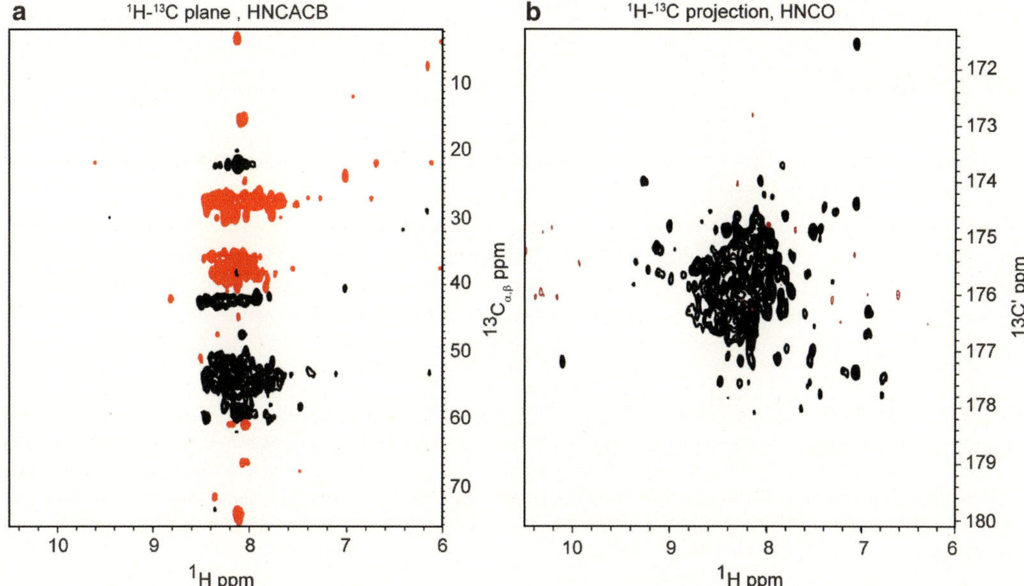

Fig. 11.4 HIV-1 gp120 outer domain expressed and purified from the adenoviral expression system is a functionally active and well folded protein. (a) A ^{1}H-C$_\alpha$C$_\beta$ plane of a 3D HNCACB experiment acquired on a ~400 μM ^{15}N/^{13}C gp120 outer domain sample at 900 MHz and 25°C showing correlations from the amide NH to C$_\alpha$C$_\beta$ atoms demonstrates the uniform enrichment of C$_\alpha$ and C$_\beta$. The downfield shifted amides are relatively weak and are not observed in the 2D plane of the HNCACB experiment. (b) A ^{1}H-^{13}C projection of a 3D HNCO experiment further demonstrates that the adenovirus-vectored mammalian expression system can provide the necessary isotopically enriched samples for heteronuclear NMR spectroscopy (Adapted from Sastry et al. [1]. With permission from Springer)

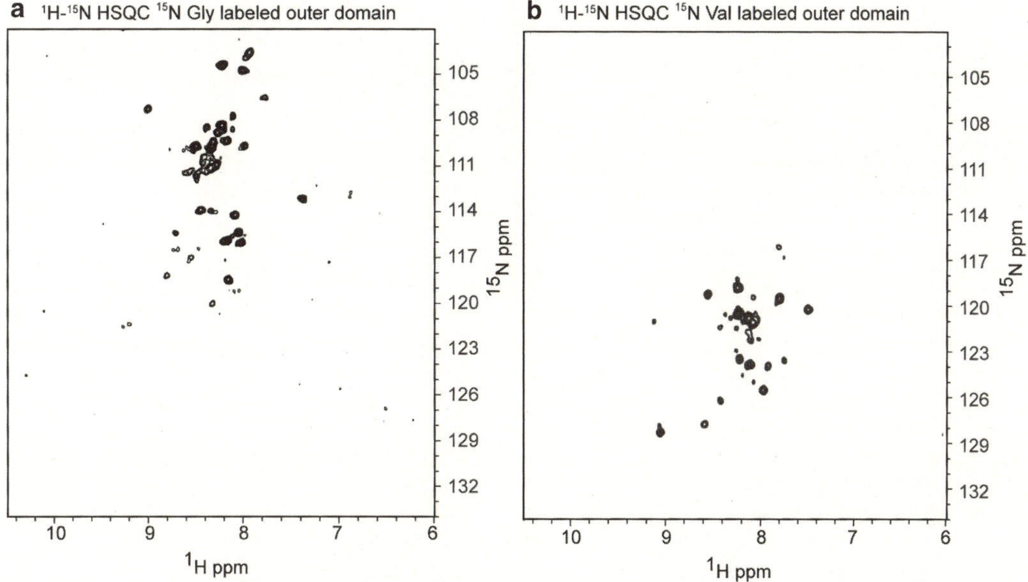

Fig. 11.5 Single amino acid labeling of HIV-1 gp120 outer domain from the adenoviral expression system is feasible. (a) ^{1}H-^{15}N HSQC spectra of HIV-1 gp120 outer domain selectively labeled with ^{15}N glycine acquired at 900 MHz and 25°C and (b) ^{1}H-^{15}N HSQC spectra of HIV-1 gp120 outer domain selectively labeled with ^{15}N valine acquired at 700 MHz and 25°C. The ^{1}H-^{15}N HSQC of selectively labeled outer domain spectra exhibit excellent signal to noise and thus demonstrate that the adenoviral system can be used to obtain selectively labeled proteins

11.6 Materials

We report materials used for selective amino-acid labeling. Materials for summarized experiments can be found in the respective primary publications [1, 40].

Specifically labeled ^{15}N Valine-CGM6750 and ^{15}N Glycine-CGM6750 with all other amino acids and components unlabeled were obtained from Cambridge Isotope Laboratories, Inc (Andover, MA). High glucose containing DMEM with Hepes and $NaHCO_3$ was obtained from Life Technologies. Kifunensine and swainsonine were obtained from Enzo Life Sciences. Nickel-NTA was obtained from Qiagen Inc.

Acknowledgments We thank the NMR staff at the New York Structural Biology Consortium for assistance with instrumentation and data acquisition. We also thank the members of the Structural Biology Section and Structural Bioinformatics Section at the Vaccine Research Center for insightful comments and discussions. Support for this work was provided by the Intramural Program of the NIH (NIAID and NIDDK). 900 MHz spectrometers were purchased with funds from NIH, USA, the Keck Foundation, New York State, and the NYC Economic Development Corporation.

11.7 Appendix

A list of reagents used in the adenoviral based mammalian expression system

Reagents	Source
DNA vectors	
pVRC1194	Available upon request from Dr. G. Nabel
pVRC1290	Available upon request from Dr. G. Nabel
Restriction enzymes	
BamHI	New England Biolabs Inc.
XbaI	New England Biolabs Inc.
SacI	New England Biolabs Inc.
ClaI	New England Biolabs Inc.
Recombinases and proteases	
Cre recombinase	Novagen-EMD Biosciences Inc.
HRV3C protease	Novagen-EMD Biosciences Inc.
Transfection kits and reagents	
Mammalian transfection system	Promega
Lipofectamine	Life Technologies
Cell lines	
HEK293	ATCC
A549	ATCC
Cell culture reagents	
DMEM with glucose, $NaHCO_3$, Hepes	Life Technologies
Fetal bovine serum	Life Technologies
Dialyzed fetal bovine serum	Life Technologies
Penicillin/Streptomycin	Life Technologies
Trypsin/EDTA	Life Technologies
Kifunensine	Enzo Life Sciences
Swainsonine	Enzo Life Sciences
Sodium pyruvate	Sigma-Aldrich
BE6000	Cambridge Isotope Laboratories, Inc.
^{15}N CGM6000	Cambridge Isotope Laboratories, Inc.

(continued)

(continued)

Reagents	Source
^{15}N/^{13}C CGM6000	Cambridge Isotope Laboratories, Inc.
^{15}N Glycine-CGM6750	Cambridge Isotope Laboratories, Inc.
^{15}N Valine-CGM6750	Cambridge Isotope Laboratories, Inc.
EDTA-d$_{12}$ (98%)	Cambridge Isotope Laboratories, Inc.
Tris-D11 (98%)	Cambridge Isotope Laboratories, Inc.
D$_2$O (99.8%)	Cambridge Isotope Laboratories, Inc.
Protein purification	
Nickel-NTA	Qiagen Inc
Protease inhibitor cocktail-EDTA free	Roche Diagnostics
Cell lysis buffer	Cell Signaling

References

1. Sastry M, Xu L, Georgiev IS, Bewley CA, Nabel GJ, Kwong PD (2011) Mammalian production of an isotopically enriched outer domain of the HIV-1 gp120 glycoprotein for NMR spectroscopy. J Biomol NMR 50:197–207
2. Nettleship JE, Assenberg R, Diprose JM, Rahman-Huq N, Owens RJ (2010) Recent advances in the production of proteins in insect and mammalian cells for structural biology. J Struct Biol 172:55–65
3. Kriz A, Schmid K, Baumgartner N, Ziegler U, Berger I, Ballmer-Hofer K, Berger P (2010) A plasmid-based multigene expression system for mammalian cells. Nat Commun 1:120
4. Trowitzsch S, Klumpp M, Thoma R, Carralot J-P, Berger I (2011) Light it up: highly efficient multigene delivery in mammalian cells. Bioessays 33:946–955
5. Kingston RE, Chen CA, Okayama H (2001) Calcium phosphate transfection. In: Current protocols in immunology, vol 31. Wiley, New York, pp 10.13.1–10.13.9
6. Potter H, Heller R (2010) Transfection by electroporation. In: Current protocols in molecular biology, vol 92. Wiley, New York, pp 9.3.1–9.3.10
7. Xia W, Bringmann P, McClary J, Jones PP, Manzana W, Zhu Y, Wang S, Liu Y, Harvey S, Madlansacay MR, McLean K, Rosser MP, MacRobbie J, Olsen CL, Cobb RR (2006) High levels of protein expression using different mammalian CMV promoters in several cell lines. Protein Expr Purif 45:115–124
8. Coleman TA, Parmelee D, Thotakura NR, Nguyen N, Bürgin M, Gentz S, Gentz R (1997) Production and purification of novel secreted human proteins. Gene 190:163–171
9. Backliwal G, Hildinger M, Chenuet S, Wulhfard S, De Jesus M, Wurm FM (2008) Rational vector design and multi-pathway modulation of HEK 293E cells yield recombinant antibody titers exceeding 1 g/l by transient transfection under serum-free conditions. Nucleic Acids Res 36:e96
10. Hopkins RF, Wall VE, Esposito D (2012) Optimizing transient recombinant protein expression in mammalian cells. In: Hartley JL (ed) Protein expression in mammalian cells: methods and protocols, vol 801, Methods in molecular biology. Humana Press, New York, pp 251–268
11. Geisse S, Fux C (2009) Recombinant protein production by transient gene transfer into mammalian cells. In: Burgess RR, Deutscher MP (eds) Methods in Enzymology, vol 463. Academic Press, San Diego, pp 223–238
12. Zhou T, Xu L, Dey B, Hessell AJ, Van Ryk D, Xiang S-H, Yang X, Zhang M-Y, Zwick MB, Arthos J, Burton DR, Dimitrov DS, Sodroski J, Wyatt R, Nabel GJ, Kwong PD (2007) Structural definition of a conserved neutralization epitope on HIV-1 gp120. Nature 445:732–737
13. Chen L, Kwon YD, Zhou T, Wu X, O'Dell S, Cavacini L, Hessell AJ, Pancera M, Tang M, Xu L, Yang Z-Y, Zhang M-Y, Arthos J, Burton DR, Dimitrov DS, Nabel GJ, Posner MR, Sodroski J, Wyatt R, Mascola JR, Kwong PD (2009) Structural basis of immune evasion at the site of CD4 attachment on HIV-1 gp120. Science 326:1123–1127
14. Kwong PD, Nabel GJ, Acharya P, Boyington JC, Chen L, Hood C, Kim A, Kong L, Kwon YD, Majeed S, McLellan J, Ofek G, Pancera M, Sastry M, Shah AC, Stuckey J, Zhou T (2011) Structural biology and the design of effective vaccines for HIV-1 and other viruses. In: Georgiev S (ed) National Institute of Allergy and Infectious Disease, NIH. Humana, New York, pp 387–402
15. Southern P, Berg P (1982) Transformation of mammalian cell to antibiotic resistance with a bacterial gene under control of the SV40 early region promoter. J Mol Appl Genet 1:327–341
16. Chen C, Okayama H (1987) High-efficiency transformation of mammalian cells by plasmid DNA. Mol Cell Biol 7:2745–2752

17. Hansen AP, Petros AM, Mazar AP, Pederson TM, Rueter A, Fesik SW (1992) A practical method for uniform isotopic labeling of recombinant proteins in mammalian cells. Biochemistry 31:12713–12718
18. Wyss DF, Dayie KT, Wagner G (1997) The counterreceptor binding site of human CD2 exhibits an extended surface patch with multiple conformations fluctuating with millisecond to microsecond motions. Protein Sci 6:534–542
19. Wyss DF, Withka JM, Knoppers MH, Sterne KA, Recny MA, Wagner G (1993) Proton resonance assignments and secondary structure of the 13.6 kDa glycosylated adhesion domain of human CD2. Biochemistry 32:10995–11006
20. Lustbader JW, Birken S, Pollak S, Pound A, Chait BT, Mirza UA, Ramnarain S, Canfield RE, Brown JM (1996) Expression of human chorionic gonadotropin uniformly labeled with NMR isotpes in Chinese hamster ovary cells: an advance toward rapid determination of glycoprotein structures. J Biomol NMR 7:295–304
21. Shindo K, Masuda K, Takahashi H, Arata Y, Shimada I (2000) Backbone ^{1}H, ^{13}C, and ^{15}N resonance assignments of the anti-dansyl antibody Fv fragment. J Biomol NMR 17:357–358
22. Hansen AP, Petros AM, Meadows RP, Nettesheim DG, Mazar AP, Olejniczak ET, Xu RX, Pederson TM, Henkin J, Fesik SW (1994) Solution structure of the amino-terminal fragment of urokinase-type plasminogen activator. Biochemistry 33:4847–4864
23. Hinck AP, Archer SJ, Qian SW, Roberts AB, Sporn MB, Weatherbee JA, Tsang MLS, Lucas R, Zhang B-L, Wenker J, Torchia DA (1996) Transforming growth factor β1: three-dimensional structure in solution and comparison with the X-ray structure of transforming growth factor β2. Biochemistry 35:8517–8534
24. Wyss D, Choi J, Li J, Knoppers M, Willis K, Arulanandam A, Smolyar A, Reinherz E, Wagner G (1995) Conformation and function of the N-linked glycan in the adhesion domain of human CD2. Science 269:1273–1278
25. Metzler WJ, Bajorath J, Fenderson W, Shaw SY, Peach R, Constantine KL, Naemura J, Leyte G, Lavoie TB, Mueller L, Linsley PS (1997) Solution structure of human CTLA 4 and delineation of a CD80/CD86 binding site conserved in CD28. Nat Struc Biol 4:527–531
26. Werner K, Richter C, Klein-Seetharaman J, Schwalbe H (2008) Isotope labeling of mammalian GPCRs in HEK293 cells and characterization of the C-terminus of bovine rhodopsin by high resolution liquid NMR spectroscopy. J Biomol NMR 40:49–53
27. Arata Y, Kato K, Takahashi H, Shimada I (1994) Nuclear magnetic resonance study of antibodies: a multinuclear approach. Method Enzymol 239:440–464
28. Klein-Seetharaman J, Yanamala NVK, Javeed F, Reeves PJ, Getmanova EV, Loewen MC, Schwalbe H, Khorana HG (2004) Differential dynamics in the G protein-coupled receptor rhodopsin revealed by solution NMR. Proc Natl Acad Sci USA 101:3409–3413
29. Klein-Seetharaman J, Reeves PJ, Loewen MC, Getmanova EV, Chung J, Schwalbe H, Wright PE, Khorana HG (2002) Solution NMR spectroscopy of [α-^{15}N]lysine-labeled rhodopsin: the single peak observed in both conventional and TROSY-type HSQC spectra is ascribed to Lys-339 in the carboxyl-terminal peptide sequence. Proc Natl Acad Sci USA 99:3452–3457
30. Mulligan RC, Howard BH, Berg P (1979) Synthesis of rabbit [β]-globin in cultured monkey kidney cells following infection with a SV40 [β]-globin recombinant genome. Nature 277:108–114
31. Warnock JN, Daigre C, Al-Rubeai M (2011) Introduction to viral vectors. In: Merten O, Al-Rubeai M (eds) Viral vectors for gene therapy: methods and protocols, vol 737, Methods in molecular biology. Humana Press, New York, pp 1–25
32. Gluzman Y, Reichl H, Scolnick D (1982) Helper-free adenovirus type-5 vectors. In: Gluzman Y (ed) Eukaryotic viral vectors. Cold Spring Harbor Laboratory, Cold Spring Harbor, New York, pp 187–192
33. Howley PM, Sarver N, Law M-F (1983) Eukaryotic cloning vectors derived from bovine papillomavirus DNA. In: Wu R, Grossman L, Moldave K (eds) Methods in Enzymology, vol 101. Academic Press, pp 387–402
34. Waehler R, Russell SJ, Curiel DT (2007) Engineering targeted viral vectors for gene therapy. Nat Rev Genet 8:573–587
35. Berk AJ (1986) Adenovirus promoters and E1A transactivation. Annu Rev Genet 20:45–79
36. Shenk T (1996) Adenoviridae: the viruses and their replication. In: Fields BN, Knipe DM, Howley PM (eds) Fields virology. Lippincott-Raven, Philadelphia, pp 2111–2148
37. Berkner K, Sharp P (1982) Preperation of adenovirus recombinant using plasmids of viral DNA. In: Gluzman Y (ed) Eukaryotic viral vectors. Cold Spring Harbor Laboratory, Cold Spring Harbor, New York, pp 193–198
38. Babich A, Feldman LT, Nevins JR, Darnell JE Jr, Weinberger C (1983) Effect of adenovirus on metabolism of specific host mRNAs: transport control and specific translational discrimination. Mol Cell Biol 3:1212–1221
39. Huang JT, Schneider RJ (1990) Adenovirus inhibition of cellular protein synthesis is prevented by the drug 2-aminopurine. Proc Natl Acad Sci USA 87:7115–7119
40. Aoki K, Barker C, Danthinne X, Imperiale MJ, Nabel GJ (1999) Efficient generation of recombinant adenoviral vectors by Cre-lox recombination in vitro. Mol Med 5:224–231
41. Moore D, Dowhan D (2002) Purification and concentration of DNA from aqueous solutions. In: Current protocol in molecular biology, vol 59. Wiley, New York, pp 2.1.1–2.1.10
42. Voytas D (2001) Agarose gel electrophoresis. In: Current protocol in molecular biology, vol 51. Wiley, New York, pp 2.5A.1–2.5A.9

43. Hall K, Blair Zajdel ME, Blair GE (2009) Defining the role of CD46, CD80 and CD86 in mediating adenovirus type 3 fiber interactions with host cells. Virology 392:222–229
44. Graham FL, Smiley J, Russell WC, Nairn R (1977) Characteristics of a human cell line transformed by DNA from human adenovirus type 5. J Gen Virol 36:59–72
45. Rich RL, Myszka DG (2010) Grading the commercial optical biosensor literature—Class of 2008: 'The Mighty Binders'. J Mol Recognit 23:1–64
46. Raghavan M, Bjorkman PJ (1995) BIAcore: a microchip-based system for analyzing the formation of macromolecular complexes. Structure 3:331–333
47. Duffy AM, O'Doherty AM, O'Brien T, Strappe PM (2005) Purification of adenovirus and adeno-associated virus: comparison of novel membrane-based technology to conventional techniques. Gene Ther 12:S62–S72
48. Tan R, Li C, Jiang S, Ma L (2006) A novel and simple method for construction of recombinant adenoviruses. Nucleic Acids Res 34:e89
49. Chillon M, Alemany R (2011) Methods to construct recombinant adenovirus vectors. In: Merten O, Al-Rubeai M (eds) Viral vectors for gene therapy: methods and protocols, vol 737, Methods in molecular biology. Humana Press, New York, pp 117–138
50. Horwitz MS (1996) Adenoviruses. In: Fields BN, Knipe DM, Howley PM (eds) Fields virology. Lippincott-Raven, Philadelphia, pp 2149–2171
51. Fisher TN, Ginsberg HS (1957) Accumulation of organic acids by HeLa cells infected with type 4 adenovirus. Proc Soc Exp Biol Med 95:47–51
52. LeMaster DM (1990) Deuterium labelling in NMR structural analysis of larger proteins. Q Rev Biophys 23:133–174
53. Murphy J, Desaive C, Giaretti W, Kendall F, Nicolini C (1977) Experimental results on mammalian cells growing in vitro in deuterated medium for neutron-scattering studies. J Cell Sci 25:87–94
54. Elbein AD, Tropea JE, Mitchell M, Kaushal GP (1990) Kifunensine, a potent inhibitor of the glycoprotein processing mannosidase I. J Biol Chem 265:15599–15605
55. Elbein AD (1991) Glycosidase inhibitors: inhibitors of N-linked oligosaccharide processing. FASEB J 5:3055–3063
56. Kong L, Sheppard NC, Stewart-Jones GBE, Robson CL, Chen H, Xu X, Krashias G, Bonomelli C, Scanlan CN, Kwong PD, Jeffs SA, Jones IM, Sattentau QJ (2010) Expression-system-dependent modulation of HIV-1 envelope glycoprotein antigenicity and immunogenicity. J Mol Biol 403:131–147
57. Magnelli P, Bielik A, Guthrie E (2012) Identification and characterization of protein glycosylation using specific endo- and exoglycosidases. In: Hartley JL (ed) Protein expression in mammalian cells: methods and protocols, vol 801, Methods in molecular biology. Humana Press, New York, pp 189–211
58. van Reis R, Brake JM, Charkoudian J, Burns DB, Zydney AL (1999) High-performance tangential flow filtration using charged membranes. J Membr Sci 159:133–142
59. Anglister J, Frey T, McConnell HM (1984) Magnetic resonance of a monoclonal anti-spin-label antibody. Biochemistry 23:1138–1142
60. Gossert A, Hinniger A, Gutmann S, Jahnke W, Strauss A, Fernández C (2011) A simple protocol for amino acid type selective isotope labeling in insect cells with improved yields and high reproducibility. J Biomol NMR 51:449–456
61. Truhlar SME, Cervantes CF, Torpey JW, Kjaergaard M, Komives EA (2008) Rapid mass spectrometric analysis of ^{15}N-Leu incorporation fidelity during preparation of specifically labeled NMR samples. Protein Sci 17:1636–1639
62. Nabel GJ (1999) Development of optimized vectors for gene therapy. Proc Natl Acad Sci USA 96:324–326

H.S. Atreya (ed.), *Isotope Labeling in Biomolecular NMR*, Advances in Experimental Medicine and Biology 992,   213
DOI 10.1007/978-94-007-4954-2, © Springer Science+Business Media Dordrecht 2012

Printed by Printforce, the Netherlands